YELLOWSTONE'S BIRDS

YELLOWSTONE'S BIRDS

DIVERSITY AND ABUNDANCE IN THE WORLD'S FIRST NATIONAL PARK

Edited by

DOUGLAS W. SMITH, LAUREN E. WALKER, and KATHARINE E. DUFFY

WITH VIDEOS BY ROBERT K. LANDIS

Princeton University Press

Princeton and Oxford

For Ruth Shea and George Melendez Wright.
Few did more for the park's iconic bird, the trumpeter swan, in Yellowstone.

Published by Princeton University Press
41 William Street, Princeton, New Jersey 08540
99 Banbury Road, Oxford OX2 6JX
press.princeton.edu
All Rights Reserved

Library of Congress Cataloging-in-Publication Data

Names: Smith, Douglas W., 1960- editor. | Duffy, Katy, editor. | Walker, Lauren E., editor.
Title: Yellowstone's birds : diversity and abundance in the world's first national park / edited by Douglas W. Smith, Lauren E. Walker, and Katharine E. Duffy ; with videos by Robert K. Landis.
Description: First edition. | Princeton : Princeton University Press, [2023] | Includes bibliographical references and index.
Identifiers: LCCN 2023006916 | ISBN 9780691217833 (hardback ; acid-free paper) | ISBN 9780691218731 (ebook)
Subjects: LCSH: Yellowstone National Park. | Birds--Yellowstone National Park. | Ornithology. | BISAC: NATURE / Birdwatching Guides | TRAVEL / United States / West / Mountain (AZ, CO, ID, MT, NM, NV, UT, WY)
Classification: LCC QL719.Y45 Y45 2023 | DDC 598.09787/52--dc23/eng/20230324
LC record available at https://lccn.loc.gov/2023006916

British Library Cataloging-in-Publication Data is available

Editorial: Robert Kirk and Megan Mendonça
Production Editorial: Mark Bellis
Typeset and designed by D & N Publishing, Wiltshire, UK
Jacket Design: Wanda España
Production: Steve Sears
Publicity: Matthew Taylor and Caitlyn Robson
Copyeditor: Patricia Fogarty

Jacket Credit: (front cover) photography by Scott Heppel; (back cover) Jake Frank, NPS; (front flap) © Tom Murphy

This book has been composed in Veneer, Museo Sans, Nexus Sans and Serif Pro

Printed on acid-free paper. ∞

Printed in Italy

10 9 8 7 6 5 4 3 2 1

CONTENTS

INTRODUCTION

BIRDING IN YELLOWSTONE

RAPTORS

CONTENTS

WATERBIRDS

PASSERINES

FOREWORD

BY DAYTON DUNCAN

WHEN YELLOWSTONE BECAME the world's first national park in 1872, the fate of wildlife did not figure much in Congress's deliberations. The focus instead centered on whether the stupendous array of thermal features—the geysers, hot springs, mudpots, and fumaroles—should be set aside and protected against private homesteading and mining. Supporters pushed the idea that the area's high elevation made it unsuitable for farming and ranching, and its volcanic origins probably precluded the likelihood of valuable mines being discovered there. In other words, it was "worthless" for the usual pursuits of a nation busily exploiting the continent's natural resources.

Driving the politics of the moment was intense lobbying by the Northern Pacific Railway, which saw the "Wonderland" of Yellowstone's scenery as a potential lure for tourists who might increase the ridership of the line it was building westward.

The bill creating "a public park or pleasuring ground for the benefit and enjoyment of the people" passed, and the new park began its slow and uneven evolution, filled with debates over what rules and regulations were necessary to ensure its survival, who would enforce them (the National Park Service didn't exist until 1916), and, equally important, whether Yellowstone's purpose was merely as a tourist attraction.

As one unexpected consequence of the park's creation (as well as its remoteness), it turned out to be a refuge for a number of iconic mammals, particularly the bison, which had been driven to near extinction during the commercial slaughter on the Great Plains in the 1870s and 1880s. Yellowstone played a crucial role in saving what is now our official national mammal from disappearing forever. Likewise, elk found a sanctuary there, as did black bears and grizzly bears (and wolves, once they were reintroduced in the 1990s).

Over time, these so-called "charismatic megafauna" became as much of Yellowstone's appeal to visitors as the geysers, and how to allow them to be both protected and wild became a major, enduring issue.

Yellowstone's birds, however, tended to be overlooked in the midst of all the thermal marvels and impressive mammals vying for attention. But the park has always been a haven for them as well, and the park has contributed to some winged species' preservation from extinction, just as much as it did the bison's.

In 1875, a young George Bird Grinnell—who would later become founder of the Audubon Society, one of Yellowstone's greatest champions, and a crucial player in the bison's salvation—served as a naturalist on a military exploration of the then three-year-old park. In addition to calling attention to the wanton slaughter of buffalo, elk, mule deer, and antelope that he witnessed on his way, and the threats to them, even within the park's boundaries, he inventoried about a hundred species of birds. (His favorite was the American dipper—the water ouzel—whose ability to dive underwater fascinated him.)

A half-century later, another visionary in the history of national parks came to Yellowstone and was captivated by its birds. George Melendez Wright, still in his twenties, had undertaken a series of field studies in the parks that ultimately would push the National Park Service to redirect its long-standing policies and emphasize the preservation of nature rather than concentrating so heavily on catering to tourists. Wright's field of vision was as broad as his impact was profound in terms of how we now envision the Park Service's mission—whether predators should be allowed to exist, whether wild animals should be treated as pets, whether management decisions should be based on science instead of popular opinion.

But in Yellowstone, he also became enthralled by its birds, especially the trumpeter swan, the nation's largest waterfowl and heaviest (26 lbs/11.8 kg) flying bird. Wright and his team spent weeks observing and studying the nesting habits of trumpeter swans.

"Days with the birds in Yellowstone are tonic to him whose spirit is bruised by reiteration of the lament that wilderness is a dying gladiator," Wright wrote of his experience in the park. "Too frequent exposure to a belief born of despair is not good for any man. But it is the birds of the water . . . that are encouragement and inspiration to the man who prays for conviction that the wilderness still lives, will always live."

"Sometimes while I am watching these birds on the water," he added, "the illusion of the untouchability of this wilderness becomes so strong that it is stronger than reality, and the polished roadway becomes the illusion, the mirage that has no substance."

At the time, in the early 1930s, there were fewer than a hundred trumpeter swans left on the continent, mostly in the Yellowstone region, and Wright's reports helped establish policies that nurtured them back from the brink and ultimately resulted in the creation of a nearby refuge that furthered their chances for survival.

This book allows Yellowstone's birds to fly out from under the large shadow of its geology and mammals, restoring them to their rightful place alongside the features that have often received all the attention. They are part of Yellowstone's story, too—part of its enduring attraction, part of its importance to our nation, and as George Melendez Wright recognized, part of the "tonic" for those who pray "that the wilderness still lives, will always live."

■ Dayton Duncan is the author of *The National Parks: America's Best Idea.*

PREFACE

DOUGLAS W. SMITH, LAUREN E. WALKER, KATHARINE E. DUFFY

WHY WRITE A book about Yellowstone's birds? In short, because there isn't one, or hasn't been one of a comprehensive nature. Precious little has been written about the birds of Yellowstone, despite a legion of books about history, geology, and large mammals—only two books (Skinner 1925, McEneaney 1988), a handful of peer-reviewed publications, and occasional popular articles. Despite Yellowstone National Park (Yellowstone; YNP) being the oldest national park in the world, established in 1872, the Yellowstone Bird Program itself is not that old, having begun only in the mid-1980s. Data on birds gathered earlier—trumpeter swans, for example—were gathered by park rangers or naturalists. Monitoring of some mammals in Yellowstone precedes the turn of the last century. Clearly, birds need more attention, and especially with the degradation of non-park environments across the continent, an assessment of avian life in an undisturbed environment is surely called for.

This is important now because many bird populations are in trouble, and some are in danger of extinction. Worldwide, 40% of avian species are declining, and North American populations have declined 29%, a loss of *3 billion birds* (Birdlife International 2018, Rosenberg et al. 2019). What is the story for YNP? Some of the leading causes for the widespread decline—habitat loss and invasive species, like cats and rats—are not issues in YNP. Or are they? Fires due to climate change, the other big cause of bird declines globally, is gobbling up forests in YNP at a worrisome rate, due to increased fire frequency and intensity (Westerling et al. 2011). But is YNP immune to other human-caused changes? Cats and rats are not an issue, but what about

increases of some species that are favored or aided by humans? Waterfowl, one such example, are on the increase in North America because of conservation attention, and YNP has seen big increases in Canada geese. But trumpeter swans have declined, and we have few insights into the status of most other waterfowl species (*see chapter 20*). Raptors are another increasing group, as exemplified by peregrine falcons, which went to extirpation in the park but are now probably at habitat saturation due to human intervention (*see chapter 10*). Golden eagles are likely declining continent-wide, and their future status in YNP is hard to predict due to low productivity (*see chapter 11*). Bald eagles and osprey are doing well, except on Yellowstone Lake with the collapse of cutthroat trout (*chapter 12*), and osprey have adapted less well than bald eagles. So raptors in YNP are not necessarily reflective of raptors elsewhere. Clark's nutcrackers (*chapter 24*) are tied to the whitebark pine, and the tree is doing poorly and has been listed as a threatened species. Other corvids (*chapter 23*) partially tied to carnivores, which have made an impressive comeback in the last few decades (Smith et al. 2020a), are also tied to people, and with park visitation (nearly 4.9 million in 2021) and human development around the park both on the increase (Hansen and Phillips 2018), their future looks good. Songbirds (*chapter 25*), with roughly half of the world's bird species, sit in the middle of all of this. Grassland birds are suffering the worst declines of any bird group (Birdlife International 2018, Rosenberg et al. 2019), mainly because of agriculture (both crops and livestock), and we are just beginning to monitor them in the park. Certainly, the full complement of grassland birds must be intact in YNP, as there is no agricultural development! But there is non-native grassland vegetation, and at this writing, we cannot give you an answer. Given this worldwide crisis, this North American crisis, should not the world's first national park assess its bird populations?

Therefore, the goal of this book is to summarize what we know and do not know about YNP birds in order to plot the next move. The next move also includes you—the park visitor interested in birds. We need your help. Somewhat unique to government is the ability to engage in long-term monitoring. Much science and university-driven research is short-term and often has a targeted purpose or poses a specific question; many park employees stay a long while, and some programs incorporate routine monitoring—in our case, bird populations—year in and year out. Our best example of this is trumpeter swans: we now have 90 years of data. What you see and report can help—properly termed, this is citizen science. Your appreciation of birds could inspire long-term monitoring projects, which contribute to science and the next edition of this book—hopefully sooner than it took to publish this one (150 years—2022 was the park's 150th birthday). Through this book, we hope you appreciate the birds found in a protected area—or an area that is thought to be protected. That distinction is precisely why your observations are important: Is YNP changing through time, given all the environmental insults occurring everywhere? Are YNP's bird populations changing? With climate change, increased visitation, and population growth, and with concomitant development in the region looming, we need to be on top of fluctuations in bird populations. Parks have long been held up as ecological baselines (Pritchard 1999), and we need to know: How well are we doing? Finally, we wrote this book with you in mind: we've offered accessible, readable language, engaging and numerous color photos, infographics

presenting complex stories simply, and an accompanying video by world-renowned cinematographer Robert K. Landis with author interviews and bird footage.

Of course, because birds are one of the best-studied animal groups on the planet, we wanted to address age-old scientific questions. YNP has about 150 breeding bird species, and dozens more that use the park or pass through but do not nest. This use, or lack of it, poses some interesting scientific questions besides population monitoring, which we emphasized above. Yellowstone is a large park, the second-largest in the continental United States (3,471 square miles/8,991 km²), with great habitat diversity across a mountainous landscape. Such a landscape is usually bird rich, yet YNP is not. Why? The park's glacial and especially its volcanic history (*see chapter 1*) sets the stage for soil and plants of vast monotony (e.g., high-elevation lodgepole pine forests), topped off by a harsh climate, which is not a recipe for bird richness (*see chapter 1*). Yet oases abound (*see chapters 3–5*). Thermal areas, unique to Yellowstone and providing year-round food and respite, and pockets of undisturbed mountain habitats can be as resplendent as anywhere in the West, and it is unspoiled (we almost wrote untouched, but that is not true). YNP is one of the few areas in the continental United States that has never been mined, logged, grazed, or otherwise developed into parking lots and plazas. How remarkable—for birds and visitors alike! We hope you will discover and appreciate Yellowstone's uniqueness and hold it dear for decades to come.

A few things that this book is not. Our goal was not to create a birding guide. Terry McEneaney, YNP's first bird biologist, did that (McEneaney 1988). We wanted a scientific treatment of all the birds that use YNP, presented in an accessible format—something a visitor to the park could use in addition to a field guide (there are many of those) and something to help you read the bird landscape. Knowing birds' natural history, populations, habits, habitats, locations, and problems are a good introduction and get you much of the way there. Further, some species have been well studied and others not at all, so presentation is uneven and sometimes many species are considered together. Waterfowl, gallinaceous birds, and shorebirds are examples, all among the least studied and least monitored. Other species, most notably trumpeter swans, have one of the longer datasets of any animal species in Yellowstone (*see chapter 17*). For many others, monitoring started with the founding of the Yellowstone Bird Program in the mid-1980s, when such species as osprey, bald eagles, and peregrine falcons became federally listed. Other programs looking at songbirds followed, nudged because of continent-wide programs (Breeding Bird Survey and Christmas Bird Count); recently the park became involved in an international songbird monitoring program, Monitoring Avian Productivity and Survivorship (MAPS). Songbirds are very niche-specific and considered to be environmental indicators, and this beefing up was long overdue.

Other important monitoring programs and research projects are worth noting, as they too are reported in this book. Beginning in 2008, the scope and intensity of Yellowstone's bird monitoring increased. Besides new songbird surveys, several other new intensive research projects were initiated. One of particular importance was the Yellowstone Raptor Initiative (YRI; Baril et al. 2017a), designed to assess what we had not—namely, raptors other than peregrine falcons, osprey, and bald eagles (although we kept monitoring them as well to keep long-term data intact). Initially, we excluded accipiters outright—we thought sharp-shinned hawks, Cooper's hawks, and northern goshawks would occupy all our time and money, and we ended tracking them only opportunistically. For everything else though, we

designed explicit targeted surveys. This took a lot of citizen science, and we developed a park-wide reporting program. Notable was a monitoring scheme for owls, never attempted before. We also embarked on targeted research that included golden eagles, common loons, osprey, Clark's nutcrackers, ravens, harlequin ducks, and trumpeter swans. Much of this was collaborative study. National Park Service (NPS) staff partnered with universities and graduate students, but also with non-governmental, private conservation organizations in a classic example of public-private partnership. Some examples are the Wyoming Wetlands Society, the Biodiversity Research Institute (BRI), the Max Planck Institute in Germany, and the Ricketts Conservation Foundation (RCF). We helped form the Greater Yellowstone Ecosystem Interagency Loon Working Group, as well as the Wyoming Golden Eagle Working Group. We had help from government partners, bringing in staff from the US Fish and Wildlife Service to help with peregrine falcons, and we worked closely with the state governments of Wyoming, Idaho, and Montana on trumpeter swans, and with Wyoming on harlequin ducks. We participated in nationwide winter eagle surveys. Our MAPS program is coordinated by the Institute for Bird Populations, and the banding station is permitted by the United States Geological Survey (USGS). We diversified and collaborated.

In the end, we want to put birds on the map in Yellowstone. Put them alongside the charismatic megafauna and geysers so long studied. But our (happy) challenge, unlike those faced by the biologists and researchers studying these other critters and resources, is that there are so many bird species in Yellowstone. So much to know, so long overlooked. We tried to be comprehensive, but there are gaps. By pointing the way, by starting the conversation, we hope to fill them—maybe even with your help.

ACKNOWLEDGMENTS

WHERE DOES ONE START? Discussions about this section had many say, of course, that we should start with the birds. We wish to acknowledge them first and foremost and say sorry it took us so long to get to you; the mother park and all, the first in-depth technical book about avian ecology and conservation took 150 years.

As always, too, we thank Yellowstone National Park, its support, infrastructure, staff, and colleagues who assist us with our work, as we are all one team. Most important are the many technicians and volunteers who have assisted with bird monitoring and protection. Notable among them are Hannah Beroske, Sarah Lindsay, and Dylan Sanborn, for their unsung efforts helping us finish this book during the summer of the 2022 bird field season. We have been lucky, too, with several graduate students who have helped us learn about and protect birds; many of them are authors in this book. In fact, many of the authors have demonstrated broad dedication to the birds of Yellowstone and the Bird Program, aiding in facets of the creation of this book that go well beyond their allotted chapters.

This book would have been markedly different without the aid of Charissa Reid, head of Science Publications for the Yellowstone Center for Resources. Charissa thanklessly helped organize, find, and submit our photos and helped in numerous other capacities whenever asked. Charissa is the perennial glue that holds so much together for the science communication in our division in Yellowstone.

We appreciate the collaboration and coordination we've had with our state and federal partners, particularly the US Fish and Wildlife Service, US Geological Survey, and the states of Idaho, Montana, and Wyoming.

Many people assisted on particular projects, represented as chapters in this book, and we wish to recognize them. Lisa Baril was the first technician under the newly revamped Bird Program in 2008, and she played a significant role in many bird-related projects. Leigh Anne Dunworth, Accessibility Coordinator for Yellowstone National Park, reviewed chapter 3. Joel (Jeep) Pagel helped re-establish peregrine falcon monitoring in Yellowstone as well as assisted in starting the Yellowstone Raptor Initiative. His expertise and companionship are appreciated and cherished. Bob Oakleaf, of the Wyoming Game and Fish Department (retired), provided invaluable advice and assistance on all things raptors. Al Harmata (eagles) and Marco Restani (raptors, especially osprey) reviewed many documents and provided valuable assistance and advice whenever asked. For swans, we appreciate the help of Cory Abrams, Emerald Gustowt, Cody Pitz, and William Rudd, and many others from the Wyoming Wetlands Society. Chuck Preston, formerly curator of the Draper Museum in Cody, Wyoming, offered as much help as he could, including helping pioneer aerial surveys for golden eagles. Jack Kirkley volunteered to survey for Swainson's hawks in the park interior after having done so during his summer maintenance job in the 1970s. His knowledge of goshawks was also appreciated. We are grateful for the essential assistance that Scott Hancock provided during numerous owl surveys. Others who assisted us with owl surveys include Amanda Boyd, Diane Renkin, Julianne Baker, Jane Olson, Lisa Strait Bowersock, and Robert Petty. Barbara Tylka, Susan Weinreis, Mary Maj, Erik Oberg, and the Sacajawea Audubon Society helped organize volunteer efforts to survey red-tailed hawks in 2020 and 2021. Jeff and Pat Lund donated significant time and effort to help survey common loons and trumpeter swans in the summer of 2021. Michael Curtis and Amanda Bramblett conducted several backcountry owl surveys in 2014; they also conducted loon surveys. Michael was a critical ranger for backcountry protection and management of bird closures. Others helping with bird protection were Dagan Klein and Ivan Kowski of the Yellowstone National Park backcountry office. We appreciate rangers who helped with transportation on Yellowstone Lake, especially Jackie Sene. The Biodiversity Research Institute (BRI) and Ricketts Conservation Foundation (RCF) supported loon research and monitoring in many ways. Carl Brown was instrumental in much loon and swan work. Raven research in Yellowstone received support from the European Union's Horizon 2020 research and innovation program under the Marie Sklodowska-Curie grant agreement No. 798091, the Max Planck Institute of Animal Behavior in Germany, the National Geographic Society (NGS-61630R-19), and the James W. Ridgeway professorship at the University of Washington.

Clark's nutcracker research in Yellowstone was supported by RCF with funding and participation in the study, particularly Arcata Leavitt and Adam Cupito, who assisted with point counts and nutcracker trapping. Yellowstone National Park and the Yellowstone Center for Resources provided additional project funding and logistical support. Yellowstone Forever, the Track Education Center, and Bob Landis graciously provided housing at various times for the field crew. We thank retired park wildlife biologists and Howard Weinberg for completing the roadside transect work during the 2019–2020 period. We are grateful to Kyla Luketta for field assistance in 2019, Ally Davidge for field assistance in 2020, Tara Durboraw and Blaire Caldwell for field assistance in 2021, and Kira Cassidy for occasional field assistance and for her creative infographics. We thank the University of Colorado Denver for support for this project. All research activities involving Clark's nutcrackers were proposed and approved under University of Colorado Denver IACUC protocol #01025.

We would also like to thank the synergistic enthusiasm of the Yellowstone Center for Resources staff, particularly many members of the Yellowstone Wolf Project, who also have an interest in birds. Notably, they are Hannah Beroske, Brenna Cassidy, Kira Cassidy, Dylan Sanborn, Daniel and Erin Stahler, Jeremy Sunderaj, and Nicole Tatton.

We thank the many photographers and artists who freely and enthusiastically donated their work. In alphabetical order, they are Greg Albrechtsen, Tyler Albrethsen (golden eagle maps), Colby Anton, Keegan Burke, Kira Cassidy (infographics), Lisa Culpepper, Jack DeLap (illustrations), Robert Dimilia, Ronan Donovan, Nick Ferrauolo, Jake Frank, Dana Haines (golden eagle maps), Charlie Hamilton James, Jordan Harrison, Scott Heppel, Cameron Ho, Bob Landis, Connor Meyer, Kyle Moon, Tom Murphy, Andrius Pasukonis, Jim Peaco, Diane Renkin, Dylan Schneider, Ben Silberfarb, and Howard Weinberg.

Alethea Steingisser is talented and patient and a master map maker. She made the map at the start of the book and those introducing each of the three chapters, which add enormously to picturing where the various groups of birds go and what the Yellowstone landscape is like.

Critical to the success and expansion of the Bird Program in Yellowstone, which partially led to the ability to complete this book, are our funders. The National Park Service, through Yellowstone National Park and Yellowstone Forever (all profits from the book will be donated to Yellowstone Forever for the Yellowstone Bird Program), have been our most significant supporters since 2008. We received many donations that helped us build the program past what our baseline support was capable of doing. In alphabetical order, funders over the years have included Deborah Erdman, Bob and Annie Graham, Patrick and Lynn Gurrentz, Carolyn and Scott Heppel, Tom Murphy, Marc Noel, the Meg and Bert Raynes Wildlife Fund, Joe Ricketts, and the Wyoming Wetlands Society.

To all the above and the ones we forgot to include—it is easy to lose track, since citizen science is most robust when it comes to birds—we thank you for increasing our understanding of Yellowstone birds. Know that your efforts will help expand knowledge and conservation of birds everywhere in the future.

ABBREVIATIONS

APHIS	Animal and Plant Health Inspection Service	IGBST	Interagency Grizzly Bear Study Team
ARU	Autonomous Recording Unit	KDE	Kernel Density Estimate
BBS	Breeding Bird Survey	MAPS	Monitoring Avian Productivity and Survivorship
BCC	Bird of Conservation Concern	MOCH	Mountain Chickadee
BMAA	β-methylamino-L-alanine	NBR	National Bison Range
BRI	Biodiversity Research Institute	NPS	National Park Service
CBC	Christmas Bird Count	RCF	Ricketts Conservation Foundation
DDE	dichloro- diphenyl- dichloroethylene	US	United States
DDT	dichloro- diphenyl- trichloroethane	USFWS	United States Fish and Wildlife Service
DNA	Deoxyribonucleic Acid	USGS	United States Geological Survey
ESA	Endangered Species Act	WGFD	Wyoming Game and Fish Department
GPS	Global Positioning System	YFAS	Yellowstone Fisheries and Aquatic Sciences
GYE	Greater Yellowstone Ecosystem	YNP	Yellowstone National Park; Yellowstone
IACUC	Institutional Animal Care and Use Committee	YRI	Yellowstone Raptor Initiative

OPPOSITE PAGE:

YELLOWSTONE NATIONAL PARK (3,471 square miles/8,991 km²) forms the core of the Greater Yellowstone Ecosystem (GYE)—one of the largest nearly intact temperate-zone ecosystems on Earth. Mostly in Wyoming, the park also extends into parts of Montana and Idaho. Much of the long-term research described in this book focuses on the northern range (depicted in purple), a 591-square-mile (1,531-km²) area (Lemke et al. 1998) mostly within the park and defined as the current and historical winter distribution of the northern Yellowstone elk herd (Houston 1982). The park is greatly impacted by its volcanic history (volcano caldera depicted) leading to, in many cases, poor soils. In turn, soil condition impacts vegetation growth and overall biodiversity; for example, lodgepole pine is dominant in much of the caldera, a forest type low in plant and bird diversity.
CARTOGRAPHY BY INFOGRAPHICS LAB, UNIVERSITY OF OREGON

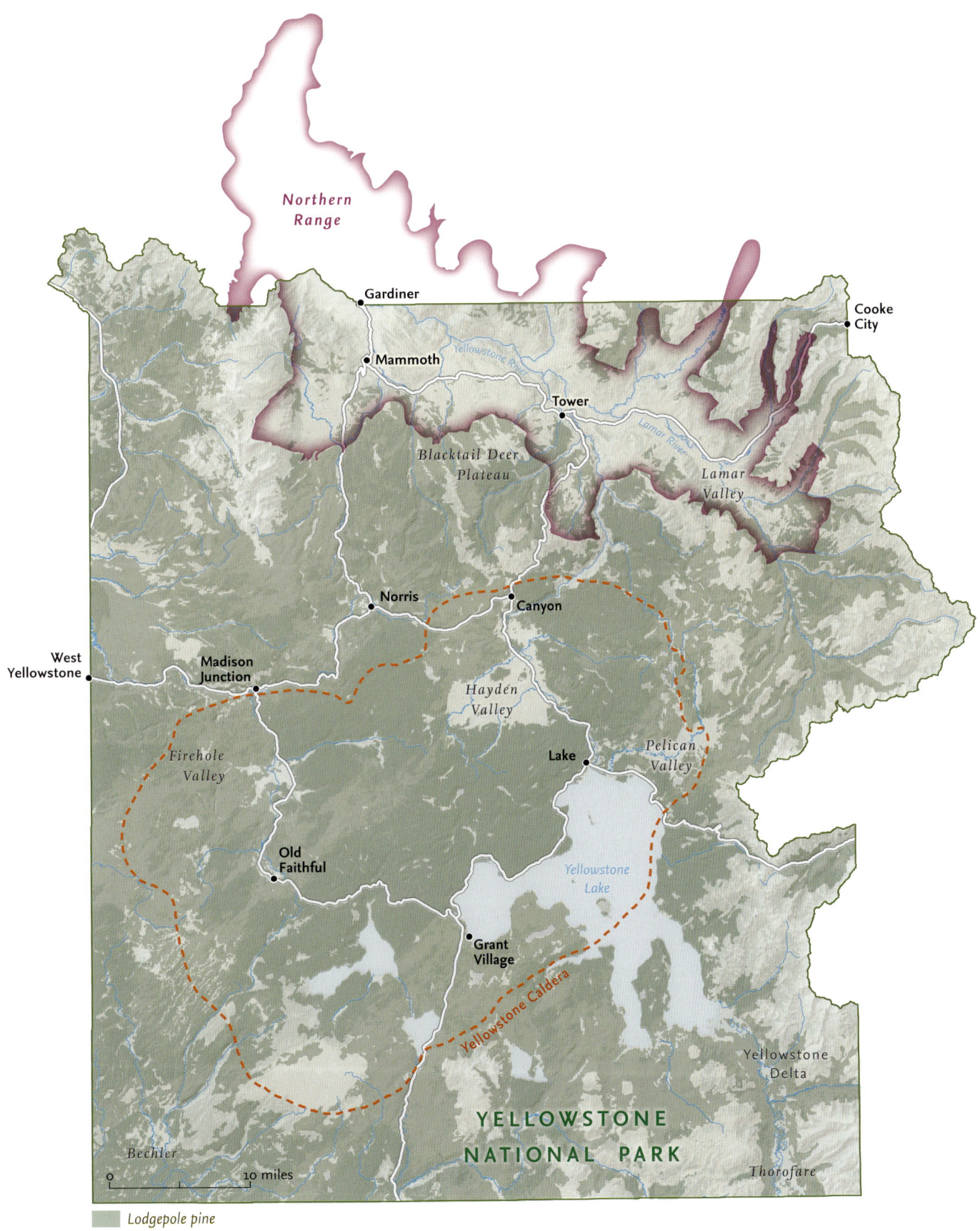

Northern Range

Gardiner

Cooke City

Mammoth

Yellowstone River

Tower

Lamar River

Blacktail Deer Plateau

Lamar Valley

Norris

Canyon

West Yellowstone

Madison Junction

Hayden Valley

Firehole Valley

Lake

Pelican Valley

Old Faithful

Yellowstone Lake

Grant Village

Yellowstone Caldera

Yellowstone Delta

YELLOWSTONE NATIONAL PARK

Bechler

0   10 miles

Thorofare

Lodgepole pine

YELLOWSTONE'S
BIRDS

1 INTRODUCTION

YELLOWSTONE'S GEOLOGICAL HISTORY CREATES DIVERSE HABITATS FOR BIRDS

KATHARINE E. DUFFY

Erupting geysers and steaming hot springs, craggy mountains with jagged summits, river valleys inhabited by bison and elk, deep canyons lined by colorful rock walls, dark conifer forests. Yellowstone National Park (Yellowstone; YNP) encompasses all of these and more. While geysers and other hydrothermal features may be the most famous and recognizable aspects of Yellowstone, it is important to understand the park's geological processes, for they created all of these landforms.

WHY IS AN introduction to the park's geology included in a book on Yellowstone birds? Geology is the bottom line: Geological processes affect soil, which determines the plants that grow on landscapes of that particular geological origin; plants form habitats that provide food, shelter, and nest sites for birds. Geological processes also resulted in lakes, ponds, and other aquatic habitats. Yellowstone's fascinating geological history accounts for a wide variety of landforms and explains the origin and location of habitats for birds and other wildlife.

Yellowstone National Park owes its designation in 1872 as the world's first national park to an immense variety of geysers, fumaroles (steam vents), hot springs, and mudpots. These hydrothermal features, a total of approximately 10,000, resulted from a supervolcano eruption 640,000 years ago. It was preceded by an even larger supervolcano eruption 2.1 million years ago, which is among the largest volcanic eruptions known. The supervolcano occurring 640,000 years ago left a crater, a caldera that occupies an area of 30 miles by 45 miles (48 km × 72 km) in the center of the park (*see page xv*). But don't look for a giant hole in the middle of Yellowstone. Stretching of Earth's crust prior to each volcanic eruption caused the center of Yellowstone to become high plateaus that remained after volcanic ash exploded violently from several vents. Massive flows of rhyolitic lava filled in much of the caldera many thousands of years ago. Gritty, nutrient-poor rhyolite today offers a relatively arid environment with little moisture retention, severely limiting the kinds of trees and other plants that now grow here.

Mountain ranges bracket two sides of the park: the Absaroka Range, along the eastern boundary of the park, and the Gallatin Range, which forms the northwestern boundary. Older volcanic rock called andesite forms the bedrock of the Absarokas and the eastern side of Yellowstone. Soil derived from this andesitic rock provides nutrients that nourish greater plant diversity than soil derived from rhyolite. Yellowstone Lake, 20 miles (32 km) north to south and 14 miles (23 km) east to west, fills much of the southeastern part of Yellowstone. Glaciers melting as recently as 14,000 years ago gouged out part of the

2

FIGURE 1.1 Yellowstone owes its existence as the first national park in the world to the presence of about 10,000 hydrothermal features, including geysers such as Lion (left) and Beehive (right) that frequently erupt, shooting hot water skyward. NPS PHOTO

basin now occupied by Yellowstone Lake and sculpted and scoured the mountains. Glaciers also left behind moraines, ridges of rocks, and other materials transported by these rivers of ice. Glaciers deposited sediments (till) in Hayden Valley and other sites and scattered numerous large boulders, known as erratics, across the northern part of the park. Glacial erratics often serve as nurse rocks by retaining moisture and soil on their leeward side, sheltering tree seedlings that become the large Douglas-fir trees seen today across Lamar Valley.

Much of the interior of Yellowstone lies above 7,500 feet (2,286 m) in elevation. The climate tends toward cold and snowy winters, while the northern tier of the park, referred to as the park's "northern range" (*see map*), has milder winters with less frequent snowfall. Bison, elk, pronghorn, deer, and bighorn sheep inhabit the northern range in winter; in summer, this lower-elevation area provides an array of habitats for nesting birds, from waterfowl in ponds and lakes, to songbirds in shrubby willows lining streams and rivers, and raptors in trees and on rock outcrops.

YELLOWSTONE'S GEOLOGICAL HISTORY CREATES DIVERSE HABITATS FOR BIRDS

LANDFORMS AND HABITATS

HYDROTHERMAL BASINS

The park's justly famous hydrothermal system of geysers (Fig. 1.1), hot springs, steam vents, and mudpots owes its existence to the Yellowstone volcano. Beneath Yellowstone, partially molten rock from below Earth's crust (magma) continues to release tremendous amounts of heat that causes groundwater from rain and melting snow to become superheated and erupt under pressure at the surface as geysers. Superheated water may reach the surface as hot springs or may release steam at the surface, forming steam vents (fumaroles). Mudpots, the fourth type of thermal feature, are acidic and contain minimal water. Hydrothermal basins, many located along the edges of the caldera, host birds year-round, but especially in migration and summer, due to the ubiquitous presence of water and insects.

FORESTS

Most (80%) of Yellowstone is forested. Vast forests of lodgepole pine cloak much of the caldera (Fig. 1.2) because these pines can tolerate drought and survive on the dry, nutrient-poor rhyolite of the lava flows (*see map*). Because of its growth as an effective monoculture, some locals refer to lodgepole pine as "*Pinus monotonous*" rather than using its real scientific name, *Pinus contorta*. Lodgepole pine forests have low diversity of tree, shrub, and ground-hugging plants. Consequently, these forests harbor fewer bird species: mountain chickadees (Fig. 1.3), brown creepers, red-breasted nuthatches, and red crossbills occur year-round, with yellow-rumped warblers, ruby-crowned kinglets, dark-eyed juncos, and American robins included among the species using these forests in summer.

Extensive forests comprised mainly of Douglas-fir thrive where sufficient moisture occurs to sustain trees in the northern part of the park. Other tree species occur with Douglas-fir, with large numbers of Engelmann spruce and subalpine fir occurring at higher elevations. Subalpine forests blanket higher elevations of the

FIGURE 1.2 Within the 640,000-year-old caldera that occupies the center of Yellowstone, lodgepole pine trees dominate. These pioneer trees can grow on the nutrient-poor rhyolite comprising the lava flows that have filled the caldera since the last eruption of the supervolcano. NPS PHOTO/DIANE RENKIN

Absarokas and the Gallatins, as well as Dunraven Pass, Craig Pass, Sylvan Pass, and areas near the Northeast Entrance. Forests of subalpine fir, Engelmann spruce, lodgepole pine, and, above 7,000 feet (2,133 m) in elevation, whitebark pine are interspersed with grassy and wildflower-filled meadows, where small stands of aspen also grow. These forests have nesting boreal owls and great gray owls, hermit thrushes and Swainson's thrushes, golden-crowned kinglets, northern goshawks, and other accipiters. In past centuries, large, lightning-caused fires rarely occurred in these higher-elevation forests due to increased moisture levels from abundant snow and rain, so these forests contain trees that are hundreds of years old. With climate change, moisture is decreasing, and summer temperatures are increasing, rendering these irreplaceable old-growth forests extremely vulnerable. Bird species dependent on forests of mature mixed conifers will have no suitable habitat when these forests burn.

ASPENS AND COTTONWOODS

Most of the trees found in Yellowstone are conifers, but aspens grow in stands, especially on the northern range and in small groups throughout the park. Woodpeckers

favor aspens because the wood is softer and older trees tend to have heart rot, so it's easier to drill nest cavities in these trees. Secondary cavity nesters (tree swallows, house wrens, mountain bluebirds, American kestrels, northern pygmy-owls, and other small owls) select old woodpecker cavities as nest sites. Black cottonwoods and narrow-leaved cottonwoods, the other deciduous trees found in the park, typically grow along streams and rivers on the northern range. A wide variety of birds nest in these deciduous trees.

AQUATIC HABITATS AND WETLANDS

Although merely 5% of the park consists of lakes, ponds, rivers, streams, and wetlands, these habitats are extremely vital to birds all year long. Yellowstone Lake and other large lakes (Lewis, Shoshone, and Heart) attract waterfowl, loons, and grebes; nesting occurs at the edges of the lakes. Waterbirds also congregate in the lakes during migration. Shorebirds find mud flats at the lakes' edges productive for foraging, especially during migration. Waterfowl seek acceptable nesting sites in numerous smaller lakes, ponds, and wetlands that dot Yellowstone. Dabbling ducks forage in shallow water, while diving ducks frequent deeper water. Trumpeter swans and common loons choose a few select lakes for nesting, but Canada geese seem to flourish everywhere. Located adjacent to the road from Mammoth, Wyoming, to Tower, Floating Island Lake, and Blacktail Ponds provide easily visible examples of waterbirds' use of park wetlands.

In Hayden Valley, glacial till and lake sediments retain moisture, nurturing growth of wetland plants adjacent to the meandering Yellowstone River. Waterfowl, soras, and Wilson's snipe nest here, while additional shorebirds stop during migration. Marshes flank other park rivers such as the Lamar, Madison, Firehole, and Gibbon. Ribbons of grasses and other marsh vegetation also grow next to the largest creeks, including Slough, Soda

FIGURE 1.3 Look for occasional mountain chickadees in the caldera; they are one of very few bird species to nest and spend time year-round in dense lodgepole pine forests. PHOTO BY CAMERON HO

FIGURE 1.4 Grasslands and sagebrush steppe of the wide Lamar Valley attract nesting songbirds, along with bison, pronghorns, and their predators, wolves and grizzly bears. NPS PHOTO/JAKE FRANK

Butte, and Blacktail. Swan Lake Flat, south of Mammoth, is an extensive marsh favored by nesting waterfowl, savannah sparrows, many other songbirds, and hunting raptors in summer. Dense willows, shrubs that can grow taller than 10 feet (3 m), thrive in some wetlands adjacent to creeks, including along the Yellowstone River in the Thorofare, near Cougar Creek south of the Gallatin Mountains, adjacent to Blacktail Deer Creek east of Mammoth, and Willow Park, south of Swan Lake Flat. These willow stands and associated marsh plants support a wide variety of nesting warblers, vireos, sparrows, flycatchers, and other birds.

SAGEBRUSH STEPPE

As elevation gradually increases above creeks and rivers, these areas typically transition to sagebrush steppe, as seen in the Hayden and Lamar valleys (Fig. 1.4), along the Madison River near the park's West Entrance, and many other areas throughout the park. Although sagebrush steppe can occur in proximity to wetlands, this habitat is quite dry. Vesper sparrows, green-tailed towhees, and sage thrashers nest here, while red-tailed and Swainson's hawks search for prey in sagebrush steppe.

CANYONS

Where large lava flows met, running water from snowmelt and precipitation eventually eroded the junction. Today, the Firehole River runs in a canyon between walls of volcanic rock. The Grand Canyon of the Yellowstone (Fig. 1.5), the most spectacular of the park's canyons, formed after hydrothermal gases and hot water weakened overlying volcanic rock that flowed after the last caldera eruption. Over thousands of years, the Yellowstone River carved a canyon that is 20 miles (32 km) long, 1,000 feet (305 m) deep, and 1,500–4,000 feet (457–1,219 m) wide through this weakened rock. Oxidation of iron compounds in the hydrothermally altered rock causes the dramatic colors seen in the canyon. In the Grand Canyon, and wherever lava flows, rocky outcrops, and ridges occur, raptors, including golden eagles, peregrine falcons, prairie falcons, and red-tailed hawks claim horizontal ledges and holes in the rock walls for nests, while osprey place their cumbersome stick nests on rock pinnacles.

The dry climate of the Black Canyon of the Yellowstone, from Mammoth east to Tower, supports open stands of Douglas-fir, especially on north-facing aspects. Rocky Mountain junipers, limber pines, and sagebrush survive

in the driest parts, usually south-facing sites. Clark's nutcrackers stash seeds of limber pines for winter food and inhabit these locations year-round. Townsend's solitaires winter where juniper berries grow profusely.

ALPINE AND SUBALPINE MEADOWS

At higher elevations in the Gallatins and Absarokas, and at the summit of Mount Washburn, forests give way to grassy meadows of colorful alpine wildflowers interspersed with rock outcrops. Above where trees can grow, black rosy finches, American pipits, horned larks, and other songbirds nest, while golden eagles, prairie falcons, and other raptors search for prey.

Today, these habitats found throughout Yellowstone's 2.2 million acres (890,308 ha) support at least 150 bird species that nest throughout the park. Numerous other bird species frequent the park during spring and fall migration, with a smaller number of species inhabiting Yellowstone in winter and year-round.

The geological history and concepts laid out above are in agreement with Yellowstone National Park, 2021, *Yellowstone Resources and Issues Handbook: 2021*, Yellowstone National Park, WY.

FIGURE 1.5 The spectacularly colorful walls of the Grand Canyon of the Yellowstone resulted from oxidation of hydrothermally altered rock. NPS PHOTO/JAKE FRANK

Visit the Yellowstone's Birds website (https://press.princeton.edu/resources/yellowstones-birds-video-collection/v1) to watch an interview with Douglas W. Smith.

YELLOWSTONE BIRDS

KATHARINE E. DUFFY, DOUGLAS W. SMITH, LAUREN E. WALKER, and KIRA A. CASSIDY

Why watch birds? They can fly! They are adorned in a rainbow of colors! They sing recognizable songs that can be beautifully melodic or strange. Many are active during the day, as we are. Birds you find in Yellowstone (Table 2.1) could also be found in other places you travel or even in your own neighborhood. Watching birds is an easy way to connect with nature; do it daily, and you're likely to feel better and view the world differently. (They say solitude is the key to a fulfilled and calm life. Throw in a morning alone with the birds, and you'll feel even better. Actually, some mornings, nothing feels better.)

REMINDER: PLEASE APPRECIATE BIRDS RESPONSIBLY AND RESPECTFULLY

- *The use of audio bird calls is illegal in the park.* Broadcasting bird territorial calls and other bird vocalizations harasses birds. These sounds cause birds to waste precious energy and time responding to a potential rival within their nest territory. Broadcasting any sound can alarm birds, even when they are not nesting, and can negatively impact other wildlife.

- *Approaching on foot within 25 yards (23 m) of wildlife, including birds, is prohibited in Yellowstone.* This includes approaching and pursuing birds to photograph them. When large groups of photographers and other people approach and surround birds, such as great gray owls, the birds may be prevented from successfully finding food or chased from areas where food is available.

- *All birds are sensitive to human disturbance, especially during the nesting season.* The presence of people often causes birds to leave an area. Birds of all species already face numerous challenges to successful nesting, and close human presence can cause nest abandonment. When an adult is chased off a nest, eggs and nestlings can die from predation or exposure to heat, cold, or precipitation. This has been documented for common loons in the park. Nestlings will die when adults are prevented from providing them with sufficient food.

- *Do not feed birds or any wildlife.* *Feeding animals is prohibited in the park.*

- *You can bird anywhere in the park.* There are four main habitat types: grassland, deciduous trees and shrubs (including riparian species), coniferous forests, and water. The number of species seen in each habitat type varies greatly; deciduous woody vegetation yields the highest number of sightings. For example, a multi-day birding trip through coniferous forest, especially within the caldera, may only produce a bird list of around 30 species or fewer. Expectations for what you will see are important. The best way to maximize your list— if that is what you are into—is to look for birds in all habitat types.

FIGURE 2.1

PLEASE APPRECIATE BIRDS RESPONSIBLY AND RESPECTFULLY

Do Not...

Do...

Stay at least 25 yards (23 m) away from wildlife, including birds. Approaching closer on foot is prohibited in YNP.

This includes approaching and pursuing birds to photograph them. When large groups of photographers and other people approach and surround birds, the birds may be prevented from successfully finding food or chased from areas where food is available.

Do not feed birds or any wildlife.

Feeding animals is prohibited in YNP. All wild animals need to learn how to find wild foods. Enjoy watching this fascinating behavior as an animal problem-solves and finds its own food.

All birds are sensitive to human disturbance, especially during the nesting season.

Birds of all species already face numerous challenges to successful nesting and close human presence can cause nest abandonment. When an adult is chased off a nest, eggs and nestlings can die from predation or exposure to heat, cold, or precipitation. Nestlings will die when adults are prevented from providing them with sufficient food.

Do not use audio bird calls.

They are illegal in YNP. Broadcasting bird territorial calls and other bird vocalizations harasses birds. These sounds cause birds to waste precious energy and time responding to a potential rival. Broadcasting any sound can alarm birds, even when they are not nesting, and can negatively impact other wildlife.

ARTWORK BY KIRA A. CASSIDY

TABLE 2.1   YELLOWSTONE NATIONAL PARK BIRD CHECKLIST, 2021

	SEASON						SEASON				
	B/b	Sp	S	F	W		B/b	Sp	S	F	W
GEESE, SWANS & DUCKS						**Hooded Merganser** *(Lophodytes cucullatus)*	B	R	O	R	O
Snow Goose *(Anser caerulescens)*		R		R		**Common Merganser** *(Mergus merganser)*	B	C	C	C	C
Ross's Goose *(Anser rossii)*		R	R	R		**Red-breasted Merganser** *(Mergus serrator)*		U		U	
Canada Goose *(Branta canadensis)*	B	A	A	A	C	**Ruddy Duck** *(Oxyura jamaicensis)*	B	C	C	C	
Trumpeter Swan *(Cygnus buccinator)*	B	U	U	U	C	**GROUSE**					
Tundra Swan *(Cygnus columbianus)*		U		U		**Ruffed Grouse** *(Bonasa umbellus)*	B	C	C	C	R
Wood Duck *(Aix sponsa)*		R		R		**Dusky Grouse** *(Dendragapus obscurus)*	B	C	C	C	R
Blue-winged Teal *(Spatula discors)*	B	U	U	U	O	**GREBES**					
Cinnamon Teal *(Spatula cyanoptera)*	B	C	C	C		**Pied-billed Grebe** *(Podilymbus podiceps)*	B	C	C	C	O
Northern Shoveler *(Spatula clypeata)*	B	C	C	C	O	**Horned Grebe** *(Podiceps auritus)*	B	R	O	R	
Gadwall *(Mareca strepera)*	B	C	C	C	U	**Red-necked Grebe** *(Podiceps grisegena)*	B	R	R	R	
American Wigeon *(Mareca americana)*	B	C	C	C	U	**Eared Grebe** *(Podiceps nigricollis)*	B	U	U	U	
Mallard *(Anas platyrhynchos)*	B	A	A	A	A	**Western Grebe** *(Aechmophorus occidentalis)*	b	R	R	R	
Northern Pintail *(Anas acuta)*	B	C	C	C	U	**Clark's Grebe** *(Aechmophorus clarkii)*		R	R	R	
Green-winged Teal *(Anas crecca)*	B	C	C	C	U	**PIGEONS & DOVES**					
Canvasback *(Aythya valisineria)*	B	U	U	U		*Rock Pigeon *(Columba livia)*	B	U	U	U	U
Redhead *(Aythya americana)*	B	U	U	U		*Eurasian Collared-Dove *(Streptopelia decaocto)*		R	R	R	
Ring-necked Duck *(Aythya collaris)*	B	C	C	C	O	Mourning Dove *(Zenaida macroura)*	B	R	R	R	
Greater Scaup *(Aythya marila)*		R		R		**NIGHTHAWKS**					
Lesser Scaup *(Aythya affinis)*	B	C	C	C	O	**Common Nighthawk** *(Chordeiles minor)*	B		C	C	
Harlequin Duck *(Histrionicus histrionicus)*	B	U	U	U							
Bufflehead *(Bucephala albeola)*	B	C	C	C	C						
Common Goldeneye *(Bucephala clangula)*		C			C						
Barrow's Goldeneye *(Bucephala islandica)*	B	A	A	A	A						

(CONTINUED OVERLEAF)

YELLOWSTONE BIRDS

TABLE 2.1 YELLOWSTONE NATIONAL PARK BIRD CHECKLIST, 2021 (CONTINUED)

	SEASON						SEASON				
	B/b	Sp	S	F	W		B/b	Sp	S	F	W
SWIFTS & HUMMINGBIRDS						Long-billed Curlew (Numenius americanus)	B	R	R	R	
White-throated Swift (Aeronautes saxatalis)	B	U	U	U		Marbled Godwit (Limosa fedoa)		R		R	
Broad-tailed Hummingbird (Selasphorus platycercus)	B		R			Ruddy Turnstone (Arenaria interpres)		R		R	
						Sanderling (Calidris alba)		R		R	
Rufous Hummingbird (Selasphorus rufus)	B		R			Baird's Sandpiper (Calidris bairdii)		U		U	
						Least Sandpiper (Calidris minutilla)		R		R	
Calliope Hummingbird (Selasphorus calliope)	B		R			White-rumped Sandpiper (Calidris fuscicollis)		R		R	
RAILS, COOTS, & CRANES						Pectoral Sandpiper (Calidris melanotos)		R		R	
Virginia Rail (Rallus limicola)	B	R	R	R	R	Semipalmated Sandpiper (Calidris pusilla)		R		R	
Sora (Porzana carolina)	B	C	C	C							
American Coot (Fulica americana)	B	A	A	A	U	Western Sandpiper (Calidris mauri)		R		R	
Sandhill Crane (Antigone canadensis)	B	C	C	C		Short-billed Dowitcher (Limnodromus griseus)		R		R	
STILTS & AVOCETS						Long-billed Dowitcher (Limnodromus scolopaceus)		U		U	
Black-necked Stilt (Himantopus mexicanus)		R		R		Wilson's Snipe (Gallinago delicata)	B	C	C	C	R
American Avocet (Recurvirostra americana)		R		R		Spotted Sandpiper (Actitis macularius)	B	C	C	C	
PLOVERS						Solitary Sandpiper (Tringa solitaria)		R		R	
Black-bellied Plover (Pluvialis squatarola)		R		R		Lesser Yellowlegs (Tringa flavipes)		R	R	R	
Killdeer (Charadrius vociferus)	B	C	C	C	U	Willet (Tringa semipalmata)		U	U	U	
Semipalmated Plover (Charadrius semipalmatus)		R		R		Greater Yellowlegs (Tringa melanoleuca)		U		U	
SANDPIPERS & ALLIES						Wilson's Phalarope (Phalaropus tricolor)	B	U	U	U	
Upland Sandpiper (Bartramia longicauda)		R	R	R		Red-necked Phalarope (Phalaropus lobatus)		R		R	

GULLS & TERNS	B/b	Sp	S	F	W
Bonaparte's Gull (*Chroicocephalus philadelphia*)		R		R	
Franklin's Gull (*Leucophaeus pipixcan*)		R	R	R	
Ring-billed Gull (*Larus delawarensis*)		C	C	C	
California Gull (*Larus californicus*)	B	C	C	C	
Herring Gull (*Larus argentatus*)		R		R	
Caspian Tern (*Hydroprogne caspia*)	B	R	R	R	
Common Tern (*Sterna hirundo*)		R		R	
Forster's Tern (*Sterna forsteri*)	B	R	R	R	

LOONS	B/b	Sp	S	F	W
Common Loon (*Gavia immer*)	B	U	U	U	

CORMORANTS	B/b	Sp	S	F	W
Double-crested Cormorant (*Phalacrocorax auritus*)	B	U	U	U	

PELICANS, HERONS, & IBISES	B/b	Sp	S	F	W
American White Pelican (*Pelecanus erythrorhynchos*)	B	C	C	C	
Great Blue Heron (*Ardea herodias*)	B	C	C	C	R
Black-crowned Night-Heron (*Nycticorax nycticorax*)		R	R	R	
White-faced Ibis (*Plegadis chihi*)		R		R	

VULTURES	B/b	Sp	S	F	W
Turkey Vulture (*Cathartes aura*)		U	U	U	

HAWKS, EAGLES, & ALLIES	B/b	Sp	S	F	W
Osprey (*Pandion haliaetus*)	B	C	C	C	

SEASON	B/b	Sp	S	F	W
Golden Eagle (*Aquila chrysaetos*)	B	U	U	U	U
Northern Harrier (*Circus hudsonius*)	B	U	U	U	R
Sharp-shinned Hawk (*Accipiter striatus*)	B	U	U	U	R
Cooper's Hawk (*Accipiter cooperii*)	B	U	U	U	R
Northern Goshawk (*Accipiter gentilis*)	B	U	U	U	R
Bald Eagle (*Haliaeetus leucocephalus*)	B	C	C	C	C
Broad-winged Hawk (*Buteo platypterus*)		R		R	
Swainson's Hawk (*Buteo swainsoni*)	B	U	C	C	
Red-tailed Hawk (*Buteo jamaicensis*)	B	C	A	C	R
Rough-legged Hawk (*Buteo lagopus*)		U		U	R
Ferruginous Hawk (*Buteo regalis*)		U		U	

OWLS	B/b	Sp	S	F	W
Great Horned Owl (*Bubo virginianus*)	B	C	C	C	C
Northern Pygmy-Owl (*Glaucidium gnoma*)	B	R	R	R	U
Great Gray Owl (*Strix nebulosa*)	B	U	U	U	R
Long-eared Owl (*Asio otus*)	B	R	R	R	
Short-eared Owl (*Asio flammeus*)	B	R	R	R	
Boreal Owl (*Aegolius funereus*)	B	R	R	R	R
Northern Saw-whet Owl (*Aegolius acadicus*)	B	R	R	R	R

KINGFISHERS	B/b	Sp	S	F	W
Belted Kingfisher (*Megaceryle alcyon*)	B	U	U	U	R

(CONTINUED OVERLEAF)

YELLOWSTONE BIRDS

TABLE 2.1 YELLOWSTONE NATIONAL PARK BIRD CHECKLIST, 2021 (CONTINUED)

	SEASON						SEASON				
	B/b	Sp	S	F	W		B/b	Sp	S	F	W
WOODPECKERS & ALLIES						Olive-sided Flycatcher (Contopus cooperi)	B	C	C	C	
Lewis's Woodpecker (Melanerpes lewis)	B	R	R	R		Western Wood-Pewee (Contopus sordidulus)	B	U	U	U	
Williamson's Sapsucker (Sphyrapicus thyroideus)	B	U	U	U		Willow Flycatcher (Empidonax traillii)	B	U	U	U	
Red-naped Sapsucker (Sphyrapicus nuchalis)	B	U	U	U		Hammond's Flycatcher (Empidonax hammondii)	B	U	U	U	
American Three-toed Woodpecker (Picoides dorsalis)	B	U	U	U	U	Dusky Flycatcher (Empidonax oberholseri)	B	C	C	C	
Black-backed Woodpecker (Picoides arcticus)	B	R	R	R	R	Cordilleran Flycatcher (Empidonax occidentalis)	B	R	R	R	
Downy Woodpecker (Dryobates pubescens)	B	U	U	U	U	**SHRIKES**					
Hairy Woodpecker (Dryobates villosus)	B	C	C	C	C	Loggerhead Shrike (Lanius ludovicianus)		R	R	R	
Northern Flicker (Colaptes auratus)	B	C	C	C	R	Northern Shrike (Lanius borealis)				U	U
Pileated Woodpecker (Dryocopus pileatus)	B	R	R	R	R	**VIREOS**					
FALCONS						Warbling Vireo (Vireo gilvus)	B	C	C	C	
American Kestrel (Falco sparverius)	B	U	U	U	R	**JAYS, MAGPIES, CROWS, & RAVENS**					
Merlin (Falco columbarius)		R		R		Canada Jay (Perisoreus canadensis)	B	U	U	U	U
Peregrine Falcon (Falco peregrinus)	B	U	U	U		Pinyon Jay (Gymnorhinus cyanocephalus)		R	R	R	R
Prairie Falcon (Falco mexicanus)	B	U	U	U	R	Steller's Jay (Cyanocitta stelleri)	B	U	U	U	U
FLYCATCHERS						Blue Jay (Cyanocitta cristata)		R		O	
Western Kingbird (Tyrannus verticalis)		R	R	R		Clark's Nutcracker (Nucifraga columbiana)	B	A	A	A	A
Eastern Kingbird (Tyrannus tyrannus)		R	R	R		Black-billed Magpie (Pica hudsonia)	B	A	A	A	A
						American Crow (Corvus brachyrhynchos)	B	U	U	U	
						Common Raven (Corvus corax)	B	A	A	A	A

LARKS	B/b	Sp	S	F	W
Horned Lark (*Eremophila alpestris*)	B	U	U	U	U

SWALLOWS	B/b	Sp	S	F	W
Bank Swallow (*Riparia riparia*)	B	R	U	R	
Tree Swallow (*Tachycineta bicolor*)	B	U	C	U	
Violet-green Swallow (*Tachycineta thalassina*)	B	U	C	U	
Northern Rough-winged Swallow (*Stelgidopteryx serripennis*)	B	R	U	R	
Barn Swallow (*Hirundo rustica*)	B	R	U	R	
Cliff Swallow (*Petrochelidon pyrrhonota*)	B	U	C	U	

CHICKADEES, NUTHATCHES & ALLIES	B/b	Sp	S	F	W
Black-capped Chickadee (*Poecile atricapillus*)	B	U	U	U	U
Mountain Chickadee (*Poecile gambeli*)	B	A	A	A	A
Red-breasted Nuthatch (*Sitta canadensis*)	B	C	C	C	C
White-breasted Nuthatch (*Sitta carolinensis*)	B	U	U	U	U
Brown Creeper (*Certhia americana*)	B	U	U	U	U

WRENS	B/b	Sp	S	F	W
Rock Wren (*Salpinctes obsoletus*)	B	U	U	U	
House Wren (*Troglodytes aedon*)	B	C	C	C	
Marsh Wren (*Cistothorus palustris*)	B	R	R	R	

DIPPERS	B/b	Sp	S	F	W
American Dipper (*Cinclus mexicanus*)	B	U	U	U	U

KINGLETS	B/b	Sp	S	F	W
Golden-crowned Kinglet (*Regulus satrapa*)	B	R	U	R	R
Ruby-crowned Kinglet (*Regulus calendula*)	B	A	A	A	

THRUSHES	B/b	Sp	S	F	W
Western Bluebird (*Sialia mexicana*)		R	R	R	
Mountain Bluebird (*Sialia currucoides*)	B	C	C	C	
Townsend's Solitaire (*Myadestes townsendi*)	B	C	C	C	C
Swainson's Thrush (*Catharus ustulatus*)	B	U	U	U	
Hermit Thrush (*Catharus guttatus*)	B	C	C	C	
American Robin (*Turdus migratorius*)	B	A	A	A	U

CATBIRDS & THRASHERS	B/b	Sp	S	F	W
Gray Catbird (*Dumetella carolinensis*)	B	R	U	U	
Sage Thrasher (*Oreoscoptes montanus*)	B	U	U	U	

STARLINGS	B/b	Sp	S	F	W
*European Starling (*Sturnus vulgaris*)	B	U	U	U	R

WAXWINGS	B/b	Sp	S	F	W
Bohemian Waxwing (*Bombycilla garrulus*)				C	C
Cedar Waxwing (*Bombycilla cedrorum*)	B	U	U	U	U

OLD WORLD SPARROWS	B/b	Sp	S	F	W
*House Sparrow (*Passer domesticus*)	B	C	C	C	U

(CONTINUED OVERLEAF)

YELLOWSTONE BIRDS

TABLE 2.1 YELLOWSTONE NATIONAL PARK BIRD CHECKLIST, 2021 (CONTINUED)

	SEASON						SEASON				
	B/b	Sp	S	F	W		B/b	Sp	S	F	W
PIPITS						Fox Sparrow (*Passerella iliaca*)	B	R	R	R	
American Pipit (*Anthus rubescens*)	B	R	R	R		American Tree Sparrow (*Spizelloides arborea*)					U
FINCHES & ALLIES						Dark-eyed Junco (*Junco hyemalis*)	B	A	A	A	O
Evening Grosbeak (*Coccothraustes vespertinus*)	B	O	O	O	O	White-crowned Sparrow (*Zonotrichia leucophrys*)	B	C	C	C	
Pine Grosbeak (*Pinicola enucleator*)	B	U	U	U	U	Vesper Sparrow (*Pooecetes gramineus*)	B	C	C	C	
Gray-crowned Rosy-Finch (*Leucosticte tephrocotis*)		O	O	O	R	Savannah Sparrow (*Passerculus sandwichensis*)	B	C	C	C	
Black-rosy Finch (*Leucosticte atrata*)	B	U	U	U	R	Song Sparrow (*Melospiza melodia*)	B	C	C	C	
House Finch (*Haemorhous mexicanus*)		R	R	R	R	Lincoln's Sparrow (*Melospiza lincolnii*)	B	C	C	C	
Cassin's Finch (*Haemorhous cassinii*)	B	C	C	C	R	Green-tailed Towhee (*Pipilo chlorurus*)	B	U	U	U	
Common Redpoll (*Acanthis flammea*)		R			U	Spotted Towhee (*Pipilo maculatus*)	B	R	R	R	
Red Crossbill (*Loxia curvirostra*)	B	U	U	U	U	**BLACKBIRDS & ALLIES**					
White-winged Crossbill (*Loxia leucoptera*)	b	R	R	R	R	Yellow-headed Blackbird (*Xanthocephalus xanthocephalus*)	B	C	C	C	
Pine Siskin (*Spinus pinus*)	B	C	C	C	R	Western Meadowlark (*Sturnella neglecta*)	B	C	C	C	
American Goldfinch (*Spinus tristis*)	B	R	R	R	R	Bullock's Oriole (*Icterus bullockii*)	B	R	R	R	
SNOW BUNTINGS						Red-winged Blackbird (*Agelaius phoeniceus*)	B	C	C	C	O
Snow Bunting (*Plectrophenax nivalis*)					O	Brown-headed Cowbird (*Molothrus ater*)	B	C	C	C	
SPARROWS						Brewer's Blackbird (*Euphagus cyanocephalus*)	B	C	C	C	
Grasshopper Sparrow (*Ammodramus savannarum*)			O								
Lark Sparrow (*Chondestes grammacus*)		O	O	O							
Chipping Sparrow (*Spizella passerina*)	B	A	A	A							
Brewer's Sparrow (*Spizella breweri*)	B	C	C	C							

WARBLERS	B/b	Sp	S	F	W
Northern Waterthrush (*Parkesia noveboracensis*)	b	R	R	R	
Black-and-white Warbler (*Mniotilta varia*)		R	O	R	
Tennessee Warbler (*Leiothlypis peregrina*)		R		R	
Orange-crowned Warbler (*Leiothlypis celata*)	B	U	U	U	
Nashville Warbler (*Leiothlypis ruficapilla*)		R		R	
MacGillivray's Warbler (*Geothlypis tolmiei*)	B	U	U	U	
Common Yellowthroat (*Geothlypis trichas*)	B	C	C	C	
American Redstart (*Setophaga ruticilla*)	B	R	R	R	

	B/b	Sp	S	F	W
Blackburnian Warbler (*Setophaga fusca*)		R		R	
Yellow Warbler (*Setophaga petechia*)	B	C	C	C	
Blackpoll Warbler (*Setophaga striata*)		R		R	
Yellow-rumped Warbler (*Setophaga coronata*)	B	A	A	A	
Townsend's Warbler (*Setophaga townsendi*)		R		R	
Wilson's Warbler (*Cardellina pusilla*)	B	U	U	U	
TANAGERS & ALLIES					
Western Tanager (*Piranga ludoviciana*)	B	C	C	C	
Black-headed Grosbeak (*Pheucticus melanocephalus*)	b	R	R	R	
Lazuli Bunting (*Passerina amoena*)	B	C	C	C	

USING THIS CHECKLIST

* Indicates non-native species.

BREEDING STATUS

(B) **Breeding:** Confirmed as breeding in Yellowstone including historic breeding.

(b) **Unconfirmed breeding:** Suspected of breeding, but not confirmed by eggs or young.

SEASONS

(Sp) **Spring:** mid-March–early June

(S) **Summer:** early June–mid-August

(F) **Fall:** mid-August–November

(W) **Winter:** December–mid-March

ABUNDANCE CATEGORIES

(A) **Abundant:** Found in moderate to large numbers and easily found in appropriate habitat.

(C) **Common:** Found in moderate numbers and usually easy to find in appropriate habitat.

(U) **Uncommon:** Found in small numbers and usually, but not always, found with some effort in appropriate habitat.

(R) **Rare:** Occurs annually in very small numbers or in a very restricted habitat. Difficult to find.

(O) **Occasional:** Occurs in some years, but not every year.

YELLOWSTONE'S BIRDS

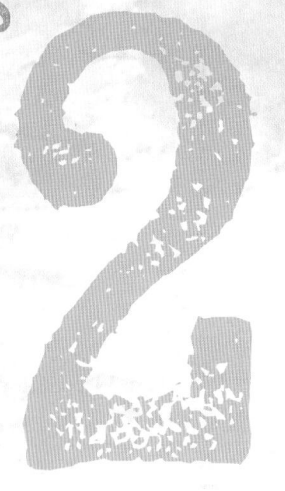

2 BIRDING IN YELLOWSTONE

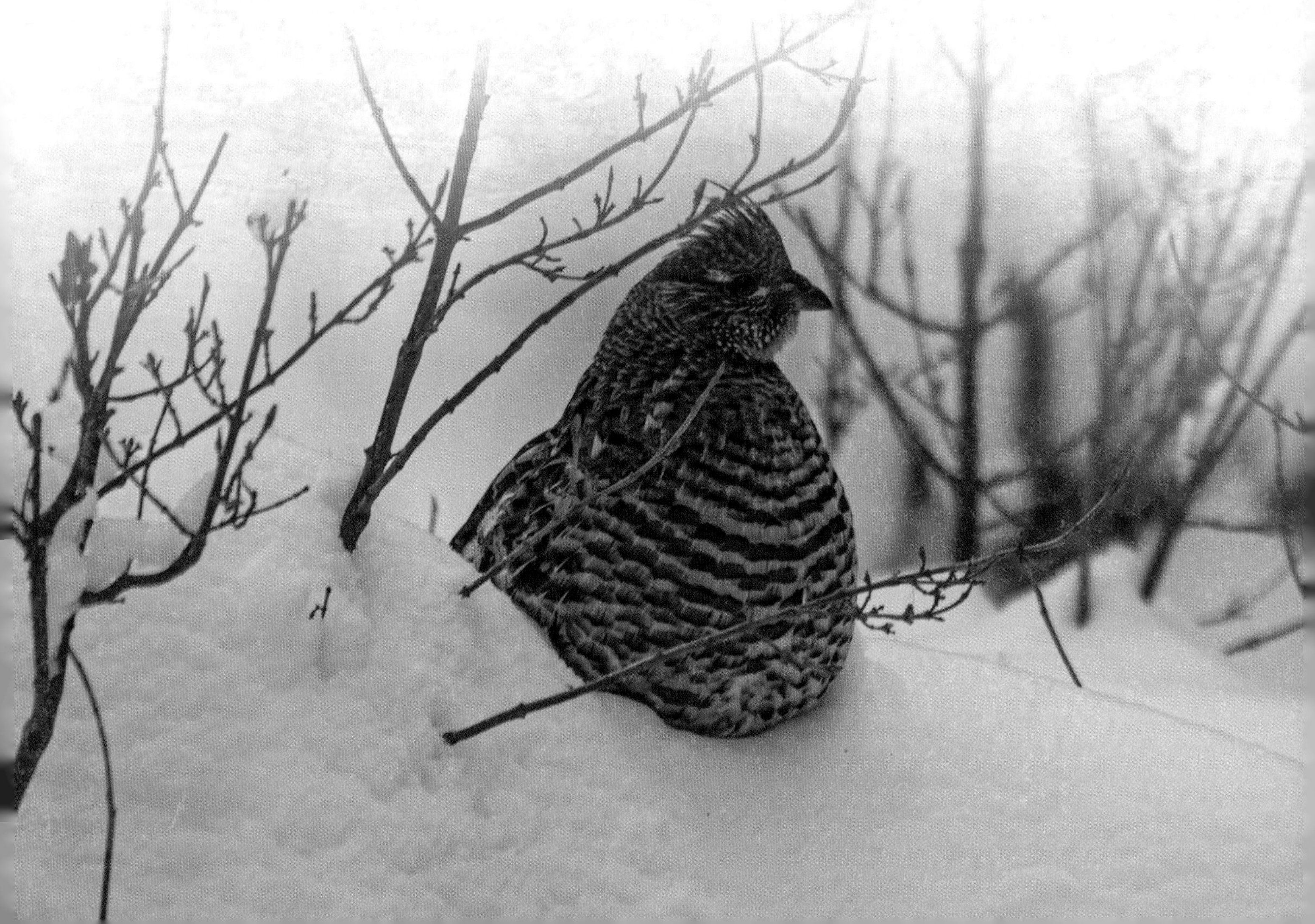

ROADSIDE BIRDING ON YELLOWSTONE'S EAST SIDE

KATHARINE E. DUFFY

The east side of Yellowstone National Park (Yellowstone; YNP) offers incredibly productive and delightfully easy birding. Numerous paved pullouts provide for birding from a vehicle or next to it. Some pullouts have accessible parking spaces, curb cuts, accessible vault toilets, and paved walks for wheelchair access and closer viewing. Use binoculars and spotting scopes to aid identification of waterfowl, other waterbirds, and shorebirds in this wildlife-rich section of Yellowstone. For current details on accessibility in Yellowstone, please see www.nps.gov/yell/planyourvisit/accessibility.htm.

HAYDEN VALLEY

From about 2 miles (3.2 km) south of Canyon Junction nearly to Fishing Bridge Junction, the Grand Loop Road lies closely adjacent to the Yellowstone River as the river winds its way gently through Hayden Valley (Fig. 3.1). Lush grasslands flourish next to the river, with sagebrush growing as elevation increases, even slightly. Numerous pullouts on both sides of the road allow views, from within or near the vehicle, of the Yellowstone River and Alum, Elk Antler, Trout, and other creeks flowing into the river. A vault toilet and accessible parking are located on the west side of the road at Trout Creek, but the path to the restroom has eroded and so isn't wheelchair accessible.

This part of the Yellowstone River has many nesting dabbling ducks (mallards, American wigeon, gadwalls, green-winged teal; Fig. 3.2), diving ducks (lesser scaup, Barrow's goldeneyes), and ubiquitous Canada geese. Trumpeter swans released by the park and other trumpeter swans often congregate near Alum Creek. Shorebirds, including greater yellowlegs, American avocets, Baird's sandpipers, and other sandpipers, seek food on mud flats where Alum Creek enters the Yellowstone River, especially from late July through the closing of interior park roads in early November. Look for western grebes, Clark's grebes, and eared grebes, also white-faced ibises in spring and fall. Bald eagles, osprey, Swainson's hawks, and great blue herons (Fig. 3.3) nest in Hayden Valley. During September and October, red-tailed hawks, Swainson's hawks, golden eagles, and other raptors migrate through Hayden Valley (*see chapter 15*); some raptors linger because the valley serves as a critical stopover site, where they hunt for small rodents and grasshoppers.

OPPOSITE PAGE:

▲ FIGURE 3.1 Wetlands lie adjacent to the Yellowstone River as it meanders north from Yellowstone Lake through Hayden Valley. These wetlands provide productive nesting habitat for an array of waterfowl. NPS PHOTO/JAKE FRANK

◢ FIGURE 3.2 Green-winged teal find sheltered nest sites at the edges of wetlands in Hayden Valley, where females raise their young. NPS PHOTO/JAKE FRANK

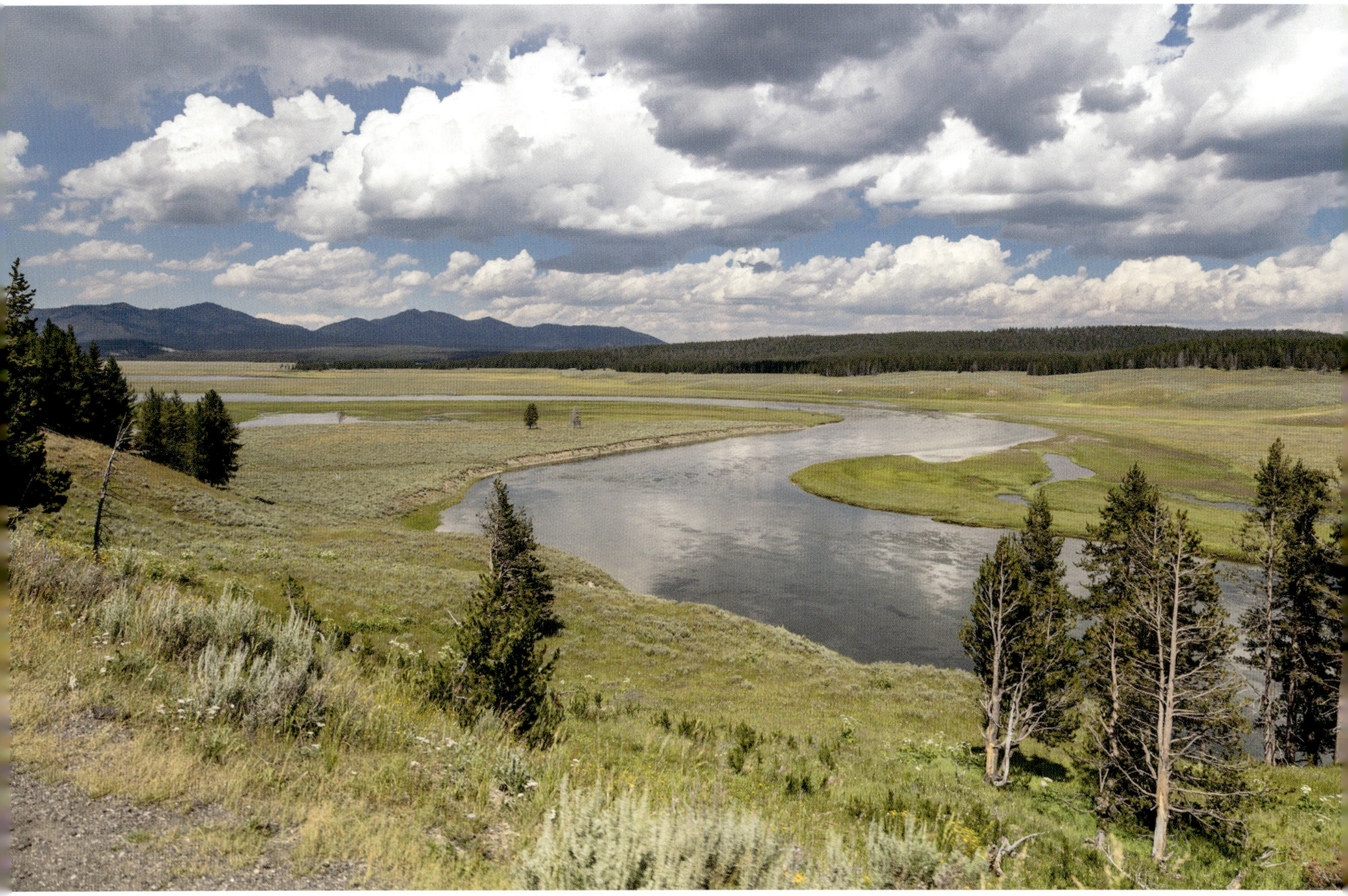

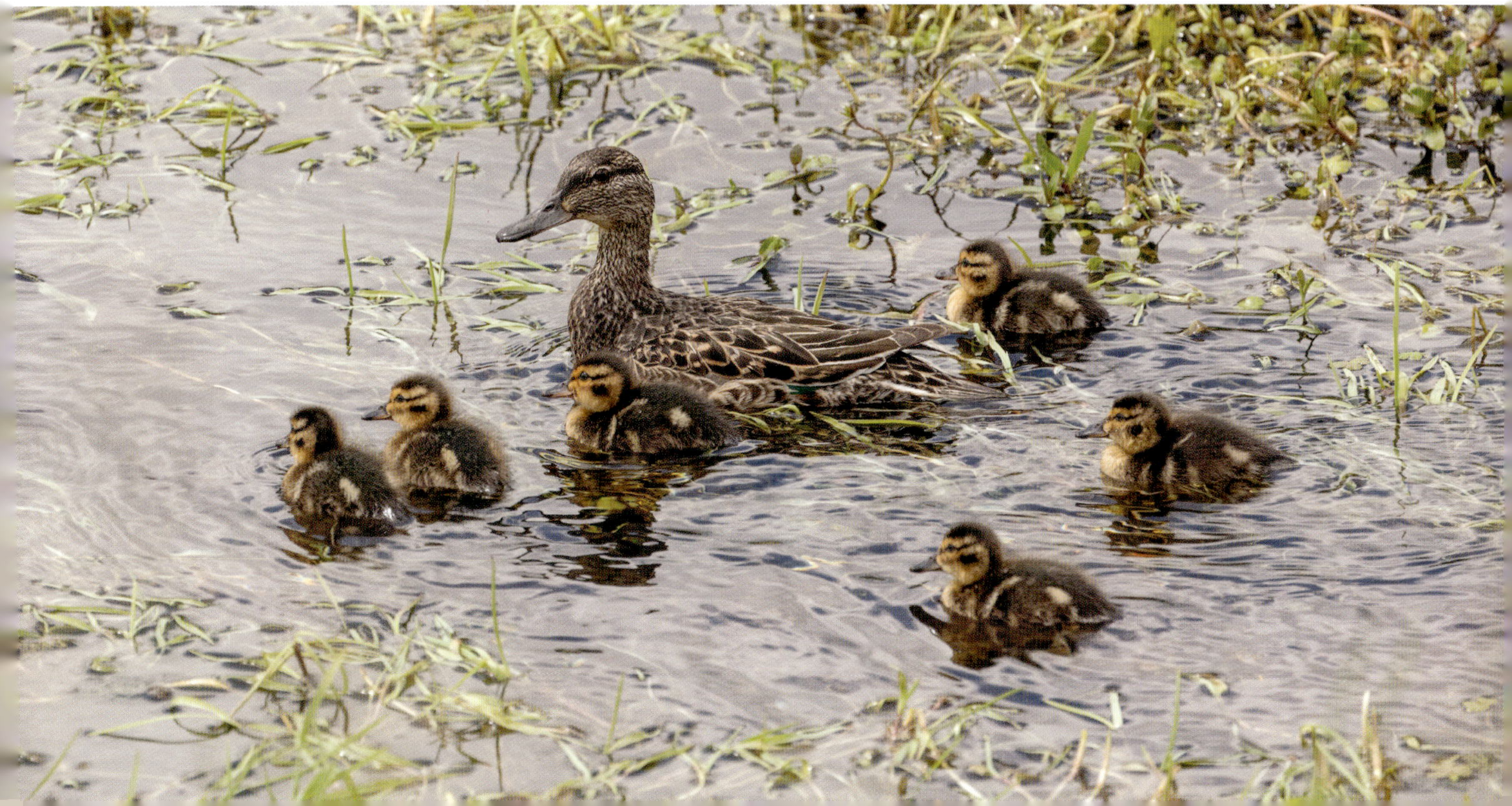

▲ FIGURE 3.3 A great blue heron flies from a heronry, a colonial nesting area, located in the southern part of Hayden Valley. NPS PHOTO/JIM PEACO

▼ FIGURE 3.4 Pelican Creek winds its way to Yellowstone Lake through a grassy marsh, offering nesting habitat for secretive birds like soras and Wilson's snipe. NPS PHOTO/JAKE FRANK

A better-known spectacle occurs when hundreds of bison congregate in Hayden Valley in July and August for their annual rut (breeding period) and cause massive traffic jams. Watch bison behavior from a distance as bulls loudly battle each other for opportunities to mate with females.

LEHARDY RAPIDS

Near the southern end of Hayden Valley, LeHardy Rapids is not wheelchair accessible, but the official pullout has accessible parking, with curb cuts and an accessible vault toilet. Viewing the rapids involves walking down a series of stairs from the parking area to a mostly unpaved path along the Yellowstone River. The first paved parking area to the north has an accessible parking space, a curb cut, and a boardwalk with a gentle grade that then leads via an unpaved path to a picnic area next to the Yellowstone River. Unpaved paths also go south to allow viewing of the rapids. Male and female harlequin ducks gather at the rapids in late spring prior to breeding. In September and October, look for females and their full-sized ducklings. Common mergansers frequent the rapids, too. American dippers nest near LeHardy Rapids. In summer, abundant salmonflies, a large species of stonefly, feed numerous

birds, as well as trout that leap out of the water to snag these tasty morsels.

FISHING BRIDGE TO INDIAN POND

The East Entrance Road crosses the Yellowstone River at Fishing Bridge. The parking area on the west side of the bridge has accessible parking spaces, curb cuts, and a paved path to the walkway across the bridge. Paved blocks, rather than pavement, cover the walkway. Birds seen on the river sometimes include American white pelicans, great blue herons, and various kinds of waterfowl, while warblers and flycatchers nest in adjacent willows and conifer forests.

The road crosses Pelican Creek (Fig. 3.4) on a causeway about 1 mile (1.6 km) past the visitor center, gas station, and store. The parking area at the west end of the causeway (the trailhead for the Pelican Creek Nature Trail) has accessible parking, curb cuts, and an accessible vault toilet. The nature trail is a 2-mile (3.2 km), unpaved path through a mature mixed conifer forest with views of Yellowstone Lake. The east end of the nature trail parking area has an excellent view of the bird-rich Pelican Bay marsh, where numerous waterfowl (dabbling and diving ducks and Canada geese), songbirds (Fig. 3.5), and other birds, including soras and Wilson's snipe, nest. At the east end of the causeway, a parking area on the north side of the road has accessible parking spaces and curb cuts, plus more views of the extensive marsh. Immediately east of the causeway, a parking area on the south side of the road offers accessible parking, curb cuts, and an accessible vault toilet.

Indian Pond, about 2 miles (3.2 km) east of Pelican Creek, has accessible parking, a curb cut, and a paved path for viewing the pond. The adjacent Storm Point trailhead parking area has accessible parking, a curb cut, and an accessible vault toilet. Indian Pond, always worth a stop, hosts a variety of interesting waterfowl and other waterbirds, especially in spring and fall.

FIGURE 3.5 A male common yellowthroat belts out its *witchity witchity witchity* territorial song from willows and other shrubs at the edge of wetlands near Pelican Creek.
PHOTO BY HOWARD WEINBERG

LAKE TO WEST THUMB—WEST SHORE OF YELLOWSTONE LAKE

After Yellowstone Lake melts in early June, numerous paved pullouts along the shore provide unobstructed views of this large and beautiful lake. Two picnic areas, located about 10 and 15 miles (16 and 24 km) south of Fishing Bridge Junction, have accessible parking, curb cuts, and accessible vault toilets, but only unpaved paths to the edge of the lake. Stop at pullouts along the lake in summer to view waterfowl, including Barrow's goldeneyes, and other waterbirds. During fall, watch for common loons, eared grebes, horned grebes, and sometimes red-necked grebes, all in non-breeding plumage, from the pullouts. Mud flats at the edge of the lake sometimes attract a variety of shorebirds during fall migration.

GULL POINT DRIVE

A 2-mile (3.2-km) paved road starts immediately south of Bridge Bay. The first mile (1.6 km) follows the edge of Yellowstone Lake and can be a good place to view grebes and loons, especially in the fall. A pond on the west side of the drive usually hosts a variety of nesting waterfowl. The Gull Point picnic area, nestled in a mixed conifer forest at the edge of the lake, affords views of the lake and has a variety of nesting forest songbirds.

BIRDS OF THE BASINS
AVIAN ACTIVITY IN YELLOWSTONE'S ICONIC HYDROTHERMAL AREAS

KATHARINE E. DUFFY

Boiling water shoots skyward from erupting geysers. Water in colorful hot springs scalds immediately. Fumaroles emit loudly hissing steam. Mats of thermophilic microbes line runoff channels from thermal features. Despite seeming to be inhospitable, Yellowstone National Park's (Yellowstone; YNP) hydrothermal areas host a surprising array of birds. A year-round abundance of food attracts birds. These locations almost always remain at least partially snow-free. Rivers and streams traversing hydrothermal areas typically never freeze due to a steady influx of hot water; in spring, rivers serve as migration corridors for birds. Grassy meadows flourish in hydrothermal areas, where trees do not grow. Although there are no bird species found exclusively in hydrothermal basins, the combination of insect prey (particularly copious during summer's flush of insects), flowing water, and grassy meadows offer appropriate conditions for many bird species during migration, nesting, and even in winter.

MOUNTAIN BLUEBIRDS HOVERING as they search for insects provide entertainment for visitors waiting for Old Faithful (Fig. 4.1) to erupt. Bluebirds nest (Fig. 4.2) in cavities excavated by northern flickers in trees and in wooden posts near boardwalks. Sometimes bluebirds find safe nest sites in the eaves of historic buildings. Violet-green swallows, tree swallows, barn swallows, and cliff swallows, occasionally joined by rough-winged swallows and bank swallows, swoop and dive over geyser basins during their aerial pursuit of flying insects. Baird's sandpipers, black-necked stilts, marbled godwits, and other migratory shorebirds sometimes stop in hydrothermal areas in April, August, and September to refuel by consuming insects and other invertebrates.

Mountain chickadees, brown creepers, and several species of woodpecker, including hairy, American three-toed, and black-backed, inhabit conifer stands within and near hydrothermal areas year-round. Clark's nutcrackers live all year in the Old Faithful and Mammoth areas because nutcrackers stash the fatty seeds of whitebark and limber pines for winter food, and these trees grow in the vicinity of the thermal basins—whitebark pine at Old Faithful and limber pine at Mammoth.

Sandhill cranes search in spring, summer, and fall for grubs and edible roots in grassy meadows within hydrothermal basins. Common yellowthroats nest in cattail marshes lining outflow streams of Mammoth Hot Springs. Yellow-rumped warblers and ruby-crowned kinglets nest

BIRDS OF THE BASINS

◢ FIGURE 4.1 Thousands of visitors congregate daily to watch as Old Faithful, Yellowstone's most renowned geyser, erupts at regular intervals. NPS PHOTO/JIM PEACO

◢ FIGURE 4.2 Meadows throughout hydrothermal basins attract insect-eating birds, such as mountain bluebirds, that forage in the basins and nest in nearby cavities in trees and buildings. PHOTO BY HOWARD WEINBERG

in lodgepole pines and other conifers in geyser basins. Marshes at Old Faithful host nesting Lincoln's sparrows, red-winged blackbirds, and Brewer's blackbirds. American pipits are regular near Old Faithful in September, during fall migration, and in May, during spring migration at the Mammoth Hot Springs terraces.

At any time of the year, golden eagles occasionally soar above hydrothermal areas. Northern harriers course over meadows during spring and fall migration. Because birds can concentrate in hydrothermal areas during migration, merlins hunt the open parts of geyser basins. Avian prey also draws sharp-shinned hawks and Cooper's hawks to hydrothermal areas during migration and while nesting. In summer, American kestrels find grasshoppers and small rodents in grassy meadows within and near hydrothermal areas, while red-tailed hawks hunt in open parts of geyser basins all summer on the west side of the park. Swainson's hawks are more numerous on the east side of the park, including in extensive meadows at Mud Volcano. Peregrine falcons nesting on ledges on lava flows near hydrothermal areas find avian prey nearby.

West Thumb Geyser Basin, immediately adjacent to Yellowstone Lake, affords late-spring, summer, and fall sightings of Barrow's goldeneyes, bufflehead, common mergansers, and other ducks, along with occasional common loons, western grebes, and red-necked grebes. Bald eagles and osprey dive for fish when Yellowstone Lake is ice-free and also hunt the Firehole, Gibbon, and Yellowstone rivers flowing through or near thermal areas. Osprey head south for winter, but bald eagles, especially territorial adults, remain within Yellowstone all year long, as do some Canada geese, common mergansers, mallards, and other kinds of ducks. Great horned owls, year-round residents, search for prey in the Upper Geyser Basin and Mud Volcano after dark. Every now and then, a great gray owl seeks rodent prey in or near hydrothermal areas.

All winter, American dippers fly underwater in rivers in hydrothermal areas as they pursue aquatic invertebrates. If a stone bridge supports sites for dippers' nests made of mosses, they might stay for the breeding season, too. When other conifer seeds become less available in winter, red crossbills use their plier-like bills to break open lodgepole pines' serotinous cones, even though these cones are extra thick and sealed with a resin. Then the crossbills, frequently found at Old Faithful in winter, gorge on the nutritious seeds within. Crossbills can nest at any time of

FIGURE 4.3 Boardwalks throughout the Upper Geyser Basin lead to close views of Beehive Geyser and numerous other geysers and hot springs. Grassy meadows near hydrothermal features attract nesting and foraging birds.
NPS PHOTO

the year when food is abundant, so it's not just in summer that you may see adults feeding dependent young.

During winter in hydrothermal areas, insect-eating birds can consume flies of the Ephydridae family that, in turn, feast on thermophiles (organisms, often one-celled, that thrive at high temperatures), with flies staying especially close to the microbial mats due to frigid air just a few inches above. It's not surprising to see killdeer and Wilson's snipe on a winter visit to Mammoth Hot Springs, the Upper Geyser Basin at Old Faithful (Fig. 4.3), and nearby Lone Star Geyser. A few winter sightings of rusty blackbirds, varied thrushes, and other rarities have happened in hydrothermal areas. Virginia rails can occur any time of the year near runoff streaming from Mammoth Hot Springs, but soras skulk through marshes at the edges of geyser basins only in summer.

Avian activity can add another dimension to the appeal of Yellowstone's iconic hydrothermal features. With many opportunities to watch birds, a pair of binoculars and the park bird checklist are essential gear for a trip to any of Yellowstone's hydrothermal basins.

5 BIRDING THE BEAVER PONDS TRAIL

JOHN PARKER

The diversity of habitats along the Beaver Ponds Trail is unmatched by any other trail of such a short length in Yellowstone National Park (Yellowstone; YNP). Only 5 miles (8 km) long and easily accessible from the park's headquarters in Mammoth, this mildly strenuous trail is a popular hike for all ages and levels of fitness. Being at a relatively low elevation, and in the driest corner of Yellowstone, the Beaver Ponds Trail is mostly free of snow from April until November. This long open season makes the trail an ideal walk during both the spring and fall migration seasons. Following a loop to the west and north of Mammoth, this trail offers continuous changes of scenery and an abundance of birds from the moment you arrive at the trailhead. The trailhead is located directly across Clematis Creek from Liberty Cap, the 37-foot (11.3-m) tall pillar of travertine that is the signature feature of the main terraces at Mammoth Hot Springs.

FIGURE 5.1 Few passerines are hardy enough to persevere through Yellowstone's winters. Mountain chickadees (A) remain ubiquitous through the park, while Townsend's solitaires (B) migrate to areas of relatively low elevation near the north entrance.
PHOTOS: (A) BY GREG ALBRECHTSEN,
(B) BY TOM MURPHY

EVEN BEFORE YOU begin up the trail, the variety of birds in the developed area in Mammoth can be quite good. Killdeer frequent the bright white travertine terraces, feeding on the brine flies that populate the run-off channel from the nearby thermal features. Even in the winter, Mammoth attracts several hardy year-round resident species, including common ravens, Clark's nutcrackers (the official Yellowstone bird), and mountain chickadees (Fig. 5.1a). Flocks of bohemian waxwings can often be seen moving around the hillsides foraging for juniper berries in early winter, much to the displeasure of the resident and very territorial Townsend's solitaires (Fig. 5.1b). If you are very lucky, gray-crowned rosy-finches might be seen along the windblown or melted areas around Mammoth. In the breeding season, however, the bird diversity in Mammoth, and particularly along the Beaver Pond trail, explodes. Leaving the parking lot and crowds behind, you begin the steady climb up Clematis Gulch and are immediately surrounded by a chorus of birds coming from all sides. Listen for lazuli buntings and MacGillivray's warblers singing from the deciduous understory and tangles of clematis vines. Shortly, warbling vireos and ruby-crowned kinglets singing from overhead will begin to dominate the soundscape and stay with you wherever the forest canopy becomes mostly continuous. This climb through semi-open Douglas-fir forest will likely be the best opportunity to see the most abundant warbler in Yellowstone, the showy "Audubon's" yellow-rumped warbler. Though a long way from the tropics, this is also the breeding grounds of the dazzlingly colored western tanager (Fig. 5.2), whose buzzy song is somewhat reminiscent of a robin's. For such a brilliantly colored bird, tanagers can be surprisingly hard to find in the canopy. Both the yellow-rumped warbler and the western tanager will sometimes flock to the open sagebrush grasslands searching for insects during spring and early-fall snowstorms.

After climbing over 400 feet (125 m) in a little more than ½ mile (1 km), the Beaver Ponds Trail splits off to

FIGURE 5.2 A male western tanager catches a tasty snack.
PHOTO BY GREG ALBRECHTSEN

the north from the Sepulcher Mountain Trail. Now the trail levels out and opens into scattered aspen groves and islands of Douglas-fir, with grasses and luxuriant wildflowers at the head of the Primrose Creek drainage. Here the mechanical trills of the chipping sparrow and the more musical trills of the pink-sided junco (a sub-species of the dark-eyed junco common to the Northern Rocky Mountains) may be drowned out by the rambling song of the house wren (Fig. 5.3). These open areas also provide an opportunity to scan for some of the resident finches, such as Cassin's finch, red crossbill (Fig. 5.4), and pine siskin.

As this section of trail meanders along the benches above Mammoth, the trees thin out, and magnificent views of Mount Everts and the Washburn and Absaroka ranges in the distance are sure to grab your attention. Now a different set of birds begin to make their appearance. Green-tailed towhees may be glimpsed dashing for cover underneath the sagebrush, and you are almost certain to continuously flush vesper sparrows as they escort the hiker through their grassland domain. At this point, you'll begin to see swallows coursing the skies above and just above the top of the sagebrush, wherever the insect hatch may be. Tree swallows nest in the aspens immediately around you (Fig. 5.5a), while the other swallows make their homes farther afield. Many of the cliff swallows nest in the eaves of the historic buildings in old Fort Yellowstone at Mammoth (Fig. 5.5b). The violet-green swallows (Fig. 5.5c) favor nesting in the high cliffs of the surrounding mountains. Barn swallows, northern rough-winged swallows, and bank swallows also nest nearby, but are present in smaller numbers.

◮ FIGURE 5.3 A house wren makes his presence known, belting out a passionate and seemingly endless solo.
PHOTO BY TOM MURPHY

◭ FIGURE 5.4 A red crossbill forages in a pine.
PHOTO BY ROBERT DIMILIA

FIGURE 5.5 (A) A pair of tree swallows nest in an old snag. (B) Cliff swallows build their own cavities out of mud along the eaves of buildings. (C) A violet-green swallow preens and shows off its brilliant iridescent coloring.
PHOTOS: (A) AND (C) BY SCOTT HEPPEL, (B) BY TOM MURPHY

The character of the walk changes as the trail begins to wind its way through a couple of small drainages. The wet and damp shaded areas here support Engelmann spruce and even an occasional subalpine fir. Hammond's flycatchers are abundant, and an attuned ear may pick up the high-pitched calls of the brown creeper, as it spirals up the trunk of a fir, or the golden-crowned kinglet's thin song, coming from high in the canopy. The thick understory and lush grasses around the forest edge are also good places to find the cryptically patterned ruffed grouse. While the feathers of the grouse may be designed for camouflage, a close examination of their feathers reveals many subtle but infinitely complex patterns. These old-growth forest remnants, interspersed with aspen glades, have just the right ingredients to make them highly attractive to woodpeckers, and careful listeners will notice the screeching calls of the Williamson's sapsucker and the gentle tapping of American three-toed woodpeckers. Northern flickers are nearly ubiquitous, while this is also the only known location in Yellowstone where pileated woodpeckers have nested. With patience, the woodpeckers' calls and drumming can lead you to a nesting cavity in one of the many snags.

Descending to the bottom of a hill, you finally arrive at the first "beaver pond." These lily-pad-covered ponds are hemmed in by low ridges, which are moraines created by the most recent period of glaciation. Throughout the year, a wide variety of waterfowl use the ponds for nesting (Fig. 5.6) and as rest stops during migration. Several species of ducks, such as mallard, ring-necked duck, lesser scaup, and Barrow's goldeneyes find enough cover to nest in the cavities or rushes and grasses that edge the ponds, and their broods of ducklings can be seen on the ponds throughout the summer months. Belligerent and loud American coots, and more circumspect birds like soras and marsh wrens, also frequent the ponds.

Between ponds, the trail undulates through the forest and past a couple of small creeks. Deep in the moist forest, where the snow lingers, the ascending, flute-like song of the Swainson's thrush can be heard, while on the opposite side of the hill, where drier conditions prevail, the

FIGURE 5.6 (A) A duckling struggles to navigate through the lily pads. (B) An American coot passes a morsel to a hungry chick. PHOTOS: (A) BY GREG ALBRECHTSEN, (B) BY TOM MURPHY

serene song of the hermit thrush can impart a sense of tranquility. Small wet meadows dot these woods and are wonderful spots to listen and watch for Lincoln's sparrows and white-crowned sparrows darting from bush to bush. Hillsides covered with the brilliant scarlet of Indian paintbrush lure rufous hummingbirds and a full palette of brightly colored butterflies.

As the trail swings around the largest and last pond, you traverse a slope with a southern exposure. This southern aspect receives intense sun and lacks snow cover; alkaline soils create the harshest growing conditions anywhere along the trail. That makes it just the place to look for the large, lovely, and highly fragrant blooms of evening primrose. Dusky flycatchers, Townsend's solitaires, and mountain bluebirds are at home in the scattered juniper and rabbit brush. This section presents an interesting juxtaposition with the dry slopes above and the wet marsh surrounding the pond below. The discordant calls of the yellow-headed blackbirds in the cattails (Fig. 5.7) are a good match for the raucous calls of the Clark's nutcrackers feeding on the limber pines above. After leaving the ponds behind, the trail meanders up and down over more ridges of rocky till, which are covered in sagebrush, grasses, and wildflowers—exactly the mix that dusky grouse prefer throughout the breeding season.

Eventually, you leave the woods behind and turn to the south, heading back toward Mammoth, through a combination of shortgrass prairie and sagebrush steppe. The immensity of the country surrounds you with spectacular scenery near and far. With wide horizons surrounding you, this is the time to scan for the raptors that make Yellowstone home. Golden eagles and prairie falcons nest in the distant cliffs, while American kestrels can be found nesting in nearby snags. These raptors often patrol this open ground for prey ranging in size from jackrabbits to grasshoppers. The birds that nest in this open country spend most of their lives hidden from view as they forage and build their nests on or near the ground. Sage thrashers, Brewer's sparrows, and western meadowlarks truly shine when they find a suitable sagebrush for broadcasting their complex songs. After a thunder shower, with bitterroot flowers at your feet, the smell of sage enveloping you, and Brewer's sparrows' endlessly entertaining songs coming from all directions, you experience the essence of the northern range.

FIGURE 5.7 A young yellow-headed blackbird fledgling clings to reeds.
PHOTO BY TOM MURPHY

OPPOSITE PAGE:
FIGURE 6.1 Kira Cassidy birding in Yellowstone National Park.
PHOTO BY KIRA CASSIDY

THE YEAR I LOST MY BIRDING MIND

KIRA A. CASSIDY

The first day of 2019 was cold—an eyelash-freezing, see-every-breath-as-vapor kind of cold; the thermometer never even reached 0° Fahrenheit. The frigid afternoon air was still and crisp as I watched a golden eagle flying back and forth across a steep hillside, looking for prey and perhaps gathering a tiny bit of warmth from the weak rays of sunlight. This was the first bird I'd seen that day, that year actually, and it gave me an idea (Fig. 6.1). What if I tried to identify every single bird I saw for the whole year?

THE YEAR I LOST MY BIRDING MIND

'VE BEEN A BIRDER my whole life, receiving my first pair of binoculars, which I dubbed "nocklers," for my fourth birthday. Most of my 4-H projects involved watching and researching backyard birds from my family's home in Illinois. I also keep Lists. Not lists, but Lists. Throughout my adult life, and with varying intensity, I've recorded the bird species I've seen in certain years or migration periods or even single days, like for Cornell's Global Big Day in the spring. I've kept track of the birds and animals I've seen or heard from my house, during graduate school, on road trips, and always during international travel. But I had never tried to keep track of the *number* of birds I saw, like many people do for reports on eBird. That first freezing day, I tallied: one golden eagle, three black-billed magpies, eight common ravens.

Throughout 2019, I also kept track of my general location and distance traveled, the morning low and afternoon high temperatures, the general weather, and how much effort I put into birding each day. I didn't want this experiment to be mistaken for a Big Year. I wasn't birding every day, or probably even more than a dozen days throughout the year. I was trying to capture, in numbers, what I normally see each year, something to which I usually gave no more than a glance. The pair of mountain chickadees that show up outside my kitchen window? That's not jotted down as "MOCH" for the year, rather "0 to 61 (!) MOCH" over 126 days, for a total of 485 MOCH throughout the year. It wasn't a Big Year. My goal was to identify *and* count every bird, even the most common ones, every day—a List Year.

FIGURE 6.2 Snow geese pass through Yellowstone during their fall migration.
PHOTO BY TOM MURPHY

FIGURE 6.3 Black-billed magpies were the second-most-observed bird during the Year of the Bird. PHOTO BY TOM MURPHY

Every day, I carried around a little pocket notebook to tally the birds I identified, species and individuals. My handwriting was barely legible as I wrote smaller and smaller, realizing that I might fill the entire book before reaching December 31. It became slightly embarrassing, as several of my friends knew about my goal to count every bird observation in 2019, and the whole plan sounded increasingly mad, especially as the spring migration began.

Part of my goal was to become a better birder. I was decent, but I often took my time with identifications, requiring a clear, steady look and a field guide. Now I wanted to be able to see a glimpse of a bird flash into the trees and identify the species. I wanted to hear a bird's song and recognize it right away. I wanted to see a flock of snow geese and know if there were 151 or 178 of them winging overhead (Fig. 6.2). I knew that increasing my

birding skills from that "pretty good" plateau would require seeing and identifying and counting a lot of birds, but that was the plan for 2019.

I live and work near the north entrance of Yellowstone National Park (Yellowstone; YNP), at 5,400 to 6,400 feet (1,650 to 1,950 m) in elevation, so winters can last for up to six months, with cold and snowy weather possible for another month before and after. The area holds only a few dozen winter resident species, so identifying and counting them was easy. There were times blizzards raged for days, sometimes making it impossible to see through the whiteout more than a few dozen feet. January 23 was the only day I came close to seeing zero birds. Despite the snow, thick clouds, and wind, I finally saw a pair of black-billed magpies as they headed for shelter in a dense juniper (Fig. 6.3).

THE YEAR I LOST MY BIRDING MIND

Spring of the List Year was exceptionally cool and wet, with a bit of snow even falling on the summer solstice. This certainly impacted the number of birds I saw, as many of the spring migrants raced through the mountainous areas to their summer breeding grounds, not delaying like they might in other years. Two neotropical migrants, however, seemed to stall out around Yellowstone in numbers I'd never seen before. Had I never seen such numbers, or had I never paid attention after that first sighting of the year? Either way, I recorded a surprising 160 western tanagers and 104 lazuli buntings from mid-May through early October.

I traveled away from the Yellowstone area four different times for a total of six weeks throughout the year, keeping track of bird observations. This was, after all, my measure of everything *I* saw and identified in one year, not what I saw in a certain place. After the big rush of bird species in the spring, the summer slows down, but it also becomes the most difficult time of year to identify birds. I'd see a flash of brown wings disappear into thick foliage or pop up briefly in a field of sagebrush before dropping to the ground—out of sight completely. Where I estimated that I successfully identified at least nine out of every 10 birds I saw in the winter, that proportion dropped dramatically in July and August to about two to four out of every 10 birds, as dozens of tiny sparrow or junco or chickadee species flicked into the corner of my eye before vanishing. Summer was still thrilling, though, especially the day I was hiking through a burned area and saw a black-backed woodpecker and a black-headed grosbeak in the same tree. By September and October, as the fall migration slowed, my identification percent climbed again.

On the very last day of the year, I went on a cross-country ski with my dogs to look for a northern pygmy-owl, which had eluded me all year. Finally, I gave in, ready to go home and total 2019's birds. I was pulling into my driveway and saw a falcon in the top of a large spruce. Hmm, I'd seen kestrels, prairie falcons, and peregrine falcons already, but most of them had migrated out of the area weeks or months ago. Could it be something else? The bird flew across a large field, so I ran inside, grabbed my scope and yelled for my sister, who is a great birder, to come see. A merlin! The only one I saw all year. It was also the last bird of the List Year, with only a few minutes of daylight left.

Throughout 2019, I ended up seeing 60,625 birds from 197 different species; 160 of those species were in the Yellowstone ecosystem. It averaged out to be 166 birds per day but ranged from the day I only saw that pair of black-billed magpies to June 5, when I saw 2,638 birds in one day. The Canada goose was by far the bird I saw in the most numbers (11,877 in total), making up most of that high-count day with 2,500 in Yellowstone's Hayden Valley alone.

Some highlights of the year ranged from seeing rare birds to interesting behavior by common birds. I was speechless when I saw six critically endangered California condors, three of them floating above me on the Marble Canyon bridge at the north end of the Grand Canyon. I watched Monty and Rose, the famous pair of endangered piping plovers who nest on a Chicago beach, as they fiercely protected their three fuzzy ping-pong-ball babies from gulls. I watched a mountain chickadee tuck away in a Douglas-fir tree and spy on a Clark's nutcracker caching seeds below it. The chickadee watched, and when the nutcracker left and I walked by, the chickadee seemed surprised to be caught as it was quietly stealing the seeds. I found I could sometimes recognize individual birds— that black-capped chickadee with the broken tail feathers, the chipping sparrow with a band of white instead of chestnut on one wing, the female Cassin's finch missing her right eye, and the male house finch with white cheek feathers.

But what bird did I see most days? This was one of my main questions driving the whole experiment. Which bird do I see so often it becomes a regular background part of my life? That bird was the common raven, which I saw on 78% of the days of 2019—some 3,900 of them over 286 days. Ravens are big birds, easy to identify, and easy to count, even when they amass in groups of 80 or more, usually at a food source like a large bison carcass—a sight I witnessed five times over the course of the year (Fig. 6.4). My favorite trail run passes underneath a raven nest,

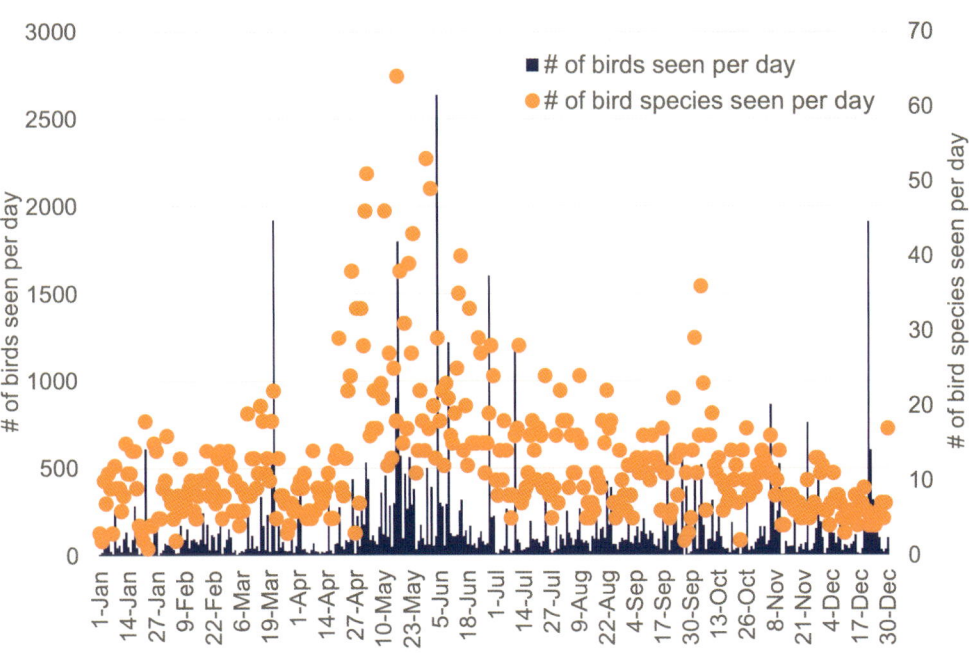

FIGURE 6.4 A raven contests a bald eagle for a meal. Ravens were viewed on more days than any other bird. PHOTO BY RONAN DONOVAN

FIGURE 6.5 Total number of species and individuals observed by Kira Cassidy per day in the Greater Yellowstone Ecosystem during 2019.

■ # of birds seen per day
● # of bird species seen per day

and while at first one of the pair flew and scolded me, once I started saying, "Hello, friend" to the pair as they flew overhead, they eventually stopped calling. After about a week, they just stayed perched as I jogged past. *There's that woman with the red jacket and the two dogs. We know her; no need to worry.*

My List Year completed, I look back on my notes and the numbers (Figs. 6.5 and 6.6). Sometimes it feels like I saw a lot, and other times I remember that I saw more sandhill cranes in one day in 2018 than all birds in 2019. But I remind myself that the total birds was not the goal. The goal was to improve my birding skills and to document my everyday, corner-of-the-eye sightings. To figure out what I normally see that used to hit my optic nerve and send an image to my occipital lobe but didn't go any further than that. To make those sightings, in a split second, grab my full attention, open a file of neurons to my past experiences, and find the matching identity, then make my arm move and hand muscles contract to create a tiny *Date: October 14, 2019/Common Raven:2.* To have a record of the experience and remember it years from now. To connect to a place and another life and say hello to a familiar creature. To help me be present wherever I am and to truly see the lives all around me.

FIGURE 6.6 Feathers from important birds observed by Kira Cassidy during 2019 in the Greater Yellowstone Ecosystem. Number of days observed is the numerator of the fraction, followed by the percentage of days observed. ARTWORK BY KIRA CASSIDY

RED-TAILED HAWK	GREAT HORNED OWL	MOUNTAIN BLUEBIRD	SANDHILL CRANE	NORTHERN FLICKER	YELLOW WARBLER	CEDAR WAXWING
99/365	27/365	36/365	33/365	78/365	32/365	68/365
27.1%	7.4%	9.9%	9.0%	21.4%	8.8%	18.6%

GEORGE BIRD GRINNELL

JOHN TALIAFERRO

First, let's establish George Bird Grinnell's birder bona fides—and we're not talking about his middle name, which, it's worth noting, he used faithfully. (He sometimes signed his name "Geo. Bird Grinnell," but never "George B. Grinnell.") If anything, his future was foretold not by a birth certificate but rather by the cosmic kismet of growing up on the estate of John James Audubon, on the wooded western shore of Manhattan Island. Grinnell's father, a New York merchant and broker, moved his family onto the Audubon's property in 1857, when Grinnell was six years old. Audubon had died in 1851, but his widow, sons, and grandchildren remained, as did the spirit of the already legendary naturalist, plus a selection of his artwork. Under the tutelage of Madame Audubon, Grinnell came to appreciate the birds that frequented what came to be known as Audubon Park, including the massive migration of passenger pigeons along the Hudson flyway.

AS GRINNELL GREW, so did his engagement with the winged world (Fig. 7.1). He hunted ducks, rails, and woodcocks in the surrounding countryside. Then, as a student at Yale, he mastered the discipline of ornithology under the mentorship of renowned bone hunter Othniel C. Marsh. Grinnell's doctoral dissertation was a morphology of the roadrunner.

He became a bit of a roadrunner himself. In 1874, he was the designated naturalist on the US Army's reconnaissance of the Black Hills, led by George Armstrong Custer—an intrusion into sovereign Native territory that triggered a stampede of white settlers and Custer's demise on the Little Bighorn River; it also honed Grinnell's talents in field observation and specimen collection. Over the next fifty years, in his wanderings across the West—pursuing big game, scaling mountains and glaciers, advocating for protection of wildlife and national parks—Grinnell always found time for his first love, birds. "Secured several birds new to me," he recorded in his Black

FIGURE 7.1 Photo of George Bird Grinnell. NPS PHOTO/YELLOWSTONE ARCHIVES

Hills notebook. "I found a nest of *Plectrophanes maccownii* [chestnut-shouldered longspurs] with four eggs." The specimens he collected that summer are still preserved in climate-controlled drawers in Yale's Peabody Museum. A year later, during a pack tour of the nation's first

GEORGE BIRD GRINNELL

national park, he compiled a thorough inventory of Yellowstone National Park's (Yellowstone; YNP) birdlife, including a species that won his particular affection, the American dipper, more familiarly known as the water ouzel (Fig. 7.2). "I must confess to the most ludicrous feeling of astonishment the first time I saw the bird walk calmly down a flat stone until its head disappeared under the water," he observed of this submarining curiosity. "When carried down a few feet by the force of the current, it would fly a few feet upstream and dive from the wing."

History tends to footnote Grinnell as a cofounder, along with a bumptious Theodore Roosevelt, of the Boone and Crockett Club, a cohort of well-heeled big-game hunters whose self-interest in "fair chase" became a contagious force for broader-gauge conservation in the late 19th century. Yet what history often overlooks is that Teddy Roosevelt was an unabashed game hog up until the time he came under the moderating influence of George Bird Grinnell; and that in 1886, two years before the Boone and Crockett Club organized, Grinnell had started the first Audubon Society, a conservation group whose influence would prove further-reaching than that of the B & C fraternity. If anybody invented a bully pulpit for conservation, it was Grinnell, who used his journal, *Forest and Stream*, to preach sensible bag limits and hunting seasons, protection of habitat, and, in general, a holistic approach to

FIGURE 7.2 The American dipper, also known as the water ouzel, a lifelong favorite species of George Bird Grinnell; he observed one in Yellowstone "walk calmly down a flat stone until its head disappeared under the water." His visit to Yellowstone in 1875, three years after the establishment of the park, produced the most thorough inventory of birds (and mammals) in the region to date. PHOTO BY CAMERON HO

nature that presaged today's environmental conscience. The tactic of shaming women—and the rookery raiders who catered to female vanity—from adorning their hats with the plumes of egrets (etc.) was, if you will, a feather in Grinnell's (and the Audubon Society's) own cap.

Let us also give Grinnell at least partial credit for the following: In 1911, when the National Association of Audubon Societies, of which Grinnell was a director, spurned a donation from Winchester, Remington, and other gun manufacturers, in support of bird protection (without birds, gun sales would suffer), Grinnell gave his blessing to a new organization, the American Game Protective and Propagation Association, whose purse and persuasions would eventually ensure passage of the Migratory Bird Treaty Act, a cornerstone of federal wildlife legislation and the precedent for the Endangered Species Act of 1973. "As long as the game is protected and increased," Grinnell reasoned, "the aim [of the gunmakers] is of no great importance." Grinnell was the prototype of a new breed of conservationist, passionate but pragmatic.

Similarly, Grinnell's strategic lobbying on behalf of national parks brought results that resound today. From the time he first gazed upon Yellowstone's wonders, in 1875, through many subsequent visits until his death in 1938, the park had no greater or more effective proponent. John Muir, the high priest of Yosemite, fought the damming of the Hetch Hetchy valley with his idealism—and lost. Grinnell did more than write livid, lachrymose editorials on behalf of his beloved Yellowstone; he stalked cabinet members, congressmen, and the White House and helped draft the laws that created a buffer of national forest around the park (thus creating the *first* national forest); stopped commercial development from defacing the geyser basins and other natural treasures; prohibited hunting inside park boundaries; prevented a railroad from penetrating the Lamar Valley; and headed off a dam at the mouth of Yellowstone Lake (Fig. 7.3). Before Grinnell, Yellowstone was America's *only* national park; thanks to Grinnell, Yellowstone came to embody what all national parks should be.

FIGURE 7.3 Grinnell's first visit to Yellowstone in 1875 included a trip to the Grand Canyon of the Yellowstone and the Upper Falls that he described as "altogether the grandest and most beautiful that I have ever seen. It altogether beggars description." The trip cemented his love of the park, and throughout his life, he tirelessly championed its beauty and wildlife, critically protecting it from development, including a dam that would have impacted Yellowstone Lake. NPS PHOTO/ YELLOWSTONE ARCHIVES, JACKSON EXPEDITION 1871

His maturation as a conservationist may not have been inevitable, but it was not hard to follow. The rapacity that violated the Black Hills led directly to his activism in Yellowstone. And finally, the lessons and methods incubated in these grand places would mold his vision for a national park on the crown of the continent, Glacier National Park.

An estimable career indeed. The boy who grew up loving birds grew mighty wings (Fig. 7.4). If only we could soar as far and high as he.

FIGURE 7.4 A prairie falcon, a species Grinnell encountered in the Black Hills in 1874 on an expedition with George Armstrong Custer. Grinnell was appointed trip naturalist and found a prairie falcon nest with three young. He took a shot with his shotgun at the female but did not kill her; apparently, the gun was loaded with "small bird seed." PHOTO BY RONAN DONOVAN

CITIZEN-SCIENCE-LED BIRD MONITORING IN YELLOWSTONE

LAUREN E. WALKER, JOHN PARKER, and KATHARINE E. DUFFY

According to a 2016 United States Fish and Wildlife Service Report, over 45 million Americans identify as birders. Embraced by new generations, birders seek both to relish the quiet calm of time spent outdoors, enjoying the company of wildlife in their natural setting, but also the secret thrill when scopes and binoculars find a new species or observe a unique behavior. For many birders, citizen science allows them the opportunity to put additional significance to their birding, combining their love of birds and their interest in contributing to local research programs by helping collect scientific data. In Yellowstone, intrepid volunteers have historically organized and conducted several broadscale programs, perhaps most notably the complementary Breeding Bird Survey (BBS) and Christmas Bird Count (CBC).

THE BBS AND CBC are both long-term and widespread surveys, conducted annually for decades at sites across the continent. Equally important, their survey target is broad, including all birds that are detected within the constraints of the survey protocol—all songbirds, but also all raptors, shorebirds, and waterfowl. Widespread and long-term datasets such as these are especially useful for documenting changes in bird communities, and across North America and in Yellowstone National Park (Yellowstone; YNP), the BBS and CBC have helped highlight decades of patterns in bird abundance and diversity. Further investigation may allow for the association between these trends in bird diversity and other broadscale ecosystem changes in the park due to changing predator and prey communities, evolving wildfire dynamics and fire management, and exponential growth in visitor presence.

BREEDING BIRD SURVEY (BBS)

Started in 1966, the BBS (www.pwrc.usgs.gov/bbs/; USGS 2018) was initiated by Chandler Robbins and the United States Geological Survey (USGS) at the Patuxent Wildlife Research Center, near Laurel, Maryland, to monitor breeding bird populations across the North American continent. Initially a means to monitor and document the impacts of the widespread effects of DDT (dichloro-diphenyl-trichloroethane) and other pesticides in the first half of the 20th century, after 55 years, BBS survey data now provide valuable insight into the impacts of changing human populations and the growing human influence on the landscape through habitat loss, fragmentation, and climate change. To date, this vast and long-term database has contributed to hundreds of scientific publications.

BBS survey routes are run each summer during the peak of the North American nesting season, anywhere from late May through early July, depending on latitude, altitude, or other climatic considerations. Birds at this time of the year are active, displaying for mates and singing to define and defend their territories, making them relatively easy to detect (Fig. 8.1). Routes are located along road corridors and are 24.5 miles (39.4 km) long, with 50 survey locations spaced every half mile (0.8 km)—a strategy designed to sample birds from a broad diversity of habitats. At each stop, the observer conducts a 3-minute point count of all birds seen or heard within a quarter-mile (0.4-km) radius. Today, over 4,000 survey routes are coordinated by the USGS, the Canadian Wildlife Service, and, more recently, the Mexican National Commission for the Knowledge and Use of Biodiversity to study the status of bird abundance, distribution, and population trends.

BBS IN YELLOWSTONE

In Yellowstone, four BBS survey routes are conducted in mid-June to help tally and track trends of the park's bird abundance and species diversity. Three routes are located

FIGURE 8.1 Birds use a variety of displays to attract mates each spring. (A) A male calliope hummingbird shows off his bright purple gorget. (B) A pair of sandhill cranes bond with a courtship "dance." (C) A male lazuli bunting shows off his bright plumage and sings to attract potential mates and deter trespassing rivals. PHOTOS: (A) BY GREG ALBRECHTSEN, (B) AND (C) BY HOWARD WEINBERG

entirely within the park (Fig. 8.2) and have been in operation since the 1980s; a fourth route that lies partially within the park boundary was added in 2017.

The Mammoth BBS route starts at the Indian Creek Campground south of Mammoth, ends at Elk Creek west of Tower Junction, and has been run continuously since 1982. Overall, the Mammoth BBS route is the lowest in elevation and the driest of the three routes inside the park. Observers conducting this route have recorded 135 different species of birds, the greatest diversity of any route in Yellowstone. Rock wren and green-tailed towhee are two of the birds commonly found on this route but rarely seen elsewhere in Yellowstone.

Conducted continuously since 1988, the Northeast Entrance BBS route starts at Tower Junction and ends across from Barronette Peak near the northeast entrance. The first 20 miles (32 km) of this route travel through the Lamar River Valley, a habitat representative of Yellowstone's northern range. Observers have tallied 125 species of birds on the Northeast Entrance route, with high numbers of sage thrashers and Brewer's sparrows in the extensive grasslands and sagebrush.

Initiated in 1987, the Yellowstone BBS route starts near the top of Dunraven Pass, at 8,800 feet (2,682 m), and continues south, following the Yellowstone River upstream to Yellowstone Lake and ending at Mary Bay along the East

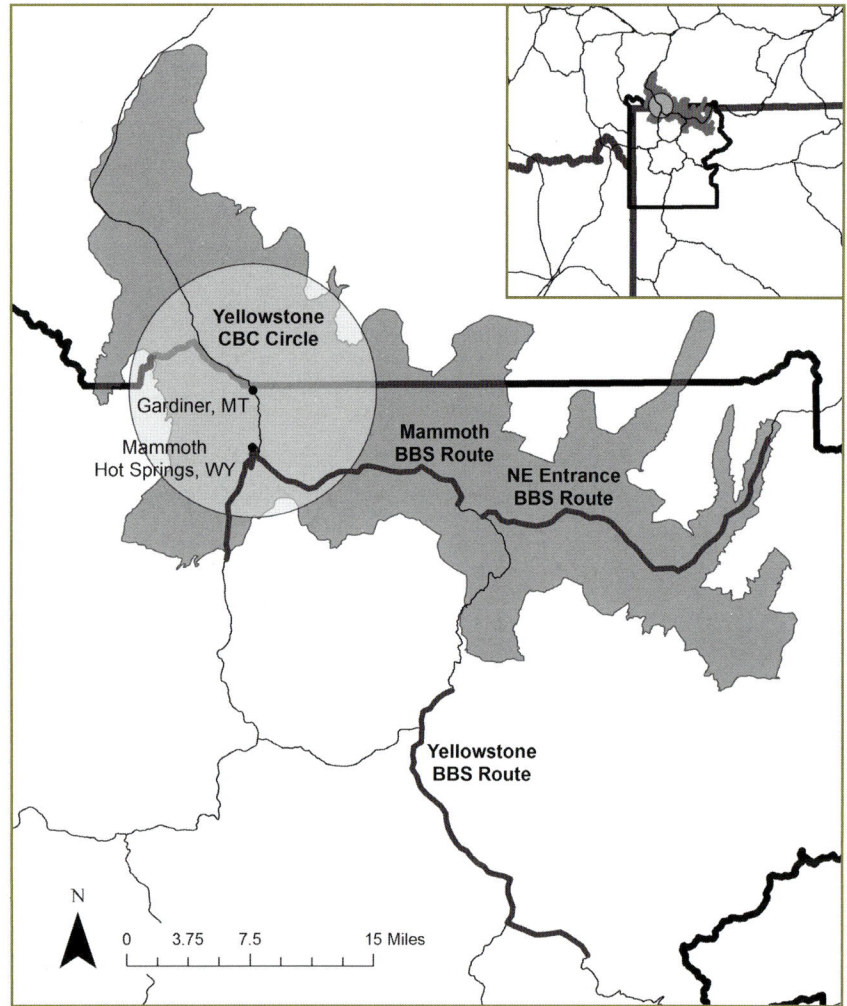

FIGURE 8.2 Long-running avian citizen-science studies in Yellowstone include the Breeding Bird Survey (BBS; three 24.5-mile/39.4-m road transects, highlighted in dark gray) and the Christmas Bird Count (CBC; a 7.5-mile/12-km radius circle, centered around Gardiner, Montana). The park boundary is delineated by a thick black line; the area shaded in gray indicates Yellowstone's northern range.

Entrance Road (Fig. 8.3). Surveyors have recorded 128 species of birds along this high-elevation route; waterfowl are both diverse and abundant, particularly Canada geese.

A fourth route, the Hebgen Lake route, was initiated relatively recently in 2017 and is located partially within the park, stretching along Yellowstone's western edge. While in Yellowstone, the Hebgen Lake route follows Grayling Creek, sampling a long stretch of high-elevation willow riparian habitat that features obligate species that only use willow, like willow flycatcher and Wilson's warbler. In total, 87 species of birds have been recorded during the short time this route has been conducted.

CHRISTMAS BIRD COUNT (CBC)

The winter season counterpart to the BBS is the CBC. Once each winter, in late December or early January, birders, naturalists, and budding scientists gather to document all the birds within survey circles 15 miles (24 km) in diameter. In the first CBC, in the winter of 1900–1901, citizen-scientists surveyed 25 circles, counting around 18,500 birds of 89 species (Chapman 1901). More recently, in the 120th CBC count during the winter of 2019–2020, volunteers surveyed 2,646 circles across the North American continent, Caribbean, and Pacific Islands, documenting

FIGURE 8.3 Heavy fog is a common impediment to early-morning Breeding Bird Surveys on Dunraven Pass and along the Yellowstone River through Hayden Valley. NPS PHOTO

the presence of 2,566 species and over 42.7 million individual birds.

The Yellowstone Christmas Bird Count circle is centered on the town of Gardiner, Montana, located along the northern boundary of Yellowstone National Park (Fig. 8.2). The count circle crosses both Montana and Wyoming and covers the central portion of Yellowstone's northern range, extending from the community of Jardine, Montana, to the northeast, Blacktail Deer Plateau to the southeast, Swan Lake Flat, Wyoming, to the southwest, and Aldridge Lake, Montana, to the northwest. The circle covers an elevational range from 5,118 to 10,958 feet (1,560 to 3,340 m) and represents a microcosm of the wide variety of Yellowstone's habitats, including rugged mountains, talus slopes, open grasslands and sagebrush, mixed-conifer

forests, hydrothermal areas and hot springs, creeks and wetlands, and the Gardner and Yellowstone rivers. Each winter, an average of 14 citizen scientists (range: 2–50) conduct the Yellowstone CBC survey, and despite the logistical constraints of wintertime access that limit most surveyors to the roadways, they document an average of 40 species (range: 28–102).

While the CBC survey method is not precisely standardized, there is wide applicability for the collected data, particularly over broad spatial or temporal scales. Although Yellowstone National Park is represented by only a single CBC circle, the survey has been conducted annually since the winter of 1975–1976, covering a 45-year period of significant ecosystem change both local to the region and continent-wide.

The eBird (www.ebird.org) system is a relatively new database for bird observations that was first launched in 2002 by the Cornell Laboratory of Ornithology, allowing bird-watchers and professional ornithologists alike to document their observations into a single system that is readily searchable and available to the public. By 2010, eBird was expanded to include worldwide bird observations and began to be used by a rapidly increasing number of active bird enthusiasts across the globe.

Primarily, the eBird system is a way for people to keep lists of the birds they have seen and heard, tracking where, when, and how the observations were made, and allowing the individual birder to store their personal observations and lists for posterity. Observers' photographs and sound recordings can also be uploaded into the individual lists. There are many portals into the database so that information can be accessed in numerous different ways, and along with the value for science and research, eBird data are also being used in numerous ways for bird education and conservation.

Currently, eBird is the best means for visiting birders to contribute to our knowledge of birds in and around Yellowstone National Park. The reported bird observations are continually reviewed by people with local knowledge and expertise using a system of filters. These filters alert the reviewers to unusually high numbers of birds, out-of-season birds, and rare birds that are out of their usual range or habitat. When an unusual observation is flagged for review, the reviewer can then query the observer, so that the reviewers can make the decision whether to accept the record.

Before the advent of eBird, the list of birds occurring in Yellowstone was limited to personal observations from a small number of park employees and the very few rare bird reports turned in to park staff and verified by the Yellowstone Bird Observation Committee. Prior to 2015, there were few observers using the eBird system in Yellowstone National Park, but in the last few years, there has been explosive growth in the number of people entering data from the region and beyond. As the system has expanded, the volume of data has vastly increased our understanding of bird distribution and abundance. While park biologists conduct focused studies and surveys of Yellowstone's birds, the eBird database is a treasure trove for a broader understanding of the birds in Yellowstone.

DOCUMENTING ECOSYSTEM CHANGE

In comparison with much of the country, Yellowstone National Park is protected from significant development, and the park's land use and habitats have seen relatively little change over the past several decades. Yellowstone is not immune, however, from broad ecosystem changes, many of which (e.g., wildfires, invasive species, increasing human visitation, climate change) have the potential to profoundly impact the park's bird community. These changes have been well-documented in the BBS and CBC survey datasets.

Over the past 100 years, wildfires in Yellowstone and throughout the West have changed in frequency and intensity due to evolving strategies surrounding land use, water rights, and fire management (Zimmerman 2009). Under likely climate change scenarios, future fires in the park may occur more frequently and at greater intensity (Westerling et al. 2011), potentially causing long-term shifts in forest composition and subsequent changes in wildlife habitat quality. The large wildfires in Yellowstone in 1988 provide an excellent example. Immediately following those fires, BBS surveys along the Mammoth and Yellowstone routes reflected a decrease in overall songbird abundance (Fig. 8.4). While some cavity nesters and insectivores may have benefited from the burned snags

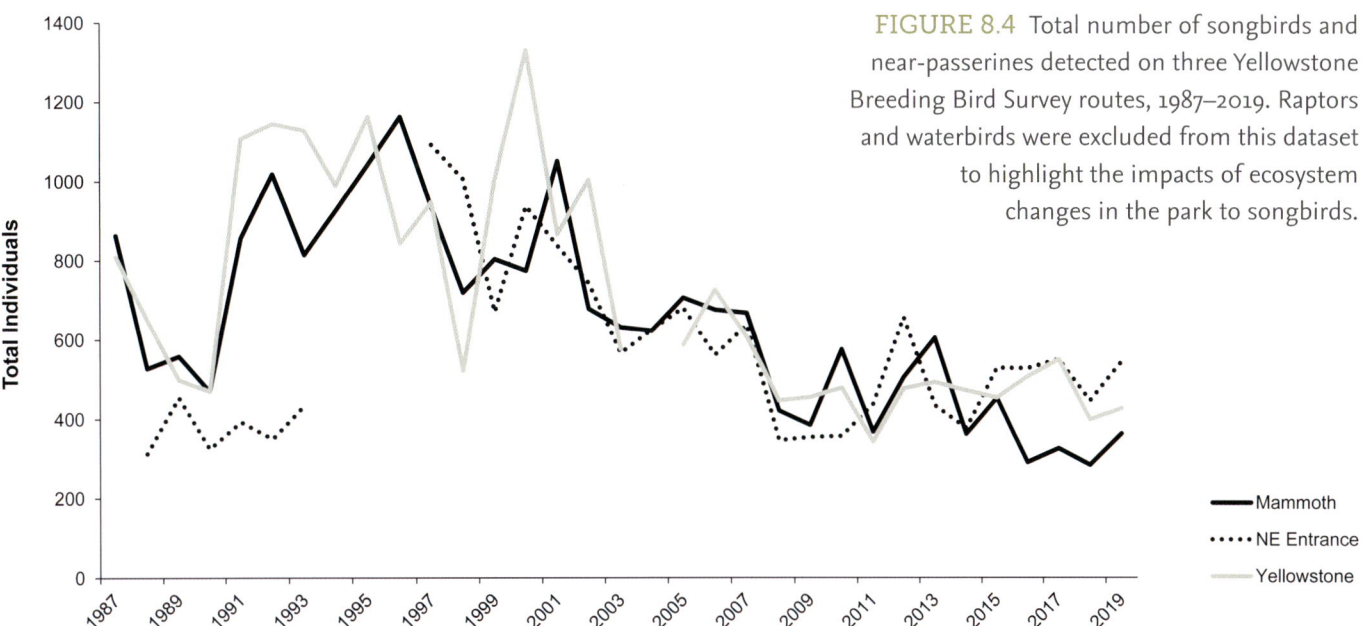

FIGURE 8.4 Total number of songbirds and near-passerines detected on three Yellowstone Breeding Bird Survey routes, 1987–2019. Raptors and waterbirds were excluded from this dataset to highlight the impacts of ecosystem changes in the park to songbirds.

FIGURE 8.5 Cavity nesters—including swallows, chickadees, and woodpeckers, like this northern flicker—readily utilize snags to nest and raise young. PHOTOS BY KEEGAN BURKE

left behind (Fig. 8.5), at least in the short term, most others would have needed to relocate to find suitable habitat. Even after the forests regrew, it was not the same as before the fires; many areas regrew as lodgepole-dominant forest, with relatively little tree or understory diversity. Ultimately, this observation helped motivate the introduction of Bird Program surveys in both recently burned and unburned, mature forests—surveys that are discussed further in chapter 25.

While the extent of non-native bird species in the park is relatively limited, both house sparrows and European starlings are present. These widespread species are highly associated with people and developed areas, and in the Yellowstone area, they are most abundant in the northern range, particularly in the areas around Gardiner, Montana, and Mammoth Hot Springs (Fig. 8.6). During the winter, European starlings have been detected in the CBC in small numbers going back to 1976, while house sparrows did not appear in the counts until 1994 and have since averaged 77 birds per survey. In contrast, house sparrows have not been detected on BBS routes during the summer, while starlings are regularly counted, averaging 23 birds a year since 1988 and peaking in 1997 with 122 starlings. These patterns in invasive species abundance could be correlated with increasing human presence on the landscape, both in surrounding communities and within the park itself. Although year-round park residents have not changed significantly in the recent decades, visitation has increased dramatically. In 1982, the year in which the first BBS route was conducted in Yellowstone, approximately 2 million people visited the park. By 2021, however, annual visitation had increased to nearly 4.9 million visitors, the majority of visits occurring in the summer months. For some species, like swans and loons, more people may mean more nest disturbance and lower reproduction. For other, anthropogenically adapted species, including non-natives like house sparrows and starlings, the increase in the human presence on the landscape may provide opportunities for range expansion. In addition to impacts to the habitat and the park's bird diversity, the increase in visitation to the park also impacts the ability

of observers, especially during the BBS, to effectively conduct their surveys. Parking spaces from which to conduct counts are hard to come by in the summer, even in the early morning, and traffic noise can make it difficult to detect the quieter chip and call notes.

Although the Canada goose is a native species, many ecologists consider these birds invasive, with a locally growing population that may ultimately interfere with regular ecosystem processes. Most Canada geese recorded in the park during the BBS surveys occur on the Yellowstone route (since 1987, between 87% and 100% of the park's total), and the total number of geese observed there has been increasing (Fig. 8.7). Anecdotally, Canada geese are common and numerous on all the park's water bodies in the summer, and the Bird Program staff suspects they may compete with other waterfowl (e.g., trumpeter swans) for resources. Additionally, the abundance of goslings, combined with the decrease in cutthroat trout availability, may have inspired bald eagles to switch primary prey targets to include young waterbirds. In contrast to the summer, Canada geese are detected sporadically during the CBC (Fig. 8.7), with numbers that likely reflect some variability in weather conditions and ice on the available water bodies (YNP Bird Program, unpublished data).

And that leads us to the elephant in the room—climate change. At the broad, continent-wide scale, climate change is already leading to range shifts for some bird species. More specific to Yellowstone, climate change has the potential to amplify the effects of wildfire (as discussed above) and invasive species, and may contribute to significant habitat change. Inevitably, these changes will be reflected in the bird communities detected in the BBS and CBC, potentially reducing the observations of some species, increasing observations of currently "rare" species, or expanding the seasons in which birds may be expected to be encountered.

For example, in the winter months, we expect to see the climate change toward shorter, warmer winters, with less snow and ice and more open water, especially in the northern range. Geese and other winter residents, like mallards, may become more abundant in the winter months and in

▲ FIGURE 8.6 Non-native European starlings gather on the backs of sleeping bison near Yellowstone's north entrance station in Gardiner, Montana. NPS PHOTO

▼ FIGURE 8.7 Canada geese detected on the Yellowstone Breeding Bird Survey route and during the Christmas Bird Count.

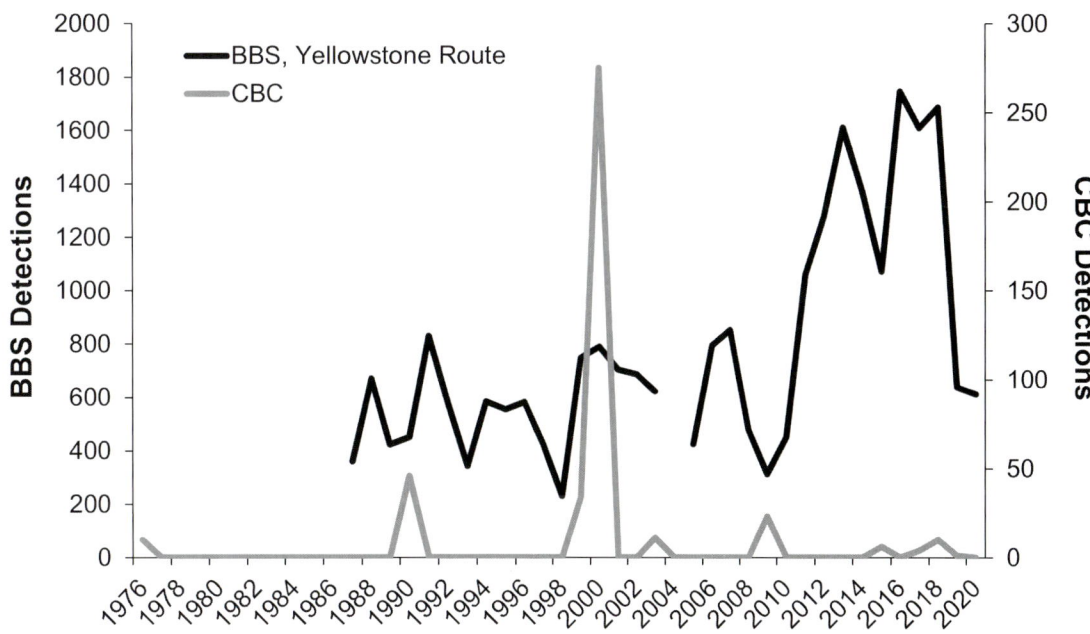

The Yellowstone Bird Observation Committee, the park's rare-bird committee, reviews sightings of species in the *Accidentals* section of the bird checklist and species not on the checklist at all. These sightings contribute to advancing knowledge of bird occurrence and distribution in Yellowstone and provide essential information when park checklists are updated. Rare-bird sighting forms are available to the public at park visitor centers and online: www.nps.gov/yell/learn/nature/wildlife-sightings.htm

the CBC. Even some songbirds (e.g., American robins) that are now only occasionally detected by the CBC may become more regular winter visitors as the warming climate allows for the availability of more food. Contrarily, species that rely on the cold winter months, like Canada jays (Fig. 8.8), are already uncommon on the CBC and may decrease further in abundance if our winters become shorter and warmer.

THE IMPORTANCE OF CITIZEN SCIENCE

According to the *Cambridge Dictionary*, citizen science is "scientific work, for example collecting information, that is done by ordinary people without special qualifications, in order to help the work of scientists." This definition, however, is both wholly inadequate and, in our opinion, insulting. Citizen scientists are not ordinary, and while they may come from a variety of backgrounds, they have many of the qualifications that are most important in conducting science: motivation to learn, willingness to invest their time, and an appreciation for the scientific pursuit of a better understanding of the world around us. Much research that Yellowstone relies on to document trends, make predictions about the future, and guide management decisions would be financially and logistically impossible without the enthusiastic participation of citizen scientists. For Yellowstone National Park and the Yellowstone Bird Program, citizen science plays a vital and diverse role that goes well beyond the Breeding Bird Survey and the Christmas Bird Count. Every visitor (including you!) can contribute to a better understanding of the park's avian ecology simply by paying attention to the birds you see, observing how the bird community changes between habitats and seasons, and noting your observations. Your notes can be as simple as an internal acknowledgment of what you've seen or as elaborate as a formal report to the Yellowstone Bird Program or the Yellowstone Bird Observation Committee. If you're the "listing" type, you might record your observations on a citizen-science platform like eBird. Regardless of how you contribute, we thank all our volunteers and citizen scientists for their continuing efforts and support.

FIGURE 8.8 A Canada jay fluffs up its feathers to keep warm. Canada jays rely on cold winters to preserve stashes of food intended for their nestlings, which hatch in early spring. PHOTO BY GREG ALBRECHTSEN

THE GROUSE OR GALLIFORMES OF YELLOWSTONE NATIONAL PARK

DAVID J. DELEHANTY

The taxonomic order Galliformes are the chicken-like birds of the world, birds such as grouse, quail, pheasants, and partridge. These are economically and culturally important birds worldwide. The chicken, perhaps the most important food-producing animal on Earth, is a galliform bird domesticated from an Asian forest pheasant, just as the turkey, another galliform bird, is the domesticated form of North America's wild turkey. As a group, galliform birds are often referred to as "game birds" because they are so widely hunted for sport and for food.

THE ORDER GALLIFORMES IS a diverse and ancient lineage of birds distributed worldwide. Distinct taxonomic forms occupy different major land masses or climate zones. Yellowstone National Park (Yellowstone; YNP) has two species of grouse, both abundant year-round. They are the dusky grouse and the ruffed grouse, both members of the pheasant family. Colloquially, both dusky and ruffed grouse are called "forest grouse." This is understandable considering that both rely heavily on forests during much of the year, but evolutionarily the dusky grouse aligns closely with North American prairie grouse, such as the greater prairie chicken and the sharp-tailed grouse (Persons et al. 2016). One can think of the dusky grouse as the member of the prairie grouse group that has become most adapted to forests, living where mountain brush communities meet mountain forests. Evolutionarily, the ruffed grouse is distinctly separate from other North American grouse and is a true forest dweller, a species that thrives in the West where northern deciduous forests meet northern conifer forests (Fig. 9.1).

At one time, a third grouse, the white-tailed ptarmigan, may have resided year-round at high elevation within current park boundaries. Two other grouse, the sharp-tailed grouse and the greater sage-grouse, are regionally present and may occasionally traverse the park's boundary but are not known to breed or otherwise have a meaningful presence in the park. Lastly, the spruce grouse occurs in northern conifer forests, including the Northern Rockies of Idaho and western Montana. Its distribution meets the northwestern corner of the park, but they are not documented to be present in the park currently.

Yellowstone's two current grouse species exhibit important traits shared by others in their order. Dusky grouse and ruffed grouse, like so many galliforms, spend much time on the ground, where they forage and nest. They have well-developed legs for walking and are strong runners. These grouse have complex plumage patterns that provide camouflage, and they are expert at hiding quietly when a threat is present. Like other galliformes, dusky grouse and ruffed grouse are strong flyers, at least for short distances, and if a threat becomes too great, they burst into flight with startling wing claps to escape.

The feet of grouse and other galliformes have three forward-facing toes and a strong rearward-facing hind

THE GROUSE OR GALLIFORMES OF YELLOWSTONE NATIONAL PARK

toe. All toes have well-developed claws. Forest grouse like the dusky grouse and the ruffed grouse can readily grasp branches, allowing them to perch and roost in trees. For forest grouse, this arboreal ability importantly translates into being able to forage on buds, fruits, and leaves from trees and shrubs, so they are not restricted to foraging on the ground. This is an essential adaptation that allows them to forage in trees during the winter when heavy winter snow blankets the park. When snow does not cover the ground, grouse can use their feet to scratch the ground and sift through leaf litter for shoots, seeds, and insects that comprise terrestrial elements of their diet. All galliformes possess a robust, decurved bill that is well suited for clipping vegetation for consumption.

Importantly, Yellowstone's grouse possess a very well-developed crop and gizzard. The crop is a stretchable out-pocketing of the esophagus, near the base of the neck, where the throat enters the body cavity. The crop serves as a storage compartment for food prior to any digestion. When a grouse picks up and swallows a food item, it can swallow the food directly into the stomach, or it can divert the food into the crop for temporary storage. Essentially, grouse and other galliformes fill their stomach at good foraging sites but then can continue to forage, storing one or more additional meals in their crop to be taken with them to roosting sites, where the food progressively is advanced to the stomach for digestion. In Yellowstone, this means that grouse can carry substantial food back to nocturnal roosts and essentially continue to eat while roosting through the long, cold winter nights.

The avian stomach is composed of two parts, a forward chamber known as the proventriculus and a rearward chamber known as the gizzard. Food swallowed to the stomach moves quickly through the proventriculus, where it is treated with digestive enzymes, and then it enters the gizzard, the muscular portion of the avian stomach. The gizzard crushes food in a manner analogous to mammals chewing food in the mouth. A powerful gizzard allows grouse to consume hard or tough foods, like seeds and fruit, and efficiently crush them for effective digestion. This mechanical pulverization of hard foods in the gizzard is aided by the grouse swallowing small stones that are held in the gizzard, where they act like free-floating teeth in the grinding process. The scientific term for gizzard stones is gastrolith. The use of gastroliths is an ancient digestive trick used by birds as

FIGURE 9.1
Ruffed grouse in snow.
PHOTO BY RONAN DONOVAN

well as their evolutionary ancestors, the dinosaurs. In winter, when heavy snow covers the ground, grouse will use the hard pits of dried fruits as grinding "stones."

Grouse have an additional trick of digestion that is central to the ability of dusky grouse and ruffed grouse to occupy Yellowstone National Park. It is called hindgut fermentation. In hindgut fermentation, grouse use their intestines to detoxify and digest plant foods that many other animals could not use. Plants often protect exposed buds and needles from herbivores by infusing vegetative parts with unpalatable secondary compounds that are bitter or noxious. Grouse defeat the plant's anti-herbivory strategy by using hindgut fermentation to neutralize plant secondary compounds. Good examples are the needles of Douglas-fir and lodgepole pine. Pine needles are nutritious but cannot be consumed in large quantities by most herbivores due to volatile oils within the needles that can alter the gut microbiome. Yet pine needles clipped from living trees make up the primary winter food of dusky grouse. Similarly, aspen and willow buds are protected by bitter acids yet are major winter foods for ruffed grouse. In fact, it is the ability to neutralize plant secondary compounds through hindgut fermentation that allows the grouse of the world to occupy the cold regions of the Northern Hemisphere. In Yellowstone, it allows dusky grouse and ruffed grouse to stay in the park year-round.

DUSKY GROUSE

The dusky grouse as a distinct species is a recent restoration of species status. For approximately 60 years prior to 2006, the dusky grouse was conjoined with the sooty grouse as a single species known as the blue grouse. Strong genetic, morphological, and behavior evidence supports the separation of these sister species. Ironically, early naturalists recognized the distinctiveness of the dusky grouse, referring to it as Richardson's grouse (e.g., Anthony 1903). Early records and commentary on Richardson's grouse apply to what is now known as the dusky grouse.

FIGURE 9.2 Dusky grouse display.
PHOTO BY HOWARD WEINBERG

The dusky grouse has evolved to exploit both mountain brush communities and mature conifer forest on a seasonal basis. Typical dusky grouse habitat consists of mountain brush with accessible mature conifer forests in the area, often upslope. Dusky grouse use brushy mountain meadows for breeding, nesting, brood-rearing, and foraging. With the onset of spring, dusky grouse depart from high-elevation conifer forests and move down to snow-free areas with mixed brush and grass, such as south-facing slopes or the margins of chokecherry stands. There, males announce their presence with low-frequency hooting whistles and court females they encounter with displays of posture, plumage, brightly colored and inflated throat pouches, and a bright yellow-orange supraorbital comb (Fig. 9.2). Males are assumed to be polygynous (having more than one female mate at a time). Little is known of female mating patterns. Females nest alone, without assistance from males and away from other females. Nests are placed on the ground in a wide range of shrubby or open habitats, but most nests are placed under a shrub canopy and contain clutches of 1–12 eggs that are incubated for about 26 days. Females brood young chicks and lead the brood to foraging sites, where the young feed themselves, and to roosting sites on the ground. Young progressively become thermally independent and capable of rapid flight. Broods break up from the end of summer through early fall. Dusky grouse do not appear to be highly social, though small groups are sometimes seen. With the onset of winter, dusky grouse make an altitudinal migration, moving upslope into conifer forests, where they switch to a wintertime diet of conifer needles, living and roosting above the snow in the trees.

Relative to many other North American game birds, dusky grouse have been lightly studied. Much of the existing literature on the species is from the perspective of the conjoined "blue grouse" species rather than exclusively dusky grouse. Nevertheless, populations in the western United States are widely believed to be secure at this time. This is because large portions of their mountain habitat lie within public lands, especially national forests and parks, which are not subject to a high degree of habitat conversion. Barring catastrophe, dusky grouse are likely to persist in the park into the foreseeable future.

FIGURE 9.3 Ruffed grouse display.
PHOTO BY HOWARD WEINBERG

RUFFED GROUSE

Ruffed grouse are common throughout deciduous and mixed conifer/deciduous forests of northern North America. They are extremely popular with hunters as a game bird; one result is that they are the subjects of substantial scientific investigation. Typical ruffed grouse habitat consists of brushy forest edges where forbs (non-woody flowering plants other than grasses, sedges, and rushes—e.g., wildflowers), fruiting shrubs, and surface water are available. In the mountain West, ruffed grouse are found where deciduous forests and conifer forests intermingle and are intersected with riparian habitat. In brief, ruffed grouse forage on green herbaceous plants from the ground in spring, adding fruits taken from shrubs, such as chokecherry and dogwood, as they become available in summer and fall, then turning to the buds of shrubs, such as willows, and trees, such as aspen, with the onset of winter. Conifer trees serve as valuable roosting sites that provide visual cover from predators and thermal cover during winter. Under harsh winter conditions, the ruffed grouse also may burrow into soft snow on the ground to form a temporary roosting chamber that modulates severe winter weather.

The presence of ruffed grouse is easy to detect in the spring due to the loud drumming of males seeking mates (Fig. 9.3). Drumming consists of a series of about 50 sharp, snapping wing claps across an approximately 10-second interval. Wing claps start slowly, rapidly increase in tempo, and finish with claps so rapid that the sounds merge in human hearing. For drumming, males use an elevated platform, commonly a fallen log within brushy cover. Males may drum at any time of year, but drumming peaks during spring, with a lesser peak during autumn.

For breeding, females visit drumming males and select their mates. Females will mate with more than one male, just as males will mate with more than one female.

Females select a nest site, always on the ground and frequently at the base of a tree or stump, where they incubate 9–14 eggs without assistance from males. Typical incubation is 23–24 days, after which females lead their broods away from the nest. As is typical of grouse, females brood young chicks as needed during the day and through the night until the young become thermally independent. Females lead their young to foraging sites on the ground, where young must feed themselves, first with insects and then, increasingly, with herbaceous matter clipped from plants.

Barring unforeseen catastrophe, ruffed grouse are highly secure in the mountain West, including the park, as long as brushy riparian zones with surface water occur with aspen and other deciduous tree stands, especially if conifers are present for secure roosting.

OTHER GROUSE

Several additional grouse species should be considered as possible for occurrence in Yellowstone National Park, though intensive survey might be necessary to confirm any presence or to have confidence in true absence. Wild turkeys, native to North America but not to the northern mountain West, and three non-native galliformes (the gray partridge, chukar, and ring-necked pheasant) have all been widely introduced to the American West as game birds over the last century, an activity still being conducted by state wildlife agencies. Currently all four species occupy private and public lands in proximity to the park. Additionally, three native species of grouse may, at some point, colonize or be present within park boundaries, including the spruce grouse, the white-tailed ptarmigan, and the greater sage-grouse. Of these, the greater sage-grouse is perhaps the most interesting possibility, due to the tenuous population status of this species.

The greater sage-grouse is an iconic bird of the American West and a sagebrush obligate species—that is, it depends on sagebrush during all phases of its life history. Sage-grouse are not known to be in the park, despite being present regionally and despite the park containing sagebrush stands within its interior basins. However, sage-grouse avoid tree cover and may be unlikely to transit high-elevation forests circling the park to reach relatively small interior sagebrush communities that are tree-free. Additionally, sage-grouse rely on exposed sagebrush for food during winter, making Yellowstone's heavy snow cover a severe winter challenge. Lastly, because of the park's high elevation, the temporal opportunity for breeding is small.

Sage-grouse employ a lek breeding system in which large male grouse with bright white plumage ornaments congregate in exposed, traditional locations within sagebrush communities and conspicuously display visually and with low-frequency sounds broadcast across the landscape. This makes sage-grouse leks relatively easy to locate, yet no leks are known to exist in the park. Sage-grouse are not reported on the Breeding Bird Survey, nor are they listed as occurring within the park, although a few reports of sage-grouse sightings have been made on citizen science (eBird) sites. Unfortunately, these reports do not contain photo documentation. It is plausible that dispersing or reconnoitering sage-grouse could be observed in the park as they transit the area. Alternatively, it is also very plausible that dusky grouse, which regularly occupy sagebrush communities in the vicinity of conifer forests, could be mistaken as sage-grouse by citizens who are unfamiliar with the grouse of the mountain West, especially large, multi-colored male dusky grouse.

Yellowstone's two grouse species are widespread and frequently encountered by visitors but, despite their commonness, and perhaps in part due to their categorization as game birds, we know relatively little about their populations in and around the park. Certainly, this is one of the avian groups that deserves more attention within the park, especially with potential climate-induced changes to habitats and fire regimes. Backcountry travelers should be on the lookout for any of the two known species and possibly for a spruce grouse, whose range abruptly ends at the northwest park line. And who knows? It may be you who spots Yellowstone's first sage-grouse.

YELLOWSTONE'S BIRDS

3 RAPTORS

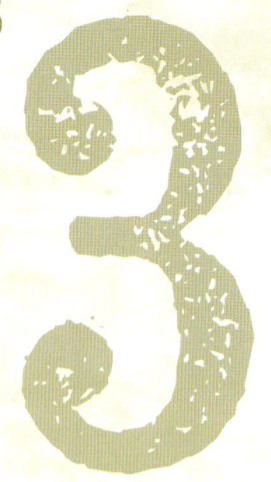

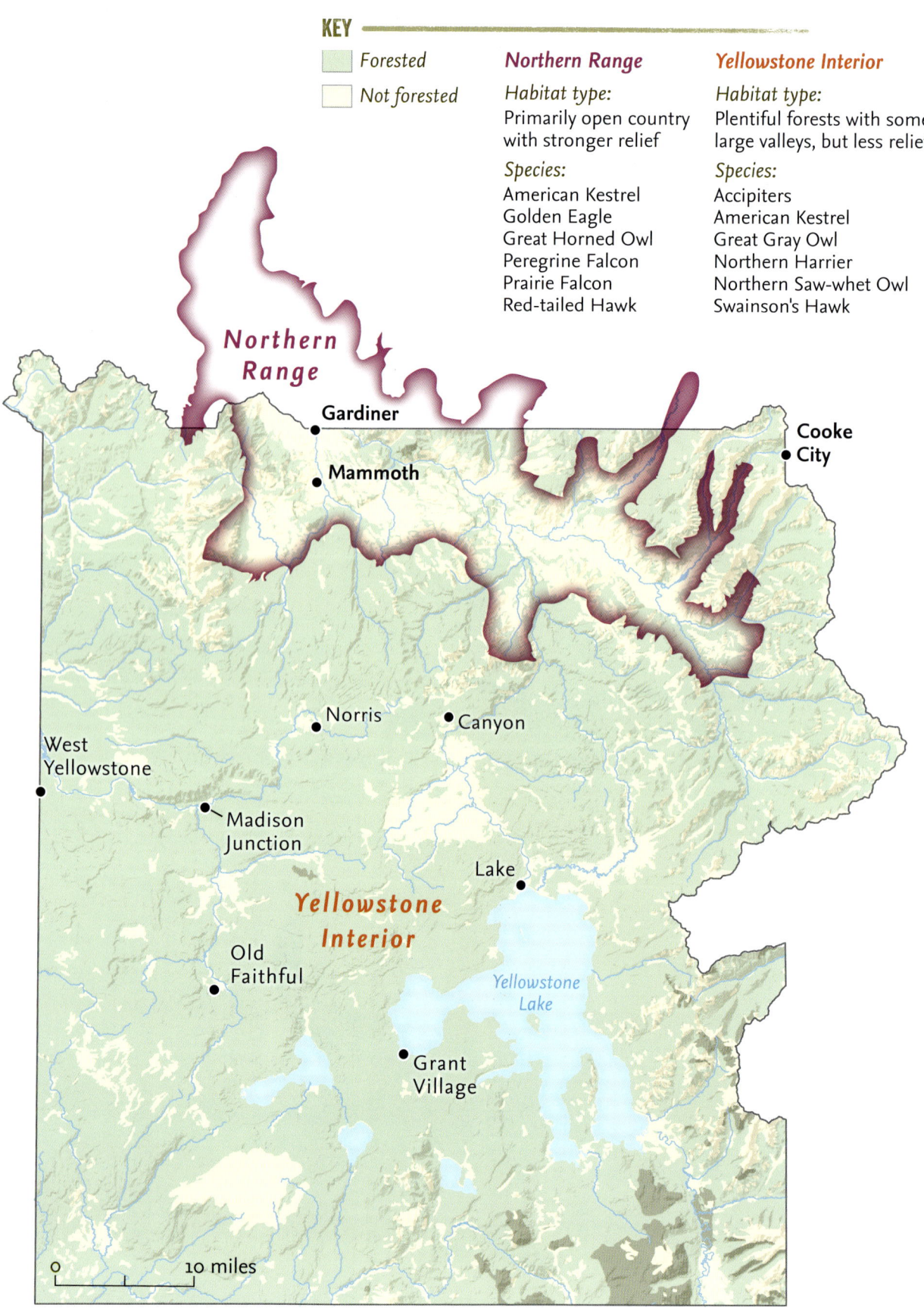

KEY

- Forested
- Not forested

Northern Range

Habitat type:
Primarily open country with stronger relief

Species:
American Kestrel
Golden Eagle
Great Horned Owl
Peregrine Falcon
Prairie Falcon
Red-tailed Hawk

Yellowstone Interior

Habitat type:
Plentiful forests with some large valleys, but less relief

Species:
Accipiters
American Kestrel
Great Gray Owl
Northern Harrier
Northern Saw-whet Owl
Swainson's Hawk

Northern Range

Gardiner

Cooke City

Mammoth

Norris

Canyon

West Yellowstone

Madison Junction

Lake

Yellowstone Interior

Old Faithful

Yellowstone Lake

Grant Village

0   10 miles

PEREGRINE FALCON
THE MOST BEAUTIFUL BIRD?

DOUGLAS W. SMITH, KATHARINE E. DUFFY, DAVID B. HAINES, and LAUREN E. WALKER

A blue-backed bullet, a male peregrine, hurtles out of the sky and surprises a common raven flying only a few feet from the nest where the peregrine's mate has just begun incubating eggs. The raven squawks as it hastily exits the area, the male peregrine on its tail. No matter that the peregrines' nest this year is a hole in a rock wall last claimed by ravens—it most definitely belongs to peregrines now!

KATHARINE E. DUFFY

NOTHING COMPARES TO a peregrine falcon. Of course, comparing anything in nature is foolhardy. Nonetheless, when beholding this bird, perched or flying, one can only think of superlatives. Strikingly beautiful, masked face, the fastest animal, and a gaze of majesty knowing the ages. Bold and powerful. Untouchable. If you have seen it, you know what we mean—the falcon epitomizes the raw power and beauty of nature all at once. Their worldwide distribution makes them observable to many and has them clinging to rocky cliffs, usually above waterways, but also city skyscrapers, which they use as cliffs. Once nearly brought to extinction, these birds have made a remarkable comeback. Nature's stunning bird has been restored, and Yellowstone National Park (Yellowstone; YNP) is no exception.

Evolution honed peregrine falcons to be unparalleled speed machines. They have long, pointed wings, enabling them to swoop and dive in flight at mind-boggling speeds as they pursue avian prey, from small birds to shorebirds to ducks, that they capture in midair (Fig. 10.1). Their bodies are tightly cloaked in sleek feathers that contribute to their streamlined aerodynamic efficiency—no

fluffy owl feathers on a peregrine. Their nasal openings have a post that baffles air so that peregrines can continue to breathe as they dive. Being struck by a diving peregrine often kills prey instantly, but if it does not, the peregrine inserts the upper part of its bill, with projections called tomial teeth, between the prey's neck vertebrae. With a quick twist, the peregrine instantly severs the spinal cord of its prey. As it flies with prey held by tightly clenched feet, each toe ending in a sharp piercing talon, the peregrine might even eat on the wing.

Peregrines are the most far-reaching terrestrial vertebrate, occupying every continent except Antarctica (White et al. 2020). This wide global reach makes them hard to characterize, as they live across a broad range of ecosystems, from desert to tundra and coasts to mountains. The species has such a broad reach that one peregrine researcher said that his boss referred to a peregrine as "a weed among hawks," as this bird adapts to almost everything—like a weed (Cade 1969). They can also be tolerant of humans. Equally broad is their diet; although they eat birds almost exclusively, they rarely specialize in a particular species (White et al. 2020). This breadth of

FIGURE 10.1 Classic peregrine falcon sighting in the Grand Canyon of the Yellowstone. NPS PHOTO

range and diet makes them resilient, a characteristic needed for life in Yellowstone because of its high elevation, unpredictable weather, and large swaths of relatively unproductive (for birds of prey) coniferous forests.

Although tough, peregrines were not spared from worldwide population declines due to organochlorine pesticide use, mainly dichloro-diphenyl-trichloroethane, commonly known as DDT. Widely used in agriculture in the 1940s and 1950s, this pesticide was used in forest management as well. Some 62 tons of DDT were applied in northern Yellowstone in 1953, 1955, and 1957 to control a spruce budworm outbreak (Furniss and Renkin 2003). This pesticide-induced decline was first identified in

Great Britain (Ratcliffe 1967) but was soon found to have global reach. Although few direct mortalities were attributed to DDT, its sub-lethal effects caused eggshell thinning, resulting in widespread reproductive failure. Across North America, peregrine populations declined during the 1950s to 1970s, depending on the level of exposure to DDT. Populations in eastern North America and southern Canada were completely extirpated, and falcons in western and northern populations declined between 10% and 75%. The portion of the population in Yellowstone National Park was lost. Some thought there would be no peregrines south of the 50th parallel, and there was nothing to be done (Cade 1969).

PEREGRINE HISTORY IN YELLOWSTONE

Piecing together peregrine history in YNP is difficult due to inconsistent records and monitoring. Likely abundant in what is now YNP for centuries, the first peregrine falcon in park archives was recorded in 1914 by naturalist Milton Skinner. Spotty records exist through the next several decades, with sightings in the Bechler area (1929; an adult with two young), the Firehole River corridor, Osprey Falls, and especially the Grand Canyon of the Yellowstone. With towering cliffs astride the Yellowstone River, the Grand Canyon was and is ideal habitat for peregrines. Others, most notably Jay Sumner and Jim Enderson, did some additional monitoring during the 1960s, knowing there was evidence of nesting in the 1950s, but the effects of DDT were already taking hold (McEneaney et al. 1998). Enderson visited the Grand Canyon of the Yellowstone in 1961, 1962, and 1964 and found no peregrines. At another site that had been active in 1960, he saw an adult in 1964 and heard a second but did not find a nest (Enderson 1969).

By the 1970s, sightings in YNP were rare and confirmation of breeding tenuous. Peregrine falcon expert Bob Oakleaf (Wyoming Game and Fish Department; WGFD) conducted surveys during this period and presumed peregrines were present, although he was unable to locate any. The last known nesting territory was vacant by 1970 (Baril et al. 2015). The combination of inconsistent monitoring and DDT-induced declines made putting together the peregrine story problematic; they were nearly extirpated before people really started looking. Because there were no estimates of population size or adequate records of known nest sites, it was impossible to know much about them or how their populations were changing within the park. Elsewhere in North America, populations reached their lowest levels by the mid-1970s. Peregrines were listed as endangered in 1970 under the Endangered Species Conservation Act of 1969, a precursor of the Endangered Species Act (ESA) of 1973 (Peakall 1976), and then listed under the ESA.

It is hard to imagine what the canyons and cliffs of Yellowstone would have been like without the cry and jet-like sound of a blurred peregrine falcon whizzing by. Having arrived post-recovery, the authors of this chapter have only known their presence; for us, the park is almost unthinkable without them. Few things can compare to pulling up binoculars and seeing that distinctive black mask and sharp pointed wings, sensing the seemingly fearless personality, or hearing the *cack cack* cry and their distinctive *ee-chup* call reverberating off canyon walls. There is something almost eerie and ancient in these sounds, and once accustomed to them, it's hard to visit these spots and not experience it.

REINTRODUCTION AND RECOVERY

Once DDT was banned and ESA protections were put in place, recovery efforts began in earnest. A massive, North America–wide reintroduction effort was undertaken. Most prominently, the Peregrine Fund was born, with a goal to captively rear peregrines for release across their decimated range. Along with others, the Peregrine Fund released nearly 7,000 peregrines across North America between 1974 and 1998 (White et al. 2020). YNP was part of this significant endeavor and was considered the center of recovery efforts for the Wyoming, Montana, and Idaho region, with 36 captively raised young released in the park at four sites between 1983 and 1988 (Oakleaf and Craig 2003). These acclimation and release areas, called hack sites, were located at Madison Junction, Slough Creek, Crown Butte, and Terrace Mountain. Regionally, 644 birds were released at 35 sites within 162 miles (260 km) of YNP from 1980 through 1997 (Baril et al. 2015). Interestingly, the last territory known to be occupied in YNP, located in the Grand Canyon of the Yellowstone, was the first to be reoccupied in 1984 by two of these released birds. The female came from a Jackson, Wyoming, release area and the male from a nearby location in Idaho (McEneaney et al. 1998).

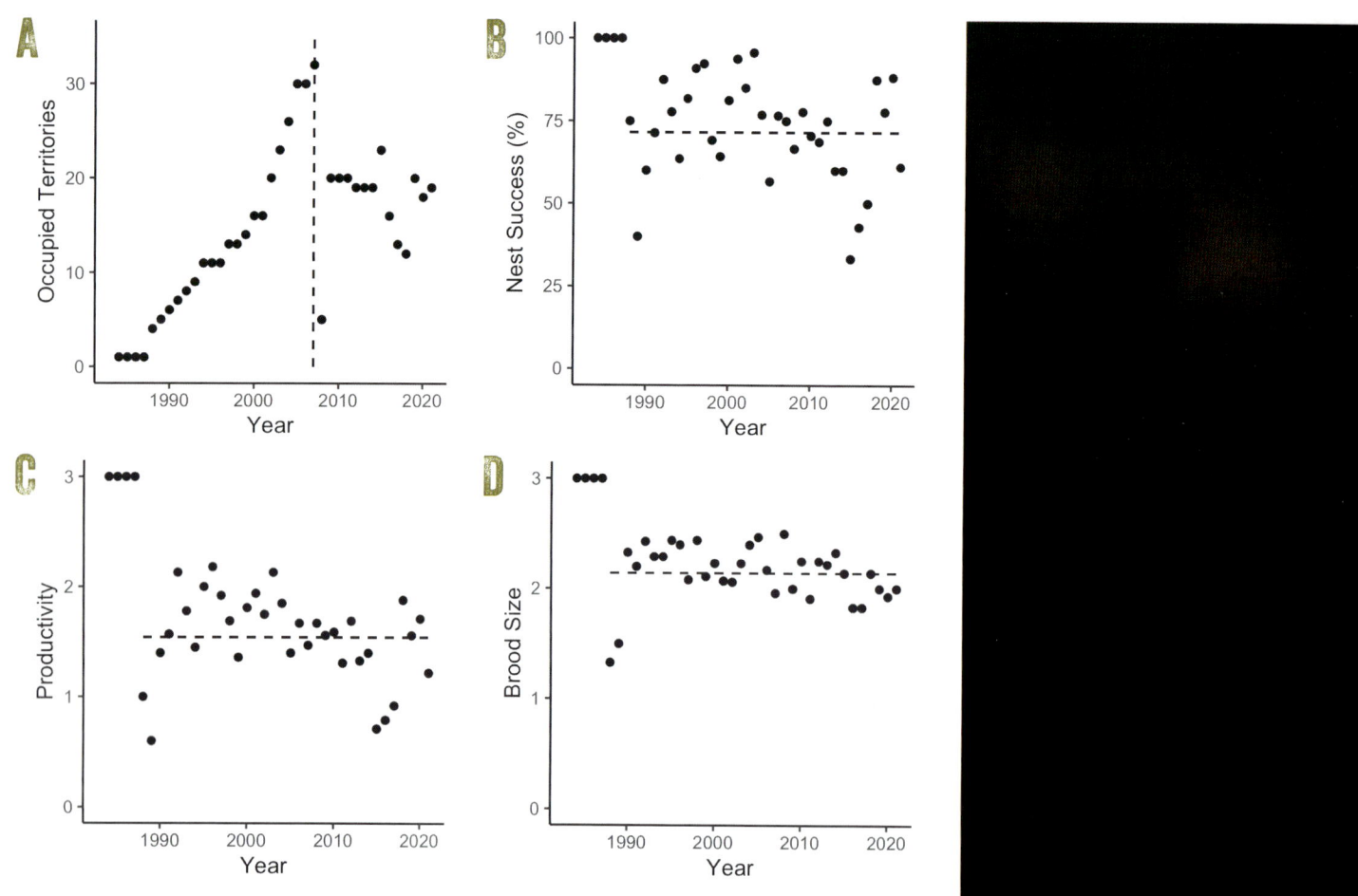

FIGURE 10.2 Population data on peregrine falcons in Yellowstone National Park, 1982–2020: (A) occupied territories, (B) nest success, (C) productivity, and (D) brood size.

After the Grand Canyon of the Yellowstone territory was occupied, reoccupation of the park was spotty and slow (Fig. 10.2). By 1994, only 11 pairs (about one new pair per year) had reoccupied YNP. Additionally, reproductive success was initially low, and through 1989, only about three young per year were produced park-wide. After 1989, however, reproduction accelerated, and in 1992, 17 young were produced. Furthermore, there was evidence, based on banding records, that most of the recruitment was from birds born to wild and not released parents. As time went on, reoccupation continued to increase, and by 2007, there were 32 known pairs across the park (Baril et al. 2015).

HOW WE MONITOR PEREGRINES

Peregrines do not winter in Yellowstone, but adult males return each year by late March or early April and claim or reclaim their nest territory, with adult females arriving soon after. Once a female is on-site, an adult male peregrine will perform spectacular aerial acrobatics as the female peregrine watches from a perch, sometimes joining the male in flight. A male peregrine rises effortlessly on updrafts along rock walls and ridges, then uses his long and pointed wings to swoop and dive. Two peregrines

FIGURE 10.3 Peregrine falcons commonly live near rocky cliffs by water and use rock outcroppings for perches. PHOTO BY RONAN DONOVAN

circling each other high in the air is a spectacular sight. While this phenomenal aerial display is breathtaking, it's not the easiest way to detect peregrine courtship activity.

A male peregrine perched in a prominent location, often at the top of a rock wall or rugged canyon, offers a definite clue to territory occupancy (Fig. 10.3). When two peregrines are perched near each other, the size difference is usually apparent. Peregrines, like many other raptors, exhibit reverse sexual dimorphism, meaning the male is considerably smaller than the female.

For an eyrie (nest site), peregrines claim a horizontal ledge, crack, or hole on a cliff or rock outcrop, formed by extensive lava flows that cloak much of Yellowstone (Fig.

10.4, 10.5, and 10.6). Male and female peregrines frequently display together on the nest ledge during early courtship; observing this behavior helps to pinpoint the chosen ledge. Vocalizations described as *ee-chup* and *ee-chip*, given by both male and female, accompany ledge displays; these calls can sometimes be heard quite far from the nest wall. During courtship, loud and strident wailing by the female can serve as solicitation for copulation. When they copulate, the male lands on the female's back, and she holds her tail to the side so that his cloaca meets hers, transferring the male's sperm to the female, with the male often flapping his wings as they copulate. The cloacal kiss, as it's called, lasts just a few seconds (Fig.

◀ FIGURE 10.4 Looking inside a nest ledge with three chicks. Although it is a natural ledge, it almost seems constructed for them. Note signs of multi-year use. Peregrines do move nest sites, but some are used repeatedly. NPS PHOTO/ DOUGLAS SMITH

▼ FIGURE 10.5 One territory with two nest ledges, each used in different years (note: the ledge on the right is the close-up in Fig. 10.4). NPS PHOTO

10.7). The female might wail after copulation, and the male might make calls, too. More than once, a male peregrine nesting near Mammoth copulated with his mate immediately after chasing a rival peregrine or other raptor from the nest area. His predictable reaction to his prowess at nest defense made it easy to find where the female was perched.

Woe to rivals and intruders daring to enter a peregrine's territory. A monitor watched as a sub-adult golden eagle flew along the top of a nest wall, right above an active peregrine falcon eyrie. As the eagle nearly reached the end of the wall, the male peregrine appeared out of nowhere and dive-bombed the eagle. With that, the hapless young eagle turned around and did the worst possible thing. It flew back over the wall, directly above the active nest. Now the male peregrine displayed his fury by swooping again and again, dancing on the back of the eagle as both flew out of sight. Peregrines can make loud "cacking" calls as

◄ FIGURE 10.6 Typical eyrie or scrape on a canyon wall. Almost no actual nest is used. This site, just off the Grand Loop Road in Yellowstone National Park, may be the most-viewed natural peregrine eyrie in North America. PHOTO BY RONAN DONOVAN

▼ FIGURE 10.7 Copulating peregrine falcons in Yellowstone National Park. Notice the size dimorphism, with the smaller male on top of the female. PHOTO BY RONAN DONOVAN

FIGURE 10.8 (A–F) A male and female peregrine falcon exchange a prey item (an eastern kingbird) in the air.
PHOTOS BY SCOTT HEPPEL

they respond to an intruder that is too near their nest ledge, another cue that strongly suggests an active nest nearby. Observation of nest-related activities provides substantiation of nest success and eventually the number of young produced.

As courtship proceeds to egg laying, the female can look "dumpy" and disheveled. At this point, she tends to stay close to the eyrie, another indication of nest location. Peregrines typically lay three or four eggs. Females do most of the incubation, including during each night. Males take brief turns at incubation and otherwise play a significant role in food procurement for the incubating female. Formerly, some referred to peregrines as the "big-footed falcons" or "great-footed falcons" because of their large toes, which they slip under their eggs while incubating. It is especially important to not abruptly disturb incubating peregrines; a hasty departure might lead to egg loss (Cade 1982, Palmer 1988; J. Pagel, pers. comm.). It should be noted that not everyone liked these monikers, as they suggest clumsiness rather than grace or agility (Bent 1961). When the female sees the male, she wails several times. The sound of wails can be a cue that the male is approaching the eyrie. Or it can reflect the female's hunger or the need for a break from incubation, called an incubation exchange.

Prey deliveries and incubation exchanges afford opportunities to determine the exact location of the eyrie. During a prey delivery, the female flies off the nest and meets the male, who transfers prey to her in the air (Fig. 10.8). Or the prey exchange can take place at a perch. During an incubation exchange, the male will proceed to the eyrie, taking a moment or two to settle on the eggs, thus pinpointing the nest site. Incubation lasts approximately 33 days. The tiny white fuzzy hatchlings grow quickly and can stand up in several days. When air temperatures are low, at night and during inclement weather, the female broods the chicks, keeping all of them safe and warm under her body. When the male delivers food to the female, she stands up to rip it apart and feed the chicks, allowing observers to count and age the chicks. Development of the chicks and growth of their body and flight feathers also permit aging of the chicks, allowing us to maintain records of breeding chronology (Cade et al. 1996).

Peregrine nestlings fledge at approximately 42 days, but usually they first meander a short distance from the nest site, if the topography allows. Watching fledgling peregrines awkwardly stumble along or away from the nest ledge, struggle to steal food from their siblings, exercise their newly grown wing feathers, and other antics can be hilarious to see, but the fledglings gain strength quickly. Within a few weeks of fledging, young peregrines begin to demonstrate the aerial abilities for which they are deservedly famous.

THE CURRENT POPULATION

With the population back on its feet, scientists in general, and in the park, utilize several standard measures to assess raptor well-being. Although we track the number of occupied territories (Fig. 10.2), this metric is most helpful if we can monitor every territory every year—a feat that logistically is impossible for us to achieve. Thus, we also monitor average annual nesting success, productivity, and brood size (Fig. 10.2). Nesting success is the *percent of occupied territories that fledge at least one young*, productivity is measured as the *average number of young fledged per occupied territory*, and brood size is the *average number of young fledged per successful pair*. These metrics can be confusing; the key is distinguishing between "occupied territory" and "successful pair." Occupied territories consist of a bonded pair; not all pairs try to nest, and not all that nest are successful in fledging young. Successful pairs, the basis for our calculations of average annual brood size, are thus a subset of all occupied territories.

Looking closely at these metrics since peregrines reoccupied the park, all appear highly variable, with no trend (Fig. 10.2). No trend is always concerning; in general, we like it when things are increasing. But peregrines are not like stocks; they can't increase indefinitely, and despite these noisy data, we believe our population is stable. Peregrines appear to be doing well in Yellowstone.

We base this conclusion on averages over long periods of time. From 1988 through 2020, nesting success averaged 72% with a range of 40% to 96%, productivity averaged 1.6 young per occupied territory, with a range from 0.6 to 2.2, and brood size averaged 2.2, with a range of 1.3 to 2.5 young per successful pair (Fig. 10.2). All these measures are considered good for peregrine falcons; they are sufficiently high to suggest that our population is in no immediate danger of decline (White et al. 2020). The highly variable annual data suggest that some years are very good, while others are not—a common pattern for many bird species. Birds, despite being incredibly tough and well adapted to a wide spectrum of temperatures and weather conditions, do need somewhat favorable conditions to raise young and find enough food to be successful. Most studies indicate they can't do this every year, and reproduction can be very responsive to this annual variability. This reproductive response to environmental conditions is more acute in high-elevation environments like Yellowstone, where climate is more extreme.

Another factor in interpreting our peregrine metrics is the population ceiling—what wildlife biologists call carrying capacity. In short, there is only so much room for peregrines in Yellowstone, and when the population becomes saturated, or meets the carrying capacity, all these measures of reproduction stabilize, decline, or vary instead of just increasing, which is what we observed in the early years of post-DDT recovery. The population is less productive overall when compared to an increasing population. Wildlife biologists call this density dependence, and it is about the only constant in the ecological world—or so it seems. In the physical sciences, the world is governed by laws; many phenomena are measurable, replicable, and found everywhere. Gravity is gravity, no matter where you go. Hardly anything is like that in ecology. Peregrines in Yellowstone are different than peregrines in Alaska. That being said, we think peregrines are at their ceiling in Yellowstone. How do we know?

First, the overall population appears stable. How does this occur when reproduction varies so much from year to year? We surmise that a non-breeding segment of the population is acting as a buffer. New birds are added to the population annually through reproduction, with a life span of about 10–15 years; most of these birds return to areas near where they were born, but they can't find a place to nest because all the good spots are occupied. So these birds wait for an opportunity, or float until they can find a place to breed, either through replacing another bird or pioneering a new territory. These floaters are key to maintaining a stable peregrine falcon population.

Another indicator of population stability is that, when peregrines were reoccupying Yellowstone after being extirpated, they steadily increased their settlement of cliff territories near water sources. Today, we rarely observe new territories; in fact, occupation of some territories will vary from year to year, while others have remained unoccupied across multiple years. For example, from 1996 through 2008, there were no unoccupied territories; yet from 2009 to 2020, there were at least two every year and sometimes as many as six (Fig. 10.2). Why? It's hard to know, but recolonization of the park probably "overshot" the number of high-quality territories, and initially, many territories that are marginal or low-quality habitat were settled and occupied. Perhaps these marginal territories are those that now come in and out of use from year to year.

Some specific examples provide support for the existence of marginal territories and population stability. One territory at Firehole Lake was, in 2009, occupied by a successful pair that produced two young, then a pair whose nests failed in 2010 and 2011, then a non-breeding territorial pair, then a single adult, and then a single subadult; it is now unoccupied. Other territories in the park interior, like Elephant Back and Fairy Falls, are similar. Besides the harsh weather at these interior locations,

another factor causing these territories to be marginal may be prey availability. All these locations are high-elevation coniferous forest—not ideal habitat for songbirds, or other birds for that matter. Food and weather may be limiting at these sites, and only in favorable years are peregrines at these locations able to successfully raise offspring.

During the early 2010s, we did a five-year, park-wide search for peregrine falcons. We confirmed 32 occupied territories (Baril et al. 2015). As we reflected on these results, one view was that there must be more; the park is so vast, and surely we had missed some. Another view was maybe not; although the park is large, optimum territories are limited. These high-elevation forests, mostly composed of lodgepole pine, harbor limited nest sites and few birds for peregrines to eat. However, we probably will never know which hypothesis is true. We have done the best we can with the resources we have to unravel this most fascinating story. Our goal now is to observe all known territories on about a three-year cycle, roughly checking in on a third of the territories each year to assess how well the population is doing. If something like DDT hits the population again, we want to catch it early.

Problems or issues like these are the beauty of studying nature. No place is like any other, so you must spend time in your place to really get what is going on. And it will not come to you in a spreadsheet full of numbers. The answers will only come through many years of observation and thought—if for no other reason than to observe and document the good/bad year cycle. Ah, nature is so beautiful. So much to find out, so little time.

PREY REMAINS AND EGGSHELL THICKNESS

There are two other important issues for Yellowstone's peregrines: food and eggshell thickness. As we discussed above, we don't know how much food is available, but we do have some idea of what they eat. A common technique for assessing food is through the collection and identification of prey remains, which is accomplished by rappelling into a nest. Coauthor David Haines and two other biologists rappelled into 24 peregrine eyries over three years (2010, 2011, and 2013) to sample what the birds had eaten, as evidenced by feathers and bones that remained. (Bird identification from a feather alone is tough, but we turned to an expert, John Schmitt, known for his expertise and attention to detail. He can often identify a species from a single feather!)

Ninety-seven percent of the remains in the eyries were birds, not surprising since peregrines almost exclusively prey on birds (White et al. 2020). American robins were the most frequently detected species (11%)—poor robins—followed by Franklin's gull (8%), unexpected since they do not nest in the park. But in early May 2017, nest monitors near Tower watched as a small flock of Franklin's gulls flew overhead with a male peregrine in hot pursuit. Gulls and peregrine disappeared out of sight over a nearby ridge, but wailing a few minutes later told the likely end of the story: the male peregrine caught and killed a gull, then dutifully delivered the gull to his mate.

Other common (about 5% of remains) species were mountain bluebirds, red-necked phalaropes (which also do not breed in Yellowstone and are considered rare), Wilson's phalaropes (which do breed in the park but are considered uncommon), and northern flickers (common). All other species were represented by six or fewer individuals (Baril et al. 2015). Clearly, although they take advantage of what is commonly available, our peregrines have a taste for certain species (e.g., phalaropes) despite them being rare. For example, we found feathers from a red phalarope, yet there are no reported records of the species in YNP.

In addition to bird remains, there were also some unforeseen findings: Two nests had fish remains, and another had a marten foot (Pagel and Schmitt 2013). Whether the peregrines hunted these prey items or stole them from another predator is unknown, but apparently, they fed on them. Life entails doing whatever is necessary to live. Or, when you move at the fastest speeds known to the animal world, you eat whatever you want.

Remains of some prey items will not be found on the nest ledge. For example, in early July 2013, an extremely voluminous hatch of salmonflies, a type of large stonefly, occurred east of Mammoth. On July 2, at about sunset, a pair of adult peregrines with an active nest nearby spent at least 35 minutes feasting on salmonflies caught on the wing. The peregrines snagged the salmonflies with their feet, lifted their feet to their bills, and quickly consumed their prey, then grabbed another and repeated the process over and over and over.

Lastly, eggshell thickness was the reason for the near extinction of this wonderful bird. Along with prey remains, our rappelling biologists also collected eggshell fragments. Thinning eggshells were the consequence of DDT and the driver for previous population declines; it was more than 20 years since they were last measured, and we wanted to confirm that today's peregrine eggshells were adequately thick. As with monitoring population demographics, we want to make sure that another environmental contaminant, another DDT, does not sneak up on us. For eggs collected from 2010 to 2013, mean eggshell thickness was 0.0137 ± 0.0006 inch (0.347 ± 0.017 mm), only 4% thinner than pre-1947 (pre-DDT), providing supporting evidence that the peregrine population has nearly recovered from the negative effects of DDT (Baril et al. 2015).

THE PEREGRINE'S FUTURE IN YELLOWSTONE

Peregrine falcons are back and doing well in Yellowstone. With continued monitoring of both birds and eggs, we hope to be able to detect any decline or new environmental contaminant before it leads to extirpation again. Protecting all the native wildlife in Yellowstone is at the heart of our most fundamental charge. Our other charge is you: we aim to keep this amazing bird visible and enjoyable for all of Yellowstone's visitors, so its sight and sound can be heard echoing off Yellowstone's canyons and cliffs (Fig. 10.9).

FIGURE 10.9 Illustration of a peregrine falcon in the Grand Canyon of the Yellowstone. This area was the last to support peregrines in the park and, after extirpation, the first place they reoccupied. ARTWORK BY JACK DELAP

Visit the Yellowstone's Birds website (https://press.princeton.edu/resources/yellowstones-birds-video-collection/v10) to watch an interview with Lauren Walker and David Haines.

THE HAUNTING RAPTOR
YELLOWSTONE'S GOLDEN EAGLES

DAVID B. HAINES, DOUGLAS W. SMITH, TODD E. KATZNER, and VICTORIA J. DREITZ

VISUAL NARRATIVE BY K. CASSIDY

When many who live in North America picture an eagle, a large, magnificent bird with a distinct white head and tail comes to mind. Rightfully, the bald eagle has garnered much attention as a national symbol of the United States, though it was nearly brought to extinction from widespread organochlorine pesticide use (e.g., DDT, dichloro-diphenyl-trichloroethane; Anderson 1972, Baril et al. 2015). Listed at the federal level as an endangered species in 1978, downlisted to threatened in 1995, and removed from the list altogether in 2007, the bald eagle has been studied extensively across its range, including 38 years of monitoring in Yellowstone National Park (Yellowstone; YNP). However, a second eagle species, equally magnificent and widely distributed throughout Earth's Northern Hemisphere, also resides in YNP but has been relatively neglected in terms of scientific study. Unlike the bald eagle, the golden eagle does not have conspicuous characteristics; instead, it is dark brown throughout, with brilliant golden feathers on the back of its head and neck and subtle gray barring in the tail (Fig. 11.1). The golden eagle is an iconic apex predator tied to human culture through spiritual beliefs, reverence, and, like many other predators, persecution as a result of misunderstanding.

N LATE MAY 2011, while sitting in Lamar Valley and viewing a distant cliff typically occupied by peregrine falcons, three of us from the Yellowstone Bird Program suddenly were distracted by the repeated cries of a red-tailed hawk. We hastily shifted our attention to the commotion, only a couple hundred meters away, and observed the hawk diving repeatedly toward something just out of view. Suddenly, a golden eagle lifted from the ground while gripping a ground squirrel in its talons, explaining the hawk's unrest. With the hawk in pursuit, the eagle forcefully flapped its large wings, powering upward from the valley floor. We followed the eagle with binoculars as it continued to gain height and to drift in the direction of another distant cliff; as the eagle reached the cliff face, it gracefully pulled upward, entering a cavity. We scrambled to grab our spotting scopes while trying our best not

FIGURE 11.1 A golden eagle flies into its nest with a prey delivery for its young.
PHOTO BY RONAN DONOVAN

to lose sight of where the eagle had landed—a difficult task! Once we had our scopes in focus, we could make out a large pile of sticks pouring over the cavity ledge. Standing there in this nest was the eagle, and at its feet were two nestlings. We were thrilled to have identified a previously unknown territory and nest site with such little effort, especially for a species of such recent interest to the Yellowstone Bird Program.

YNP initiated golden eagle territory and reproductive monitoring in 2011 as part of a five-year initiative to establish baseline data on raptor species that had not previously been monitored in the park (Yellowstone Raptor Initiative; *see chapter 14*; Baril et al. 2017b). A primary goal was to determine, for the first time, how many golden eagle territories were present park-wide, and to summarize annual reproductive rates for all known territories. The territory located in the Lamar Valley was the first found in these surveys; it was a fortuitous beginning, given the series of events that led to its discovery. However, locating other golden eagle territories would not be as straightforward; instead, they involved countless miles of hiking and hours upon hours of observation. Nonetheless, exploration of the reaches of YNP in search of golden eagles and documentation of their territories and reproduction were both overdue, given the long history of other wildlife research in the park; it was also important for raising regional awareness for the species (Fig. 11.2). Further, because the states surrounding YNP—Idaho, Montana, and Wyoming—play such a key role in supporting North American eagle populations year-round, the lack of information from Yellowstone stands out relative to the detailed knowledge of nearby areas.

Independent of the efforts set forth in 2011, written but limited documentation of the golden eagle in YNP dates

FIGURE 11.2 Yellowstone Raptor Initiative and Yellowstone Bird Program staff conducting monitoring and survey efforts for the golden eagle. (A) Lisa Baril sits on 2–3 feet (0.6–0.9 m) of snow while monitoring a golden eagle territory in early April 2014. (B) Evan Shields monitors a northern-range golden eagle territory in 2017. (C) Staff surveying the northern range in 2014. (D) Staff surveying the remote southeast corner of Yellowstone National Park in 2011. PHOTOS: (A–C) BY DAVID HAINES, (D) BY COLBY ANTON

to the 1800s. A notable record comes from Philetus W. Norris, YNP's second superintendent (1877–1882). Though he did not refer to either the bald eagle or golden eagle by name, it is easy to differentiate the species through his eloquent descriptions. For the golden eagle, he wrote: "A very large black variety haunts and raises its young upon the inaccessible and tottering pinnacles of the eroded cliffs along the Grand, Gardiner, and other cañons, and is particularly numerous and audacious in the Hoodoo labyrinths" (Whittlesey and Bone 2018). Norris went on to explain a particular observation he made in the Hoodoo Basin along the eastern boundary of the park: "In one instance it caused a lamb to fall from a towering cliff and thus secured a repast below; but from my position I was unable to observe whether the lamb was frightened from its place on the cliff, or hurled off after being crippled by the eagle's talons, breast, or wing. I incline, however, to think the latter was the case" (Whittlesey and Bone 2018).

The records from Norris are consistent with what we know about eagle nesting behavior within the modern park. Additionally, we have observed similar behaviors by eagles in YNP. On one occasion, we were fortunate to witness a golden eagle diving repeatedly toward a small group of mountain goats at Barronette Peak, apparently trying to cause one to lose its footing on the cliff. Though the eagle was unsuccessful, this observation almost 140 years later closely matches Norris's experience. This hunting strategy, though uncommonly observed, has also been documented in other portions of the species' range (Nette et al. 1984).

Norris's description, despite his unfamiliarity with the species, both literally and figuratively captures some key characteristics of the golden eagle. The species commonly, but not exclusively, builds nests on what would be considered inaccessible cliffs and is generally associated with rugged mountainous terrain. Additionally, the description of a "black variety" is appropriate given their dark color. When viewing them from the great distances at which they are normally seen, especially without modern optics, golden eagles appear black. Norris's use of the word "haunts" is more figurative and, combined with their black appearance, speaks to the uncommon and often

brief encounters people have with the species, as if they are hidden in the darkness or ghost-like. Further, their strategy for hunting and territorial defense, which is often characterized by visually searching the landscape from distant perches or soaring from high above, is reminiscent of the idea of ghosts haunting their environment.

To expand on Norris's observations, the golden eagle is a habitat generalist, capable of occupying a wide range of environments, from desert to tundra, and from sea level to high elevations above 10,000 feet (3,000 m; Katzner et al. 2020). Though widely distributed, the species is associated with some common landscape characteristics, including rugged terrain and, usually, open areas dominated by short vegetation and restricted tree cover. The presence of large, soaring raptors such as golden eagles is strongly dependent on three key factors—the availability of prey, nesting sites, and updraft (Newton 1979, Katzner et al. 2020). These dependencies are available to eagles in YNP; however, evidence suggests habitat is marginal—but more on that later.

Rugged terrain and open areas play a significant role in meeting these needs for golden eagles. An important mechanism of flight, particularly for large soaring birds, is supplemental updraft (Duerr et al. 2015), primarily achieved through orographic and thermal air movement. Orographic updraft occurs at relatively low altitudes and is driven by the horizontal deflection of winds off features of the landscape (Alerstam and Hedenstrom 1998, Duerr et al. 2015). Increased ruggedness provides the appropriate landscape structure to promote orographic updraft. In contrast, thermal updraft is created by rising columns of air formed through the warming of the ground by the sunlight. Open areas have the highest propensity for generating thermals. Further, cliffs in rugged terrain provide important nesting habitat for the species, although they do sometimes nest in trees (Katzner et al. 2020). Likewise, the food eagles eat—medium-sized mammals and birds (hares, rabbits, marmots, ground squirrels, grouse, waterfowl), young ungulates (hoofed animals), reptiles, and carrion (Watson 2010, Katzner et al. 2020)—is also associated with both rugged terrain and open areas.

Currently, golden eagles are listed as a North American species of conservation concern (USFWS 2008). Continent-wide, population trend estimates vary, providing some uncertainty in overall population trajectory. Some studies show populations declining (Kochert and Steenhof 2002, Hoffman and Smith 2003, USFWS 2016), while others suggest stable (Millsap et al. 2013, Nielson et al. 2014, USFWS 2016) or even increasing populations (Crandall et al. 2015). However, much of this work does not account for reproductive rates, and many studies do not span enough time to identify accurate long-term population trends. A detailed literature review identified only four long-term studies (i.e., longer than 20 years) of nesting golden eagles in North America (Kochert and Steenhof 2002). Except for a stable population in Alaska, declines were reported in all other areas and were attributed to the impacts of fire on prey habitat, increased change in land use, and urbanization. Further, Millsap et al. (2022) found that 74% of mortalities for golden eagles one year old and greater are human-caused. Current concerns about the population status of golden eagles across the western United States have focused on their interaction with energy development (e.g., wind, gas), human activity (Hunt 2002, Watson 2010, Pagel et al. 2013, Tack and Fedy 2015), and environmental contaminants (e.g., lead and poison for pest control; Bortolotti 1984, Newton 1998, Herring et al. 2017, Katzner et al. 2018, Slabe et al. 2022). Given the uncertainty in population trajectory across North America and the known threats acting on golden eagle survival, determining how YNP intertwines is a management necessity.

EAGLE STUDY IN YELLOWSTONE NATIONAL PARK

Perhaps more than anywhere else in the conterminous United States, YNP and the Greater Yellowstone Ecosystem (GYE) offer pristine wilderness, with large extents of the landscape conserved under some of the highest tiers of protection (Dudley 2008). However, the GYE is not immune to human-caused threats to biodiversity, with a wide array of land uses and an increasing human population (Clark 2021). Therefore, it is crucial for managers within the GYE and elsewhere to understand the role of protected areas, particularly for species such as golden eagles that are susceptible to human encroachment. Further, the ecosystem-level effects of a recently recovered large carnivore community may present a unique environment for golden eagles in YNP.

Every year since 2011, near the end of March, we throw on field packs stuffed with appropriate clothing layers to match the slow retreat of winter, and set out on foot to visit each known golden eagle territory. At the time of these initial territory visits, golden eagle pairs have already been long at work, defending their territories from other eagles through imposing flight displays (Harmata 1982, Reid et al. 2019) and sometimes violent physical encounters. Additionally, these long-term partners have been strengthening their bond through courtship, nest-site selection, and nest building. Some of the territories we visit can be viewed from the road or require only a short walk, but others are remote and require exhausting day-long commitments of 15 miles (24 km) or more and thousands of feet of elevation change. Our objectives may not always be simple to accomplish but are, at their heart, straightforward: to determine if eagles are present and, if so, whether they have laid eggs.

One challenge in determining early-season breeding status is that golden eagles commonly maintain more than one nest in their territory (Kochert et al. 2002), and not all nests may be known or viewable from one location. This means that sometimes multiple days are required to visit all known nests or to seek out new or previously unknown nests. Additionally, failure to lay eggs each year is a relatively common occurrence for golden eagles, further complicating our ability to determine early-season breeding status. If nesting has not been observed at all known nests within a territory, how do we know if the birds are simply not nesting or if there is a nest we don't know about? To answer this question, we must carefully observe

FIGURE 11.3

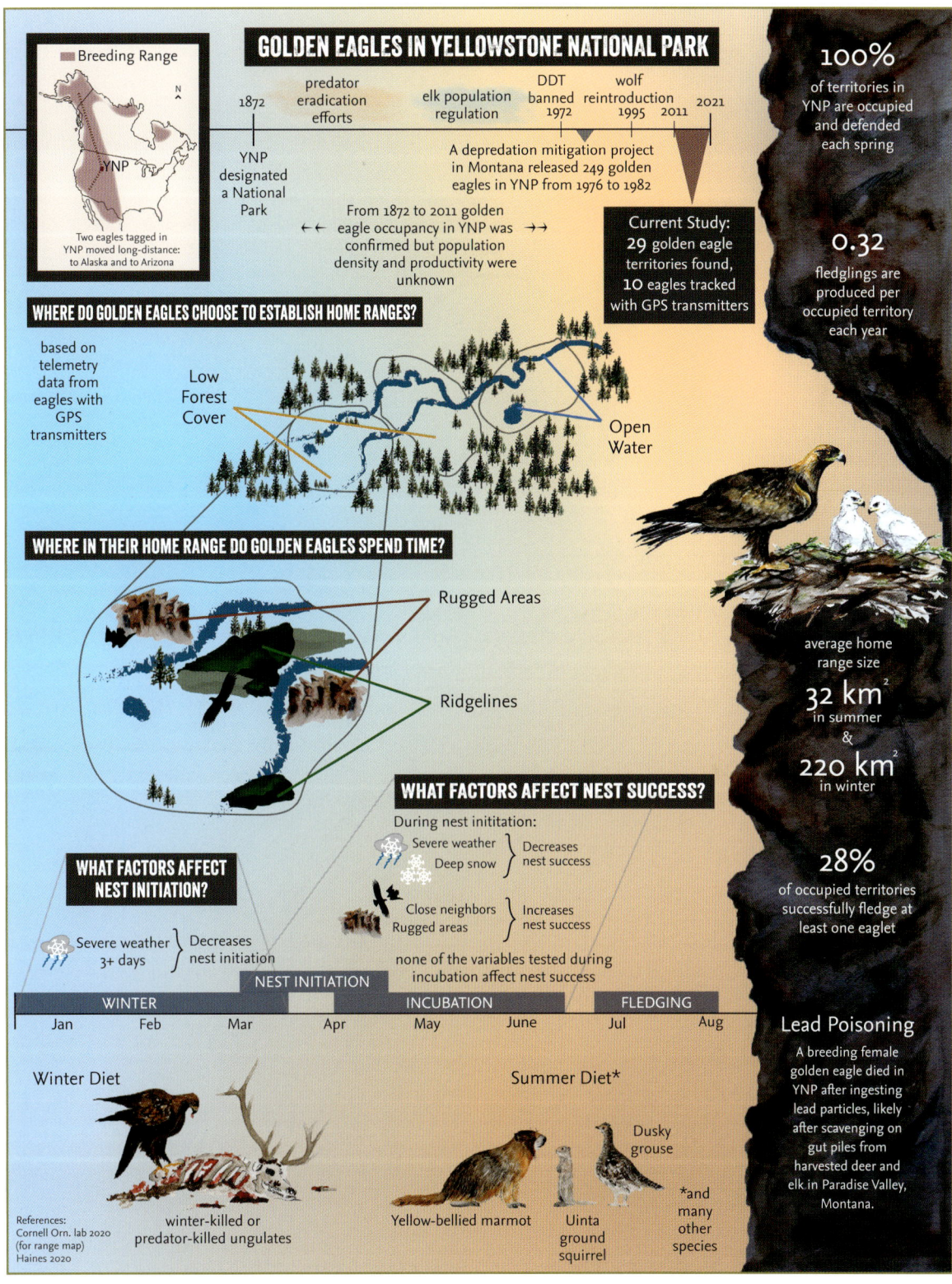

GOLDEN EAGLES IN YELLOWSTONE NATIONAL PARK

Breeding Range

N

YNP

Two eagles tagged in YNP moved long-distance: to Alaska and to Arizona

1872 — predator eradication efforts — elk population regulation — DDT banned 1972 — wolf reintroduction 1995 — 2011 — 2021

YNP designated a National Park

A depredation mitigation project in Montana released 249 golden eagles in YNP from 1976 to 1982

← ← From 1872 to 2011 golden eagle occupancy in YNP was confirmed but population density and productivity were unknown → →

Current Study: 29 golden eagle territories found, 10 eagles tracked with GPS transmitters

100% of territories in YNP are occupied and defended each spring

0.32 fledglings are produced per occupied territory each year

WHERE DO GOLDEN EAGLES CHOOSE TO ESTABLISH HOME RANGES?

based on telemetry data from eagles with GPS transmitters

Low Forest Cover

Open Water

WHERE IN THEIR HOME RANGE DO GOLDEN EAGLES SPEND TIME?

Rugged Areas

Ridgelines

average home range size
32 km² in summer
&
220 km² in winter

WHAT FACTORS AFFECT NEST SUCCESS?

During nest inititation:
Severe weather / Deep snow } Decreases nest success

Close neighbors / Rugged areas } Increases nest success

none of the variables tested during incubation affect nest success

WHAT FACTORS AFFECT NEST INITIATION?

Severe weather 3+ days } Decreases nest initiation

28% of occupied territories successfully fledge at least one eaglet

NEST INITIATION			
WINTER		INCUBATION	FLEDGING
Jan Feb Mar	Apr	May June	Jul Aug

Winter Diet

Summer Diet*

Dusky grouse

Yellow-bellied marmot

Uinta ground squirrel

*and many other species

winter-killed or predator-killed ungulates

References:
Cornell Orn. lab 2020 (for range map)
Haines 2020

Lead Poisoning

A breeding female golden eagle died in YNP after ingesting lead particles, likely after scavenging on gut piles from harvested deer and elk in Paradise Valley, Montana.

ARTWORK BY KIRA A. CASSIDY

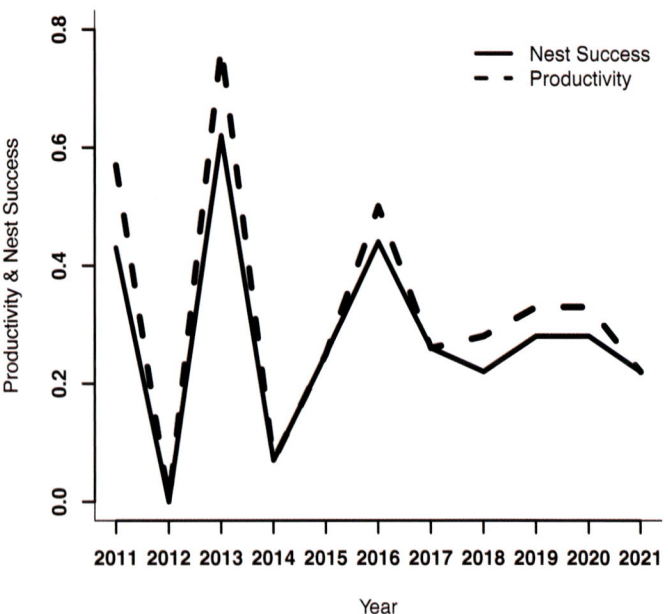

FIGURE 11.4 Northern-range golden eagle nest success (percent of occupied territories that fledge at least one young) and productivity (average number of young fledged per occupied territory), 2011–2021. Nest success is represented as a decimal for figure simplification.

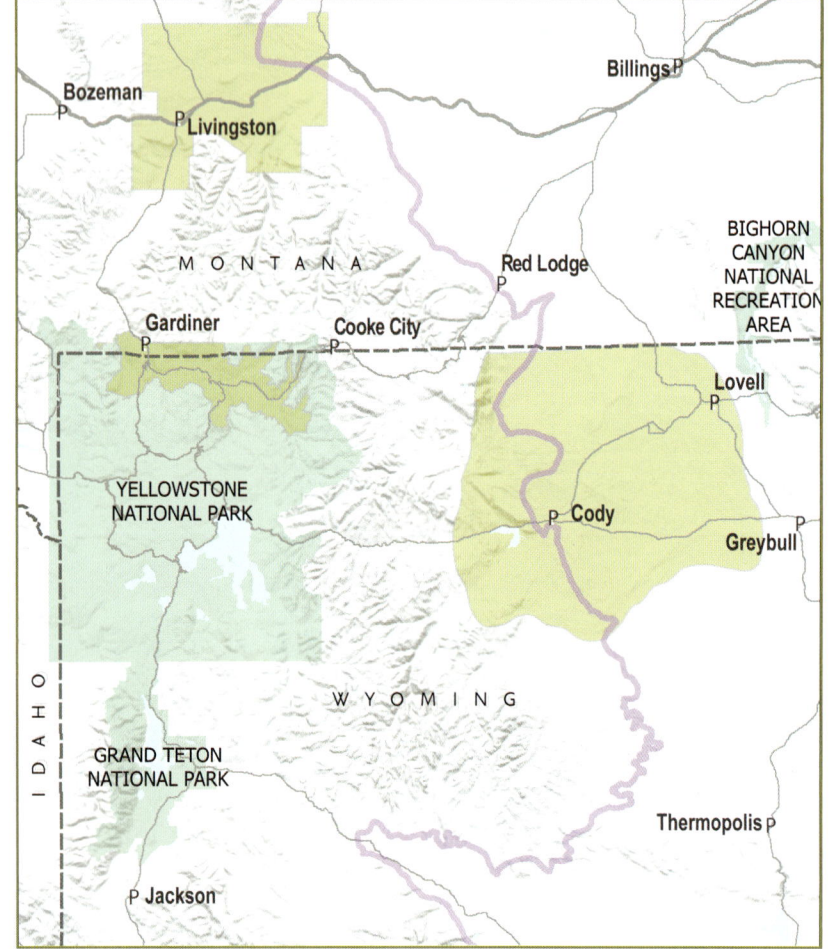

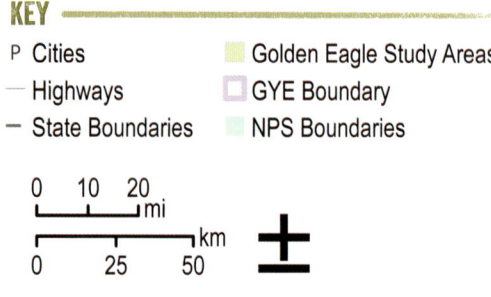

FIGURE 11.5 Three concurrent golden eagle study areas: Livingston, to the north, near Livingston, Montana; Big Horn Basin, to the east, near Cody, Wyoming; and Yellowstone, in the northern range of Yellowstone National Park.
MAP BY DANA HAINES

individual and pair behavior. For all territories where nesting is confirmed, we then return at least two more times to confirm fledging of young. With an average incubation period of 42 days, and a nestling period of approximately 70 days, our observations typically extend from the end of March to the beginning of August (Fig 11.3).

By 2015, after five years of monitoring, we had identified at least 29 territories park-wide, 21 of which are in the northern range of YNP. Density across the northern range is approximately one territory per 19 square miles (50 km²), relatively high compared with eagle densities documented elsewhere across North America (one territory per the range of 7.1–98 square miles [18.5–254 km²]; Katzner et al. 2020). Additionally, occupancy, as determined by the annual presence of a bonded pair in the territory, was 100% (Fig. 11.3). In contrast to territory occupancy rates, productivity (the average number of young fledged per occupied territory) and nest success (the percent of occupied territories that fledge at least one nestling) were relatively low, with a five-year average of only 0.35 (range 0–0.80) and 28% (range 0%–60%), respectively (Fig. 11.4). Low productivity and nest success were driven by apparent infrequent nesting attempts and high nest failure rates.

High density and occupancy rates in the northern range, contrasting with low productivity, were the impetus for continued monitoring and expanded research in YNP beyond the initial five-year effort of the YRI (Baril et al. 2017b). To give perspective, within the GYE, on unprotected private and public lands near Livingston, Montana, to the north (Crandall et al. 2015) and the Big Horn Basin, Wyoming, to the east (Fig. 11.5; Preston et al. 2017), densities of golden eagles are like those in YNP's northern range, but productivity is substantially higher. In Livingston, productivity has been 0.61 (2010–2020), while in the Bighorn Basin, it has been 0.73 (2009–2021). The variation in reproduction is likely the result of different prey and weather among these areas; however, one also might expect density to vary with these differences. Therefore, in 2016, our monitoring efforts continued, focusing on the 21 golden eagle territories within YNP's northern

range. Since that time, productivity and nest success have remained low, with an 11-year average of 0.32 and 28%, respectively (Fig. 11.3 and 11.4). Interestingly, we observed strong year-to-year variation in these reproductive measures over the first six years, but since 2017, yearly estimates have hovered around the 11-year average (Fig. 11.4).

REPRODUCTION IN A FOOD-LIMITED WORLD

Typically, nesting attempts and reproductive success of golden eagles are driven by prey abundance and weather conditions (Newton 1979). As mentioned previously, failure to lay eggs is a relatively common occurrence for golden eagles; therefore, it seems reasonable that when weather conditions are poor or prey abundance is low, eagles may more commonly skip a breeding season (Watson 2010). Throughout much of the golden eagle's diverse range, populations of small to medium-sized prey commonly serve as a significant driver of eagle reproduction (Bates and Morretti 1994, Steenhof et al. 1997, Watson 2010, McIntyre and Schmidt 2012, Preston et al. 2017, Katzner et al. 2020). Eagles in Idaho are more likely to lay eggs when black-tailed jackrabbits are more abundant and winters are less severe (Steenhof et al. 1997). Similarly, in Alaska, eagle nesting success is correlated with the cyclic nature of snowshoe hares and willow ptarmigan (Schmidt et al. 2017).

In short, the diet of golden eagles in YNP is poorly understood. Through collection of prey remains from nests (Fig. 11.6) and the use of motion-triggered nest cameras (Fig. 11.7), we have identified a diverse selection of prey that varies among territories. Eagles have been recorded to eat birds ranging from mountain bluebirds to red-tailed hawks, and mammals ranging from gophers to deer and pronghorn fawns (Fig. 11.8). The prey species we detected most frequently are yellow-bellied marmots, dusky grouse, and Uinta ground squirrels (Fig. 11.3). Although four species of rabbit and hare are present in YNP, they are absent or infrequent among the collected prey remains. From our experience, mountain

FIGURE 11.6 Senior author David Haines collects prey remains from a golden eagle nest in Yellowstone National Park. All prey remains found in the nest were collected, identified, and quantified to better understand the nesting-season diet. PHOTO BY RONAN DONOVAN

FIGURE 11.7 Motion-triggered cameras are placed above the nest site to capture prey deliveries. All prey items are identified and quantified to better understand the nesting season diet. (A) Nestling and coyote pup. (B) Adult delivers gosling. (C) Adult and nestling with partially consumed prairie falcon. (D) Nestling and red squirrel. NPS PHOTO

FIGURE 11.8 (A–D) A rare or even potentially unique event: a golden eagle attempts to capture fish from the outlet of Blacktail Ponds in northern Yellowstone National Park. PHOTO BY RONAN DONOVAN

OPPOSITE PAGE:

FIGURE 11.9 In the Lamar Valley, a golden eagle feeding on a dead bighorn sheep, a typical food source during Yellowstone's winter. PHOTO BY BEN SILBERFARB

cottontails and desert cottontails in the northern range are primarily restricted to developed areas, most frequently encountered inside the park around Mammoth Hot Springs and outside the park around Gardiner, Montana. Snowshoe hare and white-tailed jackrabbit abundance in YNP is considered low, and their distributions are described as patchy and restricted (Gunther et al. 2009, Hodges et al. 2009).

Despite this information, the prey identified above is from the nests and only represents what eagles are feeding their young during the nestling stage. Knowledge of prey selection throughout the entire year is limited. Most of the prey species identified at the nest sites are unavailable prior to egg laying because they hibernate or migrate. With low abundance and restricted availability of rabbits and hares (species that do not hibernate), it is not well known what eagles eat when those species identified at the nests are unavailable, although golden eagles are well known to scavenge carrion extensively in winter (Watson 2010, Katzner et al. 2020), a resource that is largely sustained by ungulate populations in YNP (Fig. 11.9).

The recovery of large carnivores within YNP (wolf reintroduction, cougar recolonization, and grizzly bear recovery) has changed where and when carrion is available (Smith et al. 2003), and this may have influenced eagles. Prior to reintroduction of the gray wolf in the mid-1990s, the northern Yellowstone elk herd numbered about 20,000–25,000 individuals (MacNulty et al. 2020). With no primary predator to limit elk numbers, the abundant elk herd commonly experienced late winter die-offs. These common winter die-offs occurred roughly when golden eagles initiated nesting (late February to early April), providing a reliable food source for eagles at a much-needed time. Today, with an intact large-carnivore community, the northern elk herd numbers about 6,000–8,000 individuals (MacNulty et al. 2020), and abundant winter elk die-off events are rare. Thus, even though predator kill sites provide food for eagles, there may be fewer opportunities for eagles to scavenge now, particularly during the early nesting season (Fig 11.10).

Golden eagles can live for 20 to 30 years, and unlike short-lived species whose populations can respond quickly to environmental change, populations of long-lived and highly territorial species lag behind (Watson 2010). Therefore, it may be that current eagle densities in YNP's northern range are a historical relict, reflecting the tim-

FIGURE 11.10 (A–D) Many species rely on carrion in winter and often compete for the resource. A golden eagle defends an elk carcass from a coyote. PHOTO BY RONAN DONOVAN

ing of past winter die-off events, while low eagle reproductive rates may be a response to the decrease in carrion availability.

With limited live prey through winter and early spring, and the rapid change in timing and availability of carrion since the mid-1990s, YNP eagles may be negatively impacted by food limitations interacting with weather and high local population density (Newton 1979, 1998; Bradley et al. 1997, Steenhof et al. 1997, Watson 2010, Anctil et al. 2014). For example, when density reaches or exceeds the availability of high-quality habitat, individuals are forced to occupy low-quality habitat, resulting in greater reproductive variability among territories (Ferrer and Donazar 1996). Further, when resources are low, factors of weather are more influential since the physical condition of the individuals may already be poor (Newton 1998).

WHAT'S DRIVING LOCAL PATTERNS?

While the initial goals of the five-year effort to study golden eagles in the park were relatively straightforward (to determine how many were present and their overall reproductive rates), science begets more science, and our results left us with more questions about the drivers of local population patterns. Golden eagles in YNP are subject to acute seasonal variations in prey availability and extreme weather, and a reproductive response to weather, food, and density seems likely. It is well established that the climate in YNP is harsh, leading to many weather events that can influence reproductive failure, suggesting that YNP is marginal habitat, yet density is relatively high.

Natural selection should favor the ability of individuals to distinguish between high- and low-quality habitat (Johnson 2007), so what drives our problem is understanding habitat quality, which for us was hard because we were not able to directly measure prey (food), as many other studies have done. So we used indirect measures of resources and conditions—openness, topography, vegetative cover, water, carnivore kills, and weather—to identify components of golden eagle habitat in YNP and their influence on reproduction.

MEASURING HABITAT IN YELLOWSTONE NATIONAL PARK

What is habitat? This is a historically debated question about an ecological concept that has resulted in multiple definitions that sometimes conflict. For this reason, habitat is defined here as the species-specific resources and conditions present in an area that allow occupancy, survival, and reproduction (Hall et al. 1997). A key component of this definition is that habitat is species-specific. This means that, when you look across a place like the Lamar Valley, you are looking at a mosaic of habitats specific to each species of plant and animal found there. For animals, we often try to identify habitat by looking at the use of resources (used vs. available) in relation to their home range, which is the area routinely used by an animal to meet its daily needs (Fieberg and Börger 2012). The use of GPS (global positioning system) telemetry is one way to gather data that can be used to estimate an animal's home range and to identify resources that are habitat specific. This was one of our objectives in YNP, so we first needed to capture eagles and attach GPS transmitters (Fig. 11.11).

Specific to our objectives, we targeted adult eagles in the northern range of YNP that have been monitored over the last 11 years. With our telemetry data, we were interested in identifying what resources influence where a home range is in YNP, as well as what resources are important within the home range. We found that golden eagle home-range sizes varied between individuals but on average were larger during the winter, with size estimated at 85 square miles (220 km²; SD = 282), compared with an average of 12.4 square miles (32 km²; SD = 17) during spring and summer (Fig. 11.12; Haines 2020). The larger winter home range may be driven in part by the reduction of available prey, requiring more area, and because territory defense is less pronounced outside of the

FIGURE 11.11 Prior to GPS transmitter attachment, measurements, feather samples, and blood are taken from all captured golden eagles to determine sex, assess physical condition, and identify the presence of toxins. (A) Footpad measurement. (B) Jordan Harrison and coauthor David Haines take a weight measurement. (C) Harrison and Haines attach a GPS transmitter. (D) Adult female tail. (E) Volunteer Katy Goodwin releases a golden eagle with a GPS transmitter. (F) Released golden eagle in flight with GPS transmitter. PHOTOS: (A–C, D) BY DOUGLAS SMITH, (E) AND (F) BY DYLAN SCHNEIDER

breeding season. On average, we found eagles established home ranges in areas with low forest cover and in proximity to water (Fig. 11.3). Within the home range, eagles strongly select for rugged topography, upper slopes, and ridgelines above valley bottoms (Fig. 11.3). Additionally, we have found some evidence for seasonal variation in how eagles select for resources associated with prey within their home range. For example, grass and shrub

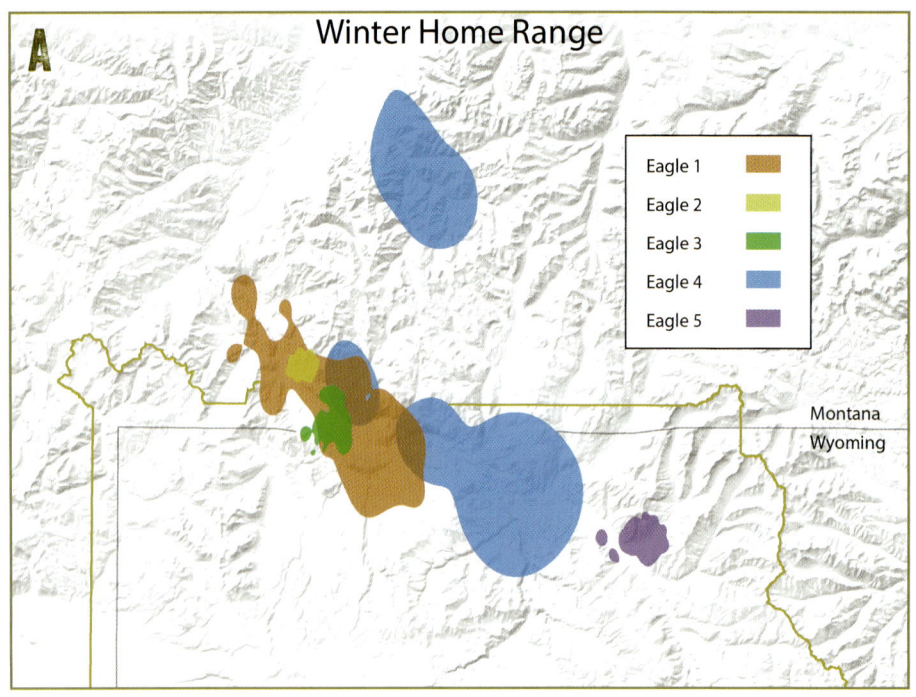

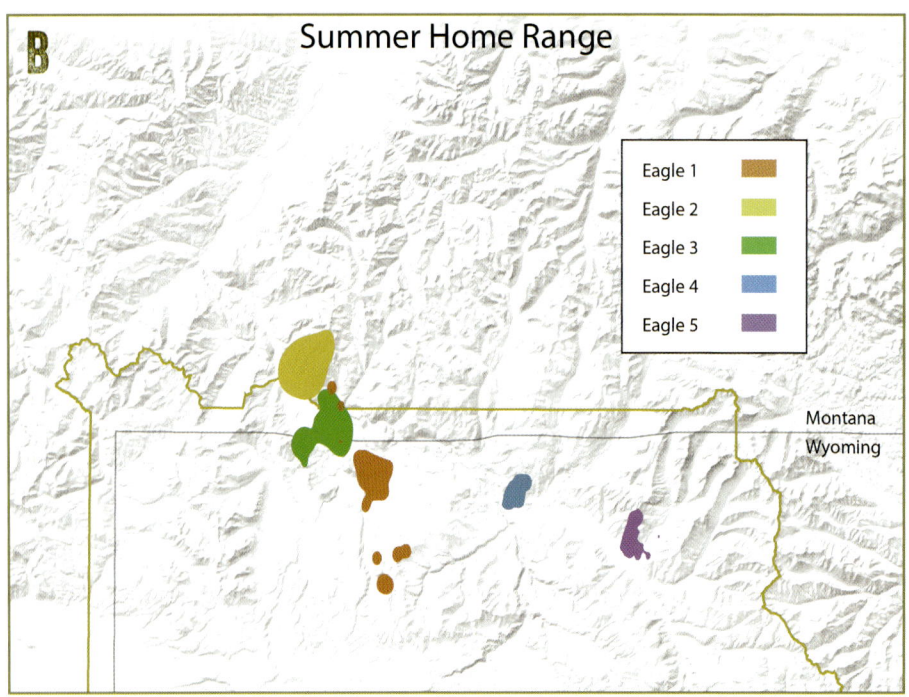

FIGURE 11.12 The home ranges of five northern-range golden eagles, calculated using 95% kernel density estimates (KDE), in (A) winter and (B) summer. MAPS BY TYLER ALBRETHSEN

communities, commonly associated with ground squirrels and other small mammals, are preferred by eagles in the spring and summer, but not in the winter (Haines 2020). Given the timing of ground squirrel emergence and the overall low abundance of rabbits and hares, this seasonal behavioral shift is not surprising.

As mentioned previously, we assumed carrion is a key food source in winter. Therefore, we were curious about how the behavior of eagles affected their probability of encountering carrion. The most reliable source of carrion through winter likely comes from predation by gray wolves and cougars (Wilmers et al. 2003). A study of wolf and cougar hunting domains in YNP found that these species hunt elk in different places. The highest probability that wolves will successfully kill elk occurs in areas of open vegetation, whereas cougars have the highest probability of successfully killing elk in moderately forested areas and rugged terrain (Kohl et al. 2019). The golden eagle's general association with open vegetation increases their probability of encountering wolf kills, and from our analysis, eagles are frequently using open areas year-round. Alternatively, we found some evidence to suggest that eagles may use moderately forested areas more regularly in the winter as opposed to summer. Our analysis did provide strong evidence that eagles are using rugged terrain; this, coupled with the winter use of moderately forested areas, may increase the likelihood of eagles encountering and benefiting from cougar kills.

These findings are supported by field data gathered by people working with wolves and cougars in YNP. Yellowstone Wolf Project staff routinely observe golden eagles on wolf kills (Wilmers et al. 2003). Surprisingly, field crews have also documented golden eagle use of cougar kills, mostly through remote cameras (D. Stahler and W. Binder, pers. comm.). As cougars are solitary, there is likely more biomass available for scavenging. Wolves, being social and living in packs, can consume a kill in a matter of hours, compared to what might be days for a cougar. Therefore, given the overlap of habitat components between these species, the abundance of cougars in this system may be a particularly important, and previously overlooked, component to golden eagle winter foraging success.

The common use of cougar kills in YNP is mirrored by research in the southern GYE, where golden eagles were found to be the third most frequent scavenger (Elbroch et al. 2017). Fortunately, the study areas for concurrent golden eagle and cougar research in northern Yellowstone strongly overlap, and data from both studies may afford the opportunity to study this relationship at a scale that could explain how breeding eagles directly benefit from cougar kills.

ENVIRONMENTAL DRIVERS OF REPRODUCTION

Since reproductive monitoring began in 2011, we have observed between zero and eight territories successfully fledge nestlings each year. Our data show that, on average, there is a 32% probability that a territorial golden eagle pair will lay eggs; if they lay, there is a 59% probability they will successfully fledge nestlings (Haines 2020). A primary objective of our research was to quantify how resources available to territorial eagles and the seasonal weather they experience influence these probabilities of egg laying and success. Some golden eagle territories in YNP are more productive than others, suggesting that differences may exist in the quality of resources available to individuals within each home range. Additionally, adverse weather may be more influential on reproduction where individuals occupy a home range with lower-quality resources (Steenhof et al. 1997, Newton 1998).

To accomplish our objectives and determine factors that influence reproductive success, we considered how the resources selected by eagles and the inter-annual weather they experience vary among home ranges. There are limited data available to estimate the abundance of live eagle prey in YNP. Therefore, we used land cover at the home-range scale to infer prey availability and perch sites that benefit foraging. Additionally, to address the influence of carnivore-contributed carrion, we considered annual estimates of edible biomass acquired per wolf per day during late winter and early spring in the northern range of YNP. Finally, we included nearest neighbor measurements (minimum distance between neighboring territories) to address the prediction that eagles nesting in closer proximity to each other may indicate areas of higher-quality habitat.

Our analysis found that winters with more prolonged precipitation events decrease the probability that eagles will lay eggs (Fig. 11.3; Haines 2020). Prolonged precipitation can limit both prey activity and the ability of the eagles to detect or hunt prey, thus negatively influencing physical condition. We found no evidence for any other factors influencing the probability of egg laying. If eggs were laid, we found that more snow on the ground and more severe weather events during late winter and early spring decrease the probability of successfully fledging nestlings (Fig. 11.3; Haines 2020). The negative influence of snow may generally indicate some lasting effect of severe winters on nest success. However, when snow persists on the ground longer into the breeding season, small mammals may delay emergence or gain increased cover from predators, but alternatively, more snow usually increases carnivore-derived carrion. Therefore, increased snow cover may limit detection as golden eagles become more reliant on live prey, but how much carrion compensates is unknown. Severe weather incorporates both low temperatures and high precipitation. However, disentangling how low temperatures and high precipitation affect prey and eagle activity requires further research related to movement and energetics of both predator and prey.

In addition to weather, though less supported, we found evidence for a positive influence of available resources within the home range. Our results suggest home ranges with more rugged terrain have a higher probability of successfully fledging nestlings (Fig. 11.3). As mentioned above, rugged terrain promotes updraft through orographic properties, which ultimately can result in increased prey acquisition and territory defense, while conserving energy required for powered flight. Finally, we found evidence that the probability of success increases where eagles nest in closer proximity (Fig. 11.3). This result suggests a connection between the quality of resources in an area, terri-

tory density, and nest success. While a positive influence for available resources was not strongly supported, results suggest that for eagles who have met the conditional requirements for egg laying, more snow cover and inclement weather during the late winter and early spring have a lasting effect on successfully fledging nestlings.

Edible biomass acquired per wolf per day appeared to have no effect on reproductive success in our analysis. Certainly, more research is warranted, but this suggests that carnivore-contributed carrion may not substitute for mass winter die-off events. Further, this annual estimate does not account for the locations of wolf kills (availability likely varies between eagle home ranges), nor does it account for what is available for scavenging animals (wolves may consume most of the kill), potentially limiting its explanatory power. Therefore, given that eagles are commonly observed at wolf kills, it remains unclear if an increased abundance of wolf-contributed carrion has a positive influence on reproductive success in YNP.

POPULATION STATUS

Relative to eagles in other areas, our probability estimates of egg laying (32%) and successfully fledging nestlings (59%) are low and are what drive our research. However, for long-lived species, slow to mature and reproduce, adult survival is particularly important to population stability (Whitfield et al. 2004). Recent broadscale demographic analyses have shown that any additive mortality posed by an increase in human-caused threats is likely to trigger population declines or exacerbate any declines that may be ongoing (Tack et al. 2017; Wiens et al. 2017, 2018).

Human-caused threats that impact survival are minimal within the boundaries of YNP. However, our research has revealed that eagles commonly cross this boundary, particularly in autumn and winter, thus exposing them to human-caused mortality factors. For example, in early December 2018, the first eagle ever outfitted with a telemetry unit in YNP, an adult female, was recovered dead after spending much of the autumn north of the park

FIGURE 11.13 (A) The first golden eagle ever captured in Yellowstone (adult female), August 2018. (B) The carcass recovery of the first golden eagle captured in Yellowstone after it died of lead poisoning, December 2018.
PHOTOS: (A) BY JORDAN HARRISON, (B) BY CONNOR MEYER.

(Fig. 11.13). Necropsy results indicated the female eagle died of lead poisoning, with liver lead levels well within the range of lethal toxicity. This poisoning was a result of lead ingestion, likely from lead particles left in a gut pile of hunter-killed game and subsequently scavenged by this eagle and perhaps others. This scenario is partially supported by the timing of her movements outside the park, which aligned with the big-game hunting

THE DENALI–YELLOWSTONE CONNECTION – CAROL L. McINTYRE

Brisk northerly winds made this late January day feel much colder than it was. It was nearly noon, and we decided to take a break after spending the morning skiing up a long ridge south of the Old Faithful area in Yellowstone National Park. We spied and then skied over to an opening in the forest that offered good views to the south and hopefully some shelter from the wind. Reaching the sheltered spot, we turned our faces southward, hoping to feel a bit of the mid-winter sun's warmth. Soon after, we spotted a very large bird soaring over a nearby ridge to the southwest. Fumbling through zippers and layers of warm clothes, I extracted my binoculars and zoomed in on the bird. I noted its long wings, held slightly upward as it circled lazily above the ridge. As it banked, we all caught the brief flash as the sun reflected off the golden feathers on its head and nape (back of the neck). Before I could tell my friends that we were watching a golden eagle, the bird stopped circling, set its wings, and headed straight for our ridge.

It didn't take the eagle long to get to us, maybe less than a minute. When it reached our ridge, it started circling right above us. From our vantage point, we could easily see the bright golden feathers on the back of its head and neck, and the long primary feathers held out and the tail feathers spread widely to capture the updrafts. We could also see the big white patches in the wings and at the base of the tail, and the similarity in the color and length of the flight and tail feathers, all indications that this was a juvenile eagle that was raised in the previous breeding season. Everyone in our group fell silent as we watched this magnificent raptor circle just a few hundred feet above us. The eagle soared overhead for a few minutes and then flew north and out of our view. Seeing the eagle made an already brilliant day even brighter.

I've been fortunate to spend most of my career as a National Park Service wildlife biologist studying golden eagles. Our study area in the northern foothills of the Alaska Range in Denali National Park and Preserve, Alaska (Denali), contains one of the highest nesting densities of golden eagles in Alaska, and perhaps North America (Katzner et al. 2020), and as such was designated an Important Bird Area by Audubon. Like many golden eagles that nest or are raised in interior and northern Alaska, Denali's golden eagles are migratory, and most overwinter thousands of miles to the south. Over the past couple of decades, we've studied the movements of Denali's younger golden eagles using lightweight satellite and GPS telemetry units. These are individuals who haven't yet entered a breeding population and whose movement patterns, migration schedules, and behavior are substantially different from members of the breeding population. We attach the telemetry units to the eagles just before they fledge (leave their natal nest). Data from the units include a highly precise GPS location that we use to track their movements. The batteries that power the telemetry units are charged by tiny solar panels on the back of the unit, thus providing us with the opportunity to study the movements and behavior of telemetered eagles for many years.

It's not hard to imagine how our excitement levels rise in late September and early October as the telemetered eagles embark on their first autumn migration, traveling independently and relying on themselves for survival. Over the next couple of months, they will migrate

season (late October through March). This seasonal shift in movement, common among most tagged eagles in YNP, illustrates the intimate connectivity of the landscape and the fact that protections in the park do not convey protection to eagles throughout their life cycle. To make this point of wide-ranging movements in the extreme, golden eagles from Alaska commonly use Yellowstone in some winters (*see below*).

FIGURE 11.14 Golden eagle migration between Alaska and the Greater Yellowstone Ecosystem. First autumn migration for three juvenile male golden eagles (red, blue, and orange lines) outfitted with GPS transmitters as nestlings in Denali National Park. Spring migration for an adult male golden eagle (purple line) captured in Lamar Valley, Yellowstone National Park, and outfitted with a GPS transmitter in December 2019. The adult male spent most of the winter inside Yellowstone before flying north to its breeding territory just outside Wrangell–St. Elias National Park. MAP BY DANA HAINES

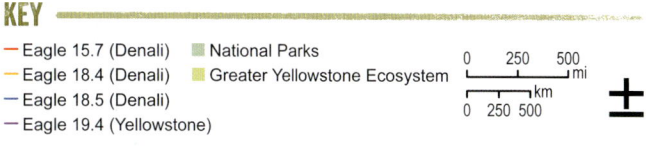

through some of the most stunning landscapes and rugged terrains in western North America. Some will spend the winter in southwestern Canada, while others will spend the winter in central Mexico. Some of the eagles that winter in the most southern wintering areas will migrate through the Greater Yellowstone Ecosystem (GYE), while others will overwinter in the GYE. Thus, my skiing buddies weren't too surprised when I mentioned that the eagle that we just saw could well be one that I saw in a nest in Denali the previous summer. If so, that young eagle had traveled at least 2,000 miles (3,200 km) since leaving its natal territory in Denali and was now making a living in the GYE.

Yellowstone and Denali are regarded as iconic national parks, and both are recognized for their wildlife and wildlife research studies. While our data show the direct connection between Denali and Yellowstone, several other golden eagle tracking studies include golden eagles from northern breeding areas that migrate or winter in the GYE, including an adult male, captured in winter during the Yellowstone-specific tracking study, that holds a territory near Chitina, Alaska, just west of Wrangell–St. Elias National Park, Alaska (Fig. 11.14). Overall, with data flowing in from multiple studies, the importance of Yellowstone and the GYE to resident and migratory golden eagles is becoming much clearer.

The period when golden eagles have been monitored in YNP does not span enough time to determine whether the local population status is stable, increasing, or decreasing. As a highly territorial and long-lived species, occupancy alone can overshadow the vital rates of the population (Van Horne 1983, Watson 2010). Thus, even with a 100% occupancy rate, our efforts to monitor reproduction are necessary for determining population trends. Estimating survival is difficult, requiring large sample sizes of data that we have only begun to collect. Ultimately, more years of reproductive monitoring and estimates of adult survival will be necessary to determine overall population status.

UNCOVERED EVENT WITH AN UNKNOWN EFFECT

A little known and undiscussed event among longtime workers in YNP, with an unknown effect on this population, involves a predator management project on two ranches near Dillon, Montana. In response to the depredation of domestic lambs by golden eagles, capture and translocation efforts were conducted from 1975 through 1983 (Matchett and O'Gara 1987). Only recently, through conversations with one of the former trappers on the project, Carter Niemeyer (APHIS and USFWS retired), have the details of this effort been made available to us. Leaving us bewildered and amazed, reports provided by Niemeyer indicated that 249 golden eagles of varying age classes were released into YNP between 1976 and 1982 (Niemeyer 1976, 1977, 1978, 1979, 1980, 1981, 1982; Fig. 11.15).

The findings from previous translocation work in Wyoming (Phillips et al. 1991) suggest that most adult eagles with previously established territories did not remain in YNP after translocation. However, far less is understood about the response of sub-adult and adult eagles with no established territories. These releases coincided with a time of the year when most of the prey we know eagles consume was available. An influx of eagles into an environment with favorable conditions may have the potential to increase overall abundance through the establishment of new territories and may in part be what contributed to northern Yellowstone's high golden eagle density. Once established, eagles will not likely relinquish these territories, given what we know about their life history. That being said, we cannot truly identify the effects of this previous management action on this local population. Nonetheless, all captured eagles were banded, and band recovery data may help to elucidate the response of some individuals after translocation.

Golden eagle translocation, as a management strategy, has been used for introductions to southern Appalachia (Touchstone 1997, Wheeler 2014) and to curtail the depredation of domestic and native wildlife in the West (Linnell et al. 1997, Latta et al. 2005). The response of golden eagles to these actions has been mixed and, in general, is poorly understood. In southern Appalachia, introductions were largely unsuccessful; however, despite an environment that did not likely support many breeding eagles historically, there is evidence of breeding, suggesting that some eagles did establish territories (T. Katzner, unpublished data). Environment most likely dictates translocation success. For example, the California Channel Islands, where the golden eagle was responsible for the catastrophic decline of the island fox, are arguably a marginal environment for the species. With all eagles translocated into an area that is known to support a sizable eagle population, none have returned (Latta et al. 2005). Alternatively, nearly all territorial adult eagles captured in Wyoming and translocated to the nearby states of Montana and North Dakota, where the environment was equal or perhaps less favored, returned to their territories (Phillips et al. 1991).

CONCLUSION

Wildlife monitoring and research in YNP has provided a suite of long-term studies that have revealed population trends, supported complex ecological theory, and, perhaps

FIGURE 11.15 (A and B) Carter Niemeyer releasing a golden eagle from Soda Butte in the Lamar Valley. The eagle was captured near Dillon, Montana, before it was relocated to Yellowstone National Park. PHOTOGRAPHER UNKNOWN

most importantly, helped guide management decisions. Golden eagle monitoring and research in YNP has laid the foundation for what one day may add to this legacy of work. Currently, our research has suggested that YNP may represent a unique golden eagle system. Perhaps it is marginal in the quality of habitat required for high reproductive rates, but it may be ideal for individual life expectancy. More years of data are necessary to confirm or deny YNP's influence on golden eagle demography.

The history of golden eagles in YNP and any changes to this local population that may have occurred over time are unknown, since monitoring did not begin until 2011.

THE HAUNTING RAPTOR

This has left us to speculate upon why eagles occur at a high density if resources and conditions throughout the full annual cycle are limiting viable reproductive rates. Therefore, our research has been an effort to establish an ecological baseline for the species today, as a means to guide management into the future. This is mission critical to the National Park Service, as restoring and protecting natural conditions is fundamental to park establishment (Organic Act 1916), yet until now, the status and condition of the golden eagle population was unknown but was likely impacted by humans. Carnivore removal, translocation of depredating golden eagles, and wolf reintroduction were all human interventions with potential long-term consequences. Unraveling this story will help us move forward as resource stewards.

Other research has recognized that golden eagles are impacted by human activity and development (Millsap et al. 2022), and our research is no exception—surprisingly. What remains unknown is how these outside-the-park threats may be impacting the status of golden eagles in YNP. The political, social, and economic boundaries that many times designate protected areas do not always correspond with the ecological boundaries that should be considered, especially with birds. In the context of what we know about golden eagles, we must look across jurisdictional boundaries to better understand the ecological relationships of golden eagles and their habitat (Fig. 11.16). Raising the awareness of this iconic symbol of wildness is imperative to the long-term conservation of the species. By taking these steps, we can ensure that Norris's characterization of golden eagles "haunting" their environment remains figurative and does not become literal.

Visit the Yellowstone's Birds website (https://press.princeton.edu/resources/yellowstones-birds-video-collection/v11) to watch an interview with David Haines.

FIGURE 11.16 On the western edge of the Greater Yellowstone Ecosystem in the Ruby River Valley area, a golden eagle feeds a nestling. Carcasses, including a rabbit and an unknown raptor nestling, can be seen at the edge of the nest. PHOTO BY RONAN DONOVAN

12

BALD EAGLES AND OSPREY

LAUREN E. WALKER and DOUGLAS W. SMITH

When biologists visited the American white pelican nesting colony on the Molly Islands in the late summer of 2019, they found destruction (*see chapter 21*). Near-intact carcasses of dead pelican chicks and full-sized fledglings dotted the islands, and additional piles of bones and feathers could be seen everywhere. The most likely culprit for this devastation? Bald eagles are the easy answer. But to more accurately reflect a complicated and interwoven ecosystem, a better answer may be bald eagles and the invasive lake trout. The trophic (food-web) effects, both bottom-up and top-down, of the invasion of lake trout into Yellowstone Lake in the 1980s are far-reaching, complex, and just beginning to be understood (Koel et al. 2019b). For the bald eagle and another common raptor of Yellowstone National Park (Yellowstone; YNP), the osprey, the lake trout invasion represents another hurdle in a century of population crash and recovery (Fig. 12.1).

WITH RECORDS THAT date back to the early 20th century, bald eagles and osprey are among the longest-studied birds in Yellowstone, surpassed only perhaps by trumpeter swans (*see chapter 17*). Both bald eagles and osprey are large, long-lived raptors found throughout Yellowstone National Park and across the United States, near bodies of water (Fig. 12.1). Adult bald eagles have completely dark brown bodies and broad wings, with a characteristic white head and tail. In contrast, juvenile eagles have bodies and underwings splotched with white but lack the all-white head and tail. With a wingspan up to 7.5 feet (2.3 m), bald eagles are one of the largest birds in North America. Osprey have a largely white body with a dark back and a distinctive dark facial stripe across a bright yellow iris. The osprey's long narrow wings span 5 feet (1.5 m) when flying but fold completely behind it to allow it to plunge feetfirst into rivers and lakes as it hunts for fish. Although eagles forage heavily on fish across their range, they are largely opportunistic and have a relatively diverse diet of fish, carrion, and waterfowl. Osprey, on the other hand, are complete piscivores, reliant on the availability of shallowly swimming fish to survive and raise young.

The history of these species in Yellowstone and across North America is tumultuous. Like many raptors, eagles and osprey both experienced dramatic continent-wide population declines in the mid-20th century due to the widespread use of organochlorine pesticides such as DDT (dichloro-diphenyl-trichloroethane). With the nationwide ban of these toxic substances in 1972, however, both species experienced significant recovery. Populations of these species across North America are today largely stable and, in many places, even thriving. Although this trend holds true throughout much of Yellowstone National Park as well, the recovery story on and around Yellowstone Lake has been complicated by the invasion of a non-native fish.

FIGURE 12.1 Bald eagles (A) and osprey (B) are formidable predators, commonly seen near Yellowstone's lakes and waterways. PHOTOS: (A) BY TOM MURPHY, (B) BY SCOTT HEPPEL

HISTORIC POPULATION TRENDS: DDT AND HUMAN DISTURBANCE

OSPREY

As far as records go back, the osprey has always been a common raptor along Yellowstone's lakes and waterways (Fig. 12.2). In 1917, M. P. Skinner reported osprey as abundant in the park, noting 60 known pairs and estimating a total population of 120 pairs. The total count included 25 in a short stretch of the Grand Canyon of the Yellowstone, and 30 along the western shore of Yellowstone Lake alone. Skinner (1917) attributed the overall abundance to the "large quantities of easily attainable fish as much as the absolute protection afforded." Osprey were still common in the park in 1929 (Kemsies 1930), but there is little information on osprey productivity or population trends in Yellowstone prior to widespread DDT

FIGURE 12.2 A juvenile osprey practices hunting for fish.
PHOTO BY GREG ALBRECHTSEN

use beginning in 1945. In the 1950s, however, declines in osprey productivity in well-studied populations in the northeastern United States helped motivate more intensive monitoring of osprey populations throughout the rest of the country. In the Greater Yellowstone Ecosystem, the first assessment of osprey reproduction and productivity did not occur until 1968 (Turner 1968). Considering populations in both Yellowstone and Grand Teton national parks, Turner (1968) observed 36% nest success and a productivity of 0.64 young per active nest. Between 1972 and 1974, the Yellowstone population was estimated as between 102 and 120 individuals; 41% of nest attempts succeeded in fledging young, and 0.64 young were produced per nesting pair (Swenson 1975). Nest success and productivity on rivers and streams were relatively high compared with reproduction on Yellowstone Lake (Fig. 12.3), a fact attributed to high levels of human disturbance on the lake (Swenson 1975). The effects of visitor disturbance on osprey and other birds on Yellowstone Lake, particularly that of fast-moving motorized boats, had been discussed by park managers as early as 1958 (Pritchard 1999), and in 1961, motorized recreation was prohibited on the arms of Yellowstone Lake. Further restrictions on motorized recreation were put in place with the passage of the Wilderness Act in 1964. To further monitor and document osprey population recovery, the park increased its osprey survey efforts, beginning in the 1980s, and a consistent monitoring regime was initiated in 1987.

BALD EAGLES

Although no true study of Yellowstone's bald eagles was conducted in the first half of the 20th century, Skinner noted in 1928 that eagles reside in the park all winter but are uncommon, and Kemsies (1930) called them occasional permanent residents (Fig. 12.4). Following the broad use of DDT and DDE (dichloro-diphenyl-dichloroethylene), beginning in 1945, bald eagle populations suffered continent-wide declines, and across the country, scientists and conservationists initiated concerted efforts to better track eagle populations and reproduction. The first efforts to

FIGURE 12.3 An adult osprey lands on a nest in the Grand Canyon of the Yellowstone River. NPS PHOTO

FIGURE 12.4 Bald eagles warm themselves in a snag on a cold winter morning. PHOTO BY TOM MURPHY

survey bald eagle populations in Yellowstone with more than casual observation were conducted in 1960 and 1961 and documented seven and nine breeding pairs, respectively, along the Yellowstone and Madison rivers and Yellowstone, Lewis, and Shoshone lakes (Murphy 1960, 1961). A few years later, in 1967, eagles were listed as federally endangered, becoming the flagship species for the Endangered Species Preservation Act of 1966 and later the Endangered Species Act of 1973. Between 1972 and 1974, Swenson (1975) found similar numbers of breeding pairs as observed in the early 1960s, results suggestive of a relatively stable population. Observations of reproductive success, however, told a troubling story: only 23% of pairs attempting to nest successfully fledged young, and no pairs fledged more than one (Swenson 1975). In 1978 and 1979, another study on eagle populations in the southern portion of the park observed that nest success and productivity had increased to approximately 40% and 0.65 young per nesting pair, respectively (Alt 1980). The overall pattern of low productivity in the early 1970s and some recovery over the following decade mirrors trends observed across North America (e.g., in Ontario; Grier 1982), as bald eagle populations began to recover after the ban of DDT in 1972. With growing interest in tracking the recovery of eagle populations, the park increased its internal monitoring efforts. While notes and observations from park rangers helped fill in the gaps, a more consistent and robust eagle monitoring program was established by the park in 1984 and continues today.

RECENT TRENDS: RESPONSES TO THE DECLINE OF A NATIVE FISH

The recent history of Yellowstone's bald eagles and osprey is closely tied to the fate of the Yellowstone cutthroat trout, and equally to the invasion of the non-native lake trout. Lake trout are a large and long-lived, piscivorous species

of freshwater char that are native to lakes of northern North America but are invasive in many lakes in the western contiguous United States. Lake trout were discovered for the first time in Yellowstone Lake in 1994 (Kaeding et al. 1996), although some data suggest the invasion likely first occurred in the mid to late 1980s (Munro et al. 2005). Although fisheries biologists initiated a gillnetting effort to remove the invasive trout, the level of effort was insufficient, and lake trout numbers quickly increased, causing subsequent declines in their native prey, the Yellowstone cutthroat trout (Kaeding et al. 1996, Koel et al. 2005). Ultimately, the decline in the cutthroat trout population has had rippling effects throughout the Yellowstone Lake ecosystem. Twenty-two species of mammals and 20 species of birds in YNP are known or suspected to use the cutthroat trout as a food source (Schullery and Varley 1995). For many of these species, including both bald eagles and osprey, lake trout do not make a suitable substitute for cutthroat because they grow much larger (and thus become too heavy) and tend to occupy deeper water than their native cousin. Thus, the decline in native trout has forced changes in diet and foraging strategies for numerous other species (Koel et al. 2019b). Osprey are highly specialized fish eaters, and their diet in Yellowstone has historically depended heavily on cutthroat trout. On Yellowstone Lake, cutthroat comprised as much as 93% of their diet (Swenson 1978). In contrast to osprey, bald eagles have a more diverse diet, which in Yellowstone was previously documented as 57% birds (largely waterfowl), 25% fish (mostly cutthroat trout), and 18% mammals (Swenson 1975). On Yellowstone Lake, however, cutthroat represented a seasonally important resource for eagles (Fig. 12.5). The bald eagle nesting season aligns well with the peak of local trout spawning runs near the lake, and historically, the cutthroat trout was noted as the sole prey species detected in eagle nests in early July (Swenson et al. 1986). After the spawning runs ended, eagles on the lake switched to targeting vulnerable, flightless waterfowl during their late-summer molt (Swenson et al. 1986). This behavioral plasticity and ability to switch between food resources allows eagles to adjust more readily when one resource becomes unexpectedly scarce.

FIGURE 12.5 Bald eagles nest early in Yellowstone, sometimes before winter conditions end.
NPS PHOTO/DOUGLAS SMITH

OSPREY

Park-wide, the osprey population grew from 56 pairs in 1987 to a peak in 1994 of 101 pairs (Fig. 12.6), including 62 pairs on Yellowstone Lake and 25 on Frank Island alone. The park population then plateaued at around 80 pairs through 2002, before declining to a relatively stable number of 30 breeding pairs (Fig. 12.6). While nest success has remained stable elsewhere in the park, nest success on Yellowstone Lake began declining in the mid-1990s, concurrent with cutthroat trout decline (Fig. 12.7; Baril et al. 2013). Furthermore, in 2003, a wildfire burned much of Frank Island, sparing only one osprey nest; the last active osprey nest on Frank Island was observed in 2006. Today, despite efforts to remove lake trout and recover cutthroat trout populations, the park has almost completely lost its osprey population on Yellowstone Lake. In recent years, few pairs have attempted to nest on the Lake, and those nests are rarely successful. The sole nesting pair in 2019 failed to fledge any young.

BALD EAGLES

From 1984, park-wide tallies of bald eagles rose steadily through 2007, after which numbers appear to decline (Fig. 12.6). Some of the recorded fluctuations in eagle abundance likely reflect changes in survey effort, as park interest and staffing levels varied from year to year. For example, relatively limited surveys were conducted in 2008 after turnover in Bird Program management. Regardless of overall population size, other metrics of eagle population health are encouraging. From 1984 through 2019, 51% of bald eagle nests in Yellowstone succeeded in fledging at least one young (Fig. 12.8), and the average productivity

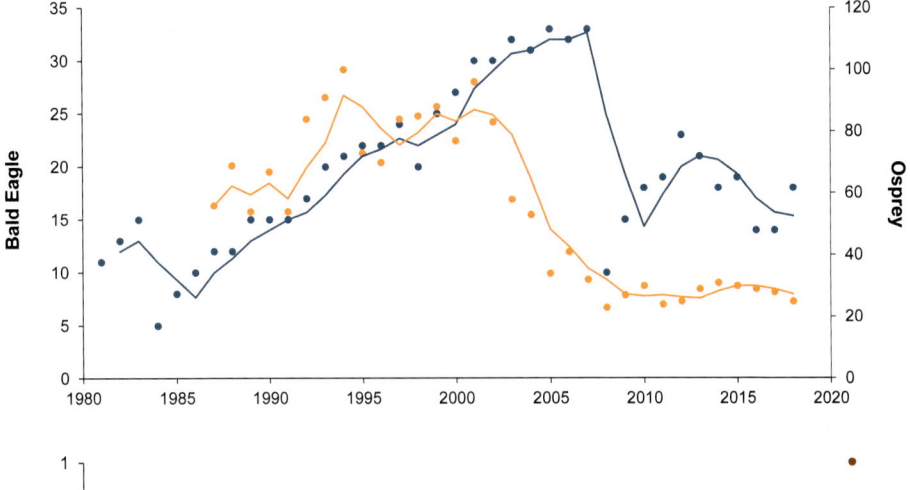

FIGURE 12.6 Estimated number of territorial pairs of bald eagles (blue) and osprey (orange) in Yellowstone National Park during the period of intensive monitoring. Trend lines are three-year moving averages.

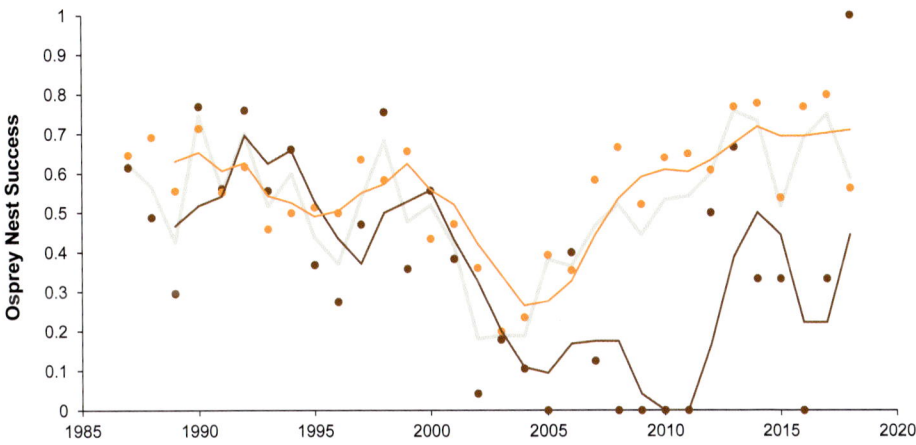

FIGURE 12.7 Observed osprey nest success per territory, with evidence of attempted nesting in Yellowstone National Park from 1987 through 2018. The osprey population is divided into those nesting on Yellowstone Lake (dark red) and those nesting throughout the rest of the park (orange). The average nest success for the park-wide osprey population is shown in gray. Trend lines are three-year moving averages.

was 0.72 young fledged per active nest. In the last 10 years, nest success and productivity across the park have averaged even higher (58% and 0.82 respectively; Fig. 12.8), suggestive of a robust population. Although nest success on Yellowstone Lake appeared to drop between 2002 and 2011, it has since recovered and appears to be in step with eagle nest success elsewhere in the park (Fig. 12.8). Without cutthroat trout, biologists believe eagles have switched to targeting waterbirds throughout their nesting season, and while osprey populations have declined, the number of breeding pairs of bald eagles on the lake has not changed significantly since 1984 (Baril et al. 2013). In recent years, Bird Program biologists have observed bald eagles depredating common loons on Yellowstone Lake, trumpeter swan cygnets from nearby Riddle Lake, and the young of American white pelicans and other colonial waterbird species nesting on the Molly Islands. Additionally, coincident with the cutthroat trout decline, the local breeding population of another large waterfowl, the Canada goose, has seemingly increased (YNP Bird Program, unpublished data). This newly available and abundant prey species may have helped to make nesting waterbirds a tempting alternative prey for hungry eagles.

Biologists believe the trophic web involving eagles and osprey on Yellowstone Lake has shifted significantly in the last three decades (Fig. 12.9; Baril et al. 2013, Koel et al. 2019b). Although lake trout numbers are down and cutthroat population estimates are slowly on the rise (Koel et al. 2019a), osprey populations have yet shown little sign of recovery, and eagles continue to kill numerous local waterbirds. This could be the new normal for Yellowstone Lake.

BALD EAGLES AND OSPREY

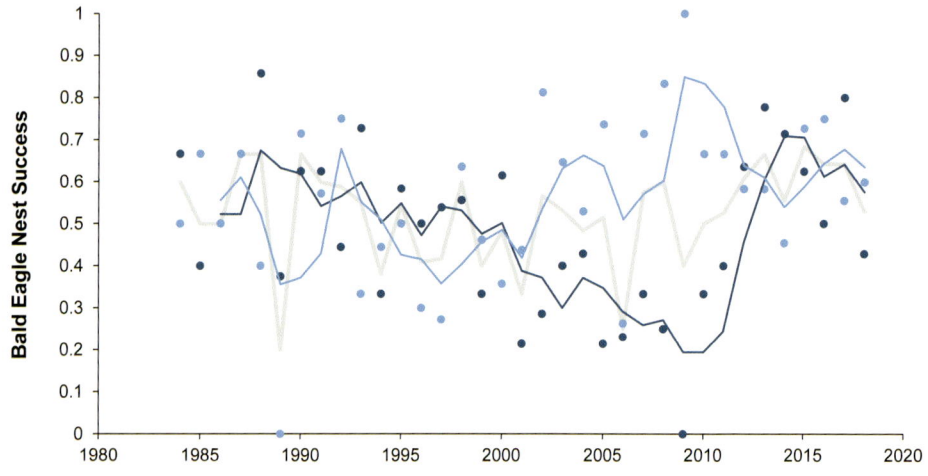

FIGURE 12.8 Observed bald eagle nest success per territory, with evidence of attempted nesting in Yellowstone National Park from 1984 through 2018. The eagle population is divided into those nesting on Yellowstone Lake (dark blue) and those nesting throughout the rest of the park (light blue). The average nest success for the park-wide eagle population is shown in gray. Trend lines are three-year moving averages.

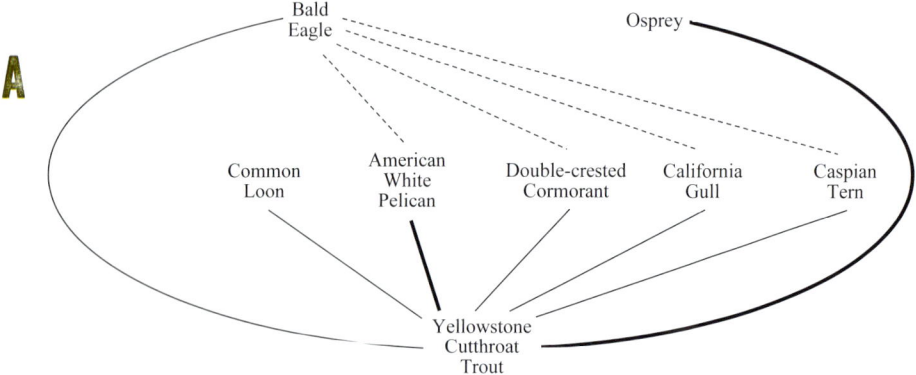

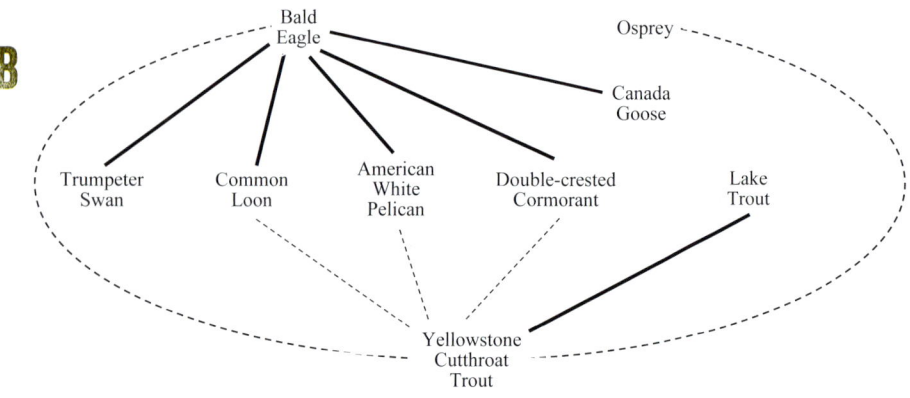

FIGURE 12.9 Yellowstone Lake trophic web before (A) and after (B) the introduction of invasive lake trout. Bold lines indicate a strong reliance of the predators on the prey species. Dashed lines indicate that predation may occur but is much diminished from historic levels.

Visit the Yellowstone's Birds website (https://press.princeton.edu/resources/yellowstones-birds-video-collection/v12-1) to watch an interview with Katy Duffy.

Visit the Yellowstone's Birds website (https://press.princeton.edu/resources/yellowstones-birds-video-collection/v12-2) to watch an interview with Douglas W. Smith.

A COMMON RAPTOR IN AN UNCOMMON PLACE

RED-TAILED HAWKS IN YELLOWSTONE

LAUREN E. WALKER, DAVID B. HAINES, and DOUGLAS W. SMITH

We make our way up the narrow canyon, lined with aspen and framed by rock walls on one side and a steep forested slope on the other. We are hoping to survey the cliffs at the terminus of the canyon, but as we turn the final corner and come out into an opening, we hear the screaming almost immediately. A loud and repetitive, impossible to ignore kee-eeee-arrr. It's a red-tailed hawk, and from its behavior, we know it's likely got a nest nearby. Although not our intended target species (we are here to survey for peregrine falcons), Yellowstone's red-tailed hawks are ubiquitous, and Bird Program biologists have recently increased monitoring efforts to better understand this relatively common raptor. Today, however, we retreat, canceling our survey to avoid unnecessary stress to this hawk and its young.

LAUREN E. WALKER

YELLOWSTONE NATIONAL PARK (Yellowstone; YNP) provides great opportunities to see raptors of all types, from the once-extirpated and still uncommonly observed peregrine falcon to the thriving bald eagle. Red-tailed hawks are a common raptor across much of North America, and Yellowstone is no exception. A familiar sight in the park, and particularly abundant on the northern range, red-tails are easy to spot in the summer as you drive the Grand Loop Road from Mammoth Hot Springs to Tower Junction, and the Northeast Entrance Road through the Lamar Valley. The plumage color of red-tailed hawks varies across their range and in

Yellowstone. First, like many raptor species, juvenile plumage color and pattern is distinct from that of their adult counterparts. Second, and more to the point, the species exhibits a variety of color morphs, typically categorized as light, rufous, and dark; however, the gradient between these categories is nearly continuous (Fig. 13.1). These color morphs can represent the variation in appearance one may expect to find within a population, like hair color, or the outward physical expression of geographically distinct subspecies. In general, however, red-tail identification can be straightforward, particularly for lighter-colored birds. In both juvenile and adult red-tails,

A COMMON RAPTOR IN AN UNCOMMON PLACE

FIGURE 13.1 Red-tailed hawks in Yellowstone may appear in a variety of color morphs, from light, depicted here, to dark. PHOTO BY GREG ALBRECHTSEN

visitors can look for the usually obvious contrast between the dark band across the belly and the white or pale breast, in combination with dark rectangular markings on the leading edge of the wings underside. For darker birds, this contrast can be partially to completely obscured by the increased pigmentation. However, except for some subspecies, the nominative red tail is a reliable identifier of adults (the juvenile tail has narrow bands and lacks red). Additionally, red-tailed hawks are unusually vocal for a

raptor and their *kee-eeee-arrr* call may be familiar to even non-birders, as it is often mistakenly used in movies and on television to represent the call of the otherwise majestic, but somewhat vocally diminutive, bald eagle.

Across the North American continent, red-tailed hawk populations are largely considered healthy and stable (Preston and Beane 2020), despite the pervasive landscape fragmentation induced by the broad human development that plagues the habitat of many other raptor species.

Truly, red-tailed hawks seem to thrive in many human-altered landscapes, commonly perching from power and fence poles, hunting in agricultural fields and along highway rights-of-way, and famously using city buildings as nesting sites (Preston and Beane 2020). Not all human infrastructure is friendly to red-tails, however; hawks are regularly hit by cars and electrocuted by power lines, and at least one recent study suggests they may be particularly vulnerable to collisions with wind turbines (Preston and Beane 2020, Diffendorfer et al. 2021).

Climate change may bring further unforeseen impacts to this widespread species. For example, changing precipitation and temperature patterns could lead to local declines in the hawk's prey populations. Alternatively, widespread drought and increasing fire frequency in the West may help create more of the open spaces preferred by red-tailed hawks (Preston and Beane 2020). While Yellowstone certainly has some human infrastructure, in general, the human impact on the landscape is minuscule relative to much of the contiguous United States. The park, therefore, presents an interesting opportunity to study this common species in a relatively wild and natural environment, allowing for comparison with both concurrent populations located in more developed areas and a red-tail population that must navigate Yellowstone's future environmental change.

A current red-tailed hawk population assessment will provide a useful baseline against red-tail populations evolving with climate change, wildfire frequency and intensity, and park management strategies. Additionally, although the park has a long-running program monitoring peregrine falcons, bald eagles, and osprey, biologists know relatively little about the populations of other raptors in the park (*see chapters 11, 14, and 16*). Red-tailed hawk populations may serve as a bellwether, warning park biologists of potential problems with a variety of less-common (and thus more difficult to study) raptor species (e.g., Swainson's hawks). Furthermore, by learning more about what drives demographic patterns in red-tailed hawk populations, park biologists hope to glean insights into the stressors faced by a broad suite of raptors.

Between 2011 and 2015, the Yellowstone Bird Program conducted an initiative focused on better understanding the wide variety of raptors that use the park, including red-tailed hawks. With little known about Yellowstone's red-tails, biologists were largely starting from scratch. During the Yellowstone Raptor Initiative (YRI), park biologists established roadside point counts from which to develop a population estimate for the northern range and additionally spent extensive effort searching for individual territories and nests to monitor reproduction. Luckily, red-tailed hawks are both numerous and relatively obvious raptors; even restricting monitoring efforts to the northern range, biologists located and monitored 17 territories in the first year of the study. That number grew, and over the course of the five-year initiative, an average of 31 red-tailed hawk territories were monitored each year. Although the YRI ended in the autumn of 2015, recently red-tailed hawk monitoring efforts were reinvigorated, utilizing the participation of a group of dedicated citizen scientists. In 2020 and 2021, volunteers visited 26 and 28 hawk territories across the northern range. What we found, both during the YRI and in following years, lays the groundwork for comparison with surveys into the future, as changing climate influences habitat.

RED-TAILED HAWKS IN YELLOWSTONE

Versatile and adaptable, red-tailed hawks breed from southern Mexico through Alaska and northern Canada; in much of the contiguous United States and Mexico, they can be found year-round (Preston and Beane 2020). In Yellowstone, however, the high elevation and harsh winters convince most red-tails to migrate south for refuge in the non-breeding season (Fig. 13.2). In the spring, red-tailed hawks return to northern Yellowstone in early to mid-March, quickly claiming breeding territories; this short migration, and relatively early return, may give them a competitive edge over other hawk species

A COMMON RAPTOR IN
AN UNCOMMON PLACE

that travel farther and arrive later in the spring, such as Swainson's hawks (*see chapter 14*).

In total, Bird Program biologists and a dedicated team of volunteers have identified 60 red-tail territories on the northern range, although we know there are many more. Based on the roadside point counts, we estimate around 184 red-tailed hawks across the area of the northern range within the park, at an average density of 0.18 birds per 0.4 square miles (1 km²; Fig. 13.3; Walker et al. 2019). Given that these surveys were conducted during the breeding season, when many individuals would have been incubating eggs or tending young (and thus would have been difficult to detect in standard surveys), this number may be a better estimate of the number of pairs rather than individual birds. We also observed local areas with a much higher density of hawks than the average. In one relatively small area of the Blacktail Deer Plateau, where we believe we were able to locate all the red-tailed hawk territories, we found them at a density of 0.59 pairs per

0.4 square miles (1 km²; Fig. 13.3; Walker et al. 2019). While the average density across the northern range is typical among other regional estimates, particularly if you consider it an estimate of the density of red-tail pairs rather than individuals, the density on the Blacktail Deer Plateau is notably higher. This region may be particularly suitable for red-tailed hawks and the habitat—open sagebrush steppe, punctuated by small stands of conifers, and home to abundant ground squirrels—provides ample foraging and nesting opportunities.

Across their range, red-tailed hawks may use a variety of nesting substrates, from power poles to building ledges (Preston and Beane 2020). In Yellowstone, most red-tail pairs nest in snags (47%) or live conifers (35%), and a relatively small percentage of nests are located on cliffs (12%) or in deciduous trees (2%; Fig. 13.4; Walker et al. 2019). In contrast to northern range golden eagles (*see chapter 11*), most territorial pairs of red-tailed hawks in the park attempt to nest each year (on average, 87%), and pairs begin

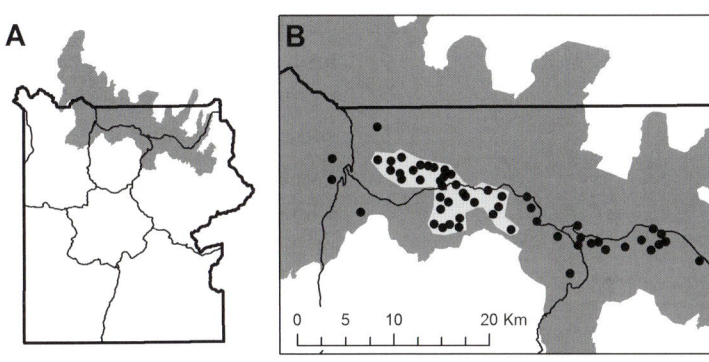

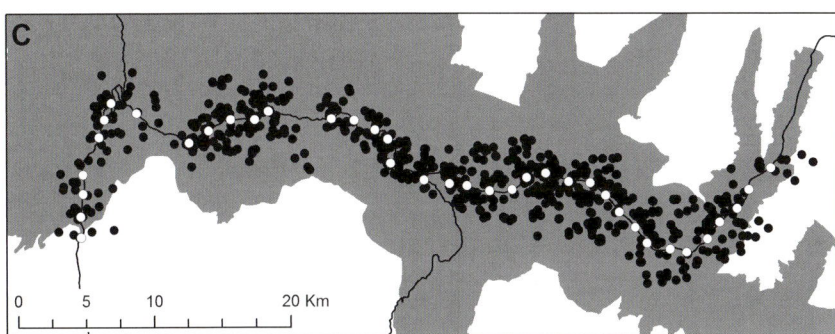

◀ FIGURE 13.3 Red-tailed hawk survey area in Yellowstone National Park's northern range (dark gray; A–C). In (B), black dots are locations of red-tailed hawk territories, monitored from 2011 and 2015; the light-gray area is the location of dense red-tailed hawk territories on the Blacktail Deer Plateau. In (C), white dots are roadside survey points, and black dots are red-tailed hawk detections from 2012 to 2015. Bold black lines are park boundaries, and thin black lines are park roads. FIGURE AND CAPTION FROM WALKER ET AL. 2019

▼ FIGURE 13.4 A classic example of a red-tailed hawk nest. PHOTO BY SCOTT HEPPEL

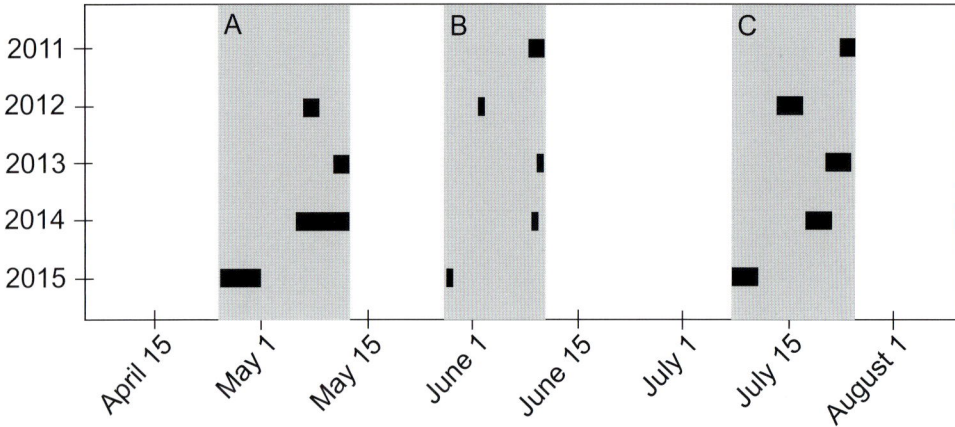

FIGURE 13.5 Nesting chronology of red-tailed hawks in Yellowstone National Park's northern range from 2011 to 2015. Gray blocks indicate estimated periods of laying (A), hatching (B), and fledging (C) across all study years. Black bars represent the observed variability within each study year. FIGURE AND CAPTION FROM WALKER ET AL. 2019

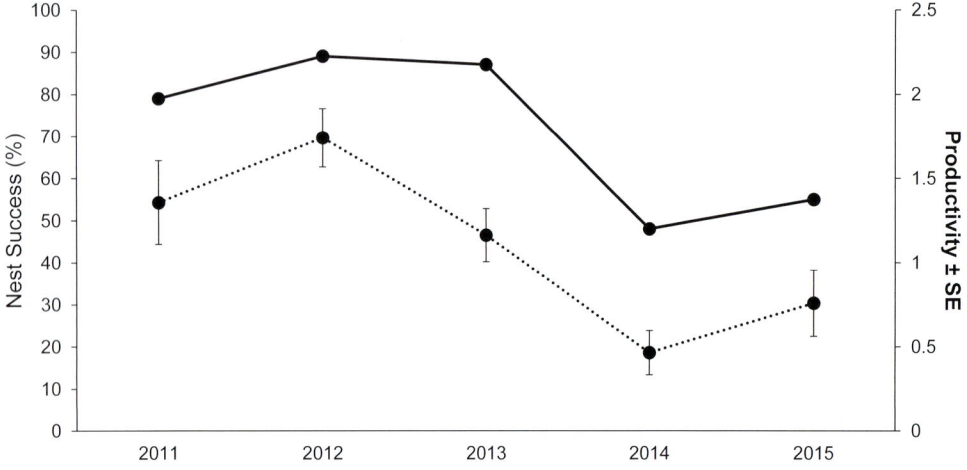

FIGURE 13.6 Nest success (percent of nests that fledge at least one young; solid black) and productivity (average number of young fledged per occupied territory ± standard error; dotted black) of red-tailed hawks on Yellowstone's northern range, 2011–2015.

laying and incubating eggs in early May. The adult hawks incubate their eggs for about 28 days, and the nestlings hatch in early June (Fig. 13.5; Walker et al. 2019). About 45 days later, by late July, most nests (63%) fledge one or two young hawks (1.7 young per successful nest on average), and observant visitors and biologists alike may watch as they awkwardly learn to fly and navigate life away from the nest (Fig. 13.5; Walker et al. 2019). During the YRI, nest success (percent of territories that successfully fledge at least one young) varied significantly between years (Fig. 13.6). In a good year like 2012, nest success was as high as 89%, while in 2014 fewer than half (48%) successfully fledged any young (Fig. 13.6; Walker et al. 2019).

In other locations, annual variation in hawk reproduction has been attributed to differences in weather or cli-mate patterns, prey availability, interspecific competition (particularly with other buteos), and even nest parasites (Schmutz et al. 1980, Stinson 1980, Janes 1984, Smith et al. 1998, Craighead and Smith 2002). Our nest-monitoring efforts do not allow for full transparency of what factors may impact reproduction, nor do we have the data to determine the effect of prey availability and abundance. Additionally, preliminary analyses for the effect of weather conditions on reproductive success have been inconclusive. What little we have observed leaves us to speculate upon some factors that negatively impact young at the nest. From our observations of nesting red-tails and our research on golden eagles, nest depredation appears to be one factor that can impact hawk reproductive success. We know golden eagles prey upon red-tailed hawk nestlings through both the collection of prey remains and the use of remote cameras at golden eagle nests. Further, we suspect that great horned owls also play a role in this form of interspecific competition. For example, over one three-day period, at a particularly easy-to-view roadside red-tail nest, we observed the sequential disappearance of three nestlings. Though we cannot confirm this was the act of a great horned owl, each evening nestlings were present, and by morning one was removed until none remained, suggesting a repeated nighttime event. We searched the area below the nest in case young had prematurely fledged, but none were found, thus partially supporting a depredation event. Another anecdotal observation suggests a second potentially important factor in red-tailed hawk reproductive success: biting flies.

The abundance of biting flies varies greatly from year to year, and their overall impact on annual productivity is unknown. Nonetheless, young birds unable to fly and driven from the refuge of their nest are extremely vulnerable, and many are unlikely to survive. Our observations suggest a potential threat in years when fly abundance is high.

Currently, it is difficult to say whether Yellowstone's red-tailed hawk population is stable. Henny and Wight (1972) estimated that northern red-tailed hawk populations must fledge, on average, between 1.33 and 1.38 young per occupied territory to maintain a stable population—well above our observed average productivity of 1.07 young (Fig. 13.6). However, local conditions may have different requirements. Additionally, the five-year period of the YRI may not be representative of long-term average productivity for the region. Ongoing red-tailed hawk monitoring could confirm our previous observations or identify significantly different patterns in reproduction or overall population size, ultimately helping the park's biologists effectively and appropriately manage for both this hawk species and the habitat it uses. While we have little concrete data to explain the variability in our observations in Yellowstone, determining what drives the differences between years could ultimately enable biologists to make predictions about the long-term stability of red-tailed hawks in the park.

DAVID B. HAINES

In 2014, while assessing the breeding success of red-tails on the Blacktail Deer Plateau, we peered through our spotting scopes, doing our best to cope with an excessive amount of biting flies. Upon our previous visit to this nest, three nestlings had been present, however, as we adjusted our focus it was clear that the nest was empty. We scanned through the tree in case the nestlings had moved into the branches, but still, none were observed. We made the decision to approach the nest tree in hopes of revealing more than what we could determine through our scopes. As we approached, the unmistakable *kee-eeee-arrr* rang out from the adult red-tails, suggesting that young may still be present. Suddenly, within 50 meters of the tree, we noticed a young red-tail lying on the ground, alive but covered by the same biting flies we were ourselves contending with. With the cries of the adults above, and the chance that we may alert other predators to this vulnerable young red-tail, we quickly retreated, surmising that the nestlings had prematurely jumped from the nest in response to the inordinate number of flies. When we returned a week later, viewing the area from across a creek drainage, we observed two juveniles soaring with the adults, indicating that at least two of the three had survived.

During August and September, Yellowstone's adult red-tails are foraging and tending juveniles, while the young hawks are perfecting their hunting skills and quickly becoming self-sufficient. By October, Yellowstone's red-tails are on the road again, migrating to warmer southern locales, with more predictable winter food sources. Adults and juveniles go their separate ways, and the young birds will face their first migration as independent birds. Red-tailed hawks that survive the trip and the long winter will return to the park the following spring to try it all again.

14 RAPTOR DIVERSITY IN YELLOWSTONE NATIONAL PARK

DAVID B. HAINES, DOUGLAS W. SMITH, and KATHARINE E. DUFFY

In Yellowstone National Park (Yellowstone; YNP), not everyone who parks in a pullout and hastily exits the car with binoculars adjusts their focus on large mammals. Hawk watchers view Yellowstone as a cornucopia of habitats where many species of raptor migrate, nest, and hunt. People gazing skyward with binoculars could be witnessing spectacular raptor behavior, perhaps a sleek prairie falcon zooming along the cliff walls of Mount Everts and disappearing into one of the Swiss-cheese-like holes where it nests. During courtship, American kestrels vocalize loudly from aspen stands with old woodpecker cavities, favored nest sites for these small falcons. Numerous red-tailed hawks find productive nest territories throughout the northern range. Powerful golden eagles often choose nest ledges high in rugged canyons. Long-winged northern harriers cruise back and forth, low to the ground over meadows, where they look and listen for small rodents and bird prey. When Swainson's hawks return from wintering areas in southern South America, some will build their stick nests in large conifers at the upland edge of Hayden Valley. Cooper's hawks hunt by speedily darting out of forest edges, taking what they captured back to their nests hidden deep in the forest. The Yellowstone bird checklist contains 24 raptors that nest, winter, or migrate through the area, diverse representation for a relatively harsh high-elevation environment.

YELLOWSTONE NATIONAL PARK'S long-term Bird Program has provided more than 30 years of data regarding the population status of three raptor species, the peregrine falcon, bald eagle, and osprey (*see chapters 10 and 12*). Some of these data reflect the population recovery of bald eagles and peregrine falcons from the negative effects of widespread organo-chlorine pesticide use (DDT; dichloro-diphenyl-trichloroethane) following its restrictions in 1972. Additionally, these monitoring efforts have revealed a population response of both the bald eagle and the osprey to the introduction of non-native lake trout around Yellowstone Lake (Baril et al. 2013). However, an additional 16 species of hawk, eagle, falcon, and owl have been confirmed to breed in YNP and have been relatively neglected in terms of monitoring.

In 2009, coauthors Smith (National Park Service, retired) and Duffy (National Park Service, retired), along with Joel (Jeep) Pagel (US Fish and Wildlife Service), recognized that little to no attention had been given to these other raptor species in YNP. So they proposed a project to establish baseline data on specific species with no previous data regarding their local status. This proposal resulted in the Yellowstone Raptor Initiative (YRI, 2011–2015; Baril et al. 2017b), a five-year program funded from private donations to the Yellowstone Park Foundation (now Yellowstone Forever). Species of interest included the golden eagle, red-tailed hawk, Swainson's hawk, American kestrel, prairie falcon, and several owl species. The results from golden eagle, red-tailed hawk, and owl monitoring are reported in their respective chapters (*chapters 11, 13, and 16*). In this chapter, we summarize some of our findings, insight, and questions regarding species with little or no data gathered since the start of the YRI.

SWAINSON'S HAWK

The Swainson's hawk may best be recognized for its gregarious long-distance migration between breeding areas of western North America and wintering areas in the pampas (lowland grassland plains) of southern South America (primarily Argentina). Nearly the entire North American population of this relatively slender and narrow-winged buteo (Fig. 14.1) exhibit this incredible migration, and for some individuals, the roundtrip can exceed 12,400 miles (20,000 km), a distance among raptor species exceeded only by the arctic peregrine falcon (Fuller et al. 1998). As opposed to many migrating raptors that travel individually or in small groups, the Swainson's hawk typically travels in flocks, or kettles. This spectacular migration converges along the gulf coast of eastern Mexico, where one can observe individual flocks of 5,000 to 10,000

individuals soaring overhead (Fuller et al. 1998, Bechard et al. 2020a). A true marvel of nature.

Swainson's hawk native environments vary regionally but are typically characterized by open grasslands, sparse shrublands, and small open woodlands at low to mid-elevations (Bechard et al. 2020a). But, today, Swainson's hawks have adapted to agricultural landscapes, particularly those with similar vegetative structure to native grassland communities (e.g., wheat and alfalfa; Bechard 1982, Estep 1989, Woodbridge 1991). These environments often support favored prey, which includes insects, ground squirrels, and pocket gophers.

Despite adaptation to farmland, there are factors that can, both immediately and over the long term, negatively impact populations. For example, in response to apparent declines observed on breeding grounds during the early 1990s, researchers found that thousands of Swainson's hawks were dying in Argentina as a direct result of an organophosphate pesticide used to control grasshoppers (Woodbridge et al. 1995, Goldstein et al. 1996). It was suggested that approximately 5% or more of the global population died as a direct result of this pesticide (Goldstein et al. 1996). Fortunately, these findings led to a complete ban of the pesticide throughout Argentina in 2000. Over the long term, conversion of agriculture to urban landscape (a common trend) destroys habitat, thus limiting population size. Consequently, much of what we know regarding Swainson's hawk ecology has often focused on populations associated with this human-induced change to the landscape. Therefore, for the purpose of conservation, it is important to better understand the species across portions of its range where resources resemble their natural state.

SWAINSON'S HAWK IN YELLOWSTONE NATIONAL PARK

A species typically associated with agricultural landscapes at lower to mid-elevations, Swainson's hawks breed across the high elevations of YNP, a remaining vestige of the native environment of the species. Surprisingly, the northern

OPPOSITE PAGE:

FIGURE 14.1 A Swainson's hawk with an intermediate color morph between rufous and dark soars overhead, easily identified by its slender and tapered wings.
PHOTO BY SCOTT HEPPEL

range of YNP, with its wide-open grassland and sagebrush communities, and less densely forested areas, supports few territorial Swainson's hawks. Instead, the abundance of territorial pairs increases as you move to the higher-elevation plateaus, valleys, and montane meadows of YNP's interior.

An initial objective of the YRI, like that for the golden eagle and the red-tailed hawk (*see chapters 11 and 13*), was to locate Swainson's hawk territories and monitor reproduction annually. However, though observing territorial individuals was not difficult, locating nest sites proved to be an arduous task with a low rate of success. Accordingly, efforts were reduced to include only repeat surveys along two park interior road segments, reproductive monitoring at a few territories, and documentation of opportunistic encounters. Here we summarize some of our observations and personal insights regarding the species in YNP.

THE CONTRIBUTIONS OF A FORMER GARBAGE COLLECTOR

In 2010, after presenting the objectives of the YRI at the annual Raptor Research Foundation conference in Fort Collins, Colorado, coauthor Smith was approached by Dr. Jack Kirkley, a raptor ecologist and professor at the University of Montana, Western. Kirkley informed Smith that he had worked as a seasonal maintenance worker in YNP, hauling garbage during the mid-1970s, and that, as an aspiring ecologist, he had documented raptors along his routes. For Smith, this was a fortuitous encounter. Kirkley volunteered to establish and conduct surveys for Swainson's hawk that would align with his observations from the mid-1970s, thereby providing a reference for

changes in Swainson's hawk occurrence between two time periods that would otherwise have been unknown.

From 2011 through 2013, Kirkley surveyed points along two road segments from the Grand Canyon of the Yellowstone south to Yellowstone Lake (15 locations) and the northern edge of Yellowstone Lake (11 locations). Kirkley surveyed the predetermined points 8–19 times per year, documenting the presence or absence of Swainson's hawk, nest sites if observable, and recent fledglings. Additionally, when possible, Kirkley photographed and recorded the color morph of territorial pairs to document the turnover of individuals. Swainson's hawk color morphs are typically grouped into three categories (light, dark, and rufous), but variation between these morphs is almost continuous. Therefore, plumage patterns are often unique enough to visually differentiate individuals, a confirmation that is rarely possible for many species outside of the use of capture and marking techniques.

Across all three years and 26 points, Kirkley observed Swainson's hawks at 19 points at least one time and identified a total of 13 territorial pairs. In 2011 and 2012, evidence of breeding (vocalizing juveniles, observed fledglings, carrying of food and nesting material) was observed along survey routes at a minimum of eight territories each year. Over the three years, Kirkley identified site reoccupancy and the turnover of individuals at five occupied territories using the unique plumage patterns he had documented. One territory maintained the same breeding pair across all three years. From 2011 to 2012, the remaining four territories all replaced a single individual, and between 2012 and 2013, two territories maintained the same individuals from the previous season, and one territory replaced a single individual. Finally, at one location, Kirkley determined that one female occupied the same territory and paired with a different male in each of the three years.

Kirkley's observations of Swainson's hawk aligned with many areas where he had observed the species during the mid-1970s. But there was an apparent absence from the Canyon horse riding complex, Indian Pond, and Mary Bay. Between 2011 and 2013, red-tailed hawks occupied these areas that were once occupied by Swainson's hawks; perhaps more importantly, large-scale forest fires during the 1980s had altered the forest, grassland, and shrub communities. The effects of these fires may have caused shifts in territorial boundaries, as increased open areas allowed for expanded foraging grounds and redistributed suitable nesting areas, potentially allowing Swainson's hawks to use areas that were previously unsuitable. A more intensive survey, including areas away from the road, would be necessary to further examine the absence or presence of Swainson's hawks throughout these areas.

Outside of Kirkley's survey points, YRI staff observed Swainson's hawks throughout many other areas of the interior, which include but are not limited to Pelican Valley and Hayden Valley, Upper Geyser Basin and Lower Geyser Basin, Mirror Plateau, various wet meadows, and the Thorofare. Composition of the vegetative communities where individuals were observed varied greatly from shrub-steppe to willow-dominated river valleys. However, all included forest communities interspersed within or surrounding these open areas, creating structural similarities that likely represent habitat preferences in this high-elevation environment. Through portions of the interior, Swainson's hawk abundance appeared to be equal to or greater than the related red-tailed hawk, a stark contrast to our observations in the northern range of YNP.

Yellowstone's northern range is largely accessible throughout the year and, as an environment that at first glance appears most suitable for open-country raptors, was central to the YRI's objectives regarding golden eagles and red-tailed hawks (*see chapters 11 and 13*). Additionally, we actively surveyed this area for Swainson's hawks during the YRI. Despite the extensive amount of time spent surveying the northern range, for Swainson's specifically and raptors in general, we confirmed only five Swainson's hawk territories. Occasional observations of individuals in other areas of the northern range suggest that there are likely more, but by no means equal the abundance levels observed in areas of the interior. Of these five northern territories, only one is located at a lower relative elevation, while all others are found along the elevated margins of the northern range. Red-tailed hawk abundance in the northern range greatly outweighs that of Swainson's hawk, an observation that raises questions regarding what factors drive this disparity (*see below*).

One observation in particular highlights this interesting distribution, documented on the near-annual late-summer horseback rides by coauthor Smith across the eastern portion of the northern range and into Yellowstone's interior. Equipped with pocket binoculars dangling around his neck, Smith tried to identify every hawk while riding along on one of his horses, Amos or Joker. Starting at a relatively low elevation in Lamar Valley, roughly 6,500 feet (1,980 m), it was all red-tailed hawks. Then, gradually, as elevation was gained to about 9,000 feet (2,740 m) on Mirror Plateau, more and more Swainson's hawks appeared. For a time across the plateau, it was only Swainson's. After camping for a night, this continued through the next day along Astringent Creek (some years Pelican Creek) and into Pelican Valley, where he surveyed near Kirkley's routes and again saw scattered red-tailed hawks.

REPRODUCTION

Between the interior and the northern range, we located nests in only five territories. In many attempts to locate other nests, individuals or mated pairs would often fly beyond the forest edge, obscuring our view of suspected nest sites. On multiple occasions, we observed individuals carrying sticks for their nest or prey we suspected was being delivered to nestlings. Nest-site characteristics vary across the breeding range of the Swainson's hawk (Bechard et al. 2020a); nests observed in YNP are commonly located in coniferous trees set back from the forest edge of open valleys and meadows, often appearing small and well camouflaged.

We monitored breeding at between one and three of the five nest sites each year and, over the five-year period of the YRI, observed 11 breeding attempts that we followed until young fledged or failed. Of the 11 nesting attempts, 10 were successful (90.1%) in fledging a minimum of 14 young. This small sample does not allow us to make inferences on the reproductive rates of Swainson's hawk in YNP or comparison among populations. Few studies have monitored reproduction in areas of low human disturbance or nearly native environments (Andersen 1995), particularly at high elevations, highlighting the need to better understand the breeding ecology of this species in a place like YNP.

WHY DOES ABUNDANCE VARY?

The relative abundance of Swainson's hawks between the interior and the northern range of YNP has raised questions regarding the species' local ecology. The low abundance in the northern range may be driven by the high density of territorial red-tailed hawks (Walker et al. 2019), who travel shorter migration distances and arrive in YNP weeks prior to Swainson's hawks. This means that red-tails occupy territories in the absence of Swainson's hawk, a species that may otherwise compete for similar resources. Given that the retreat of winter throughout the interior greatly lags behind the northern range, perhaps the abundance of the two species appears more similar in the interior than in the lower-elevation areas of the park, an interaction between the arrival of the species and the time at which conditions become suitable for occupation. That being said, in areas where the timing of suitable conditions aligns with the presence of both species and competitive abilities are similar, abundance of the two species may be similar.

Alternatively, this difference in relative abundance of the two hawks in areas of high and low elevation may be regulated through resources preferred by the respective species in each area. No analysis of habitat or prey selection by either species has been done in YNP, but both are known to exploit small mammals and birds elsewhere. Throughout many parts of YNP's interior, ground squirrels (a preferred prey of many medium to large raptors) are generally absent or in low abundance compared to the northern range, where they are ubiquitous in summer. For example, in Hayden Valley, where many Swainson's hawks are observed, small mammal species diversity consists mainly of pocket gophers and montane voles, species previously identified as prey in nearby areas (Restani 1991). Additionally, the Swainson's hawk, unlike the red-tailed hawk, is known to exploit invertebrate prey sources (e.g., grasshoppers and other insects; Bechard et al. 2020a). On multiple occasions, we have observed territorial individuals capturing and consuming insects like salmonflies, suggesting insects may play an important role for interior birds during the breeding season. In addition to differences in prey availability, the composition and structure of forest and other vegetative communities vary between the northern range and the interior. Therefore, this difference in distribution and abundance may reflect how resources are partitioned between two species that otherwise have the potential to negatively influence successful breeding through competition.

LOOKING FORWARD

There is a need to better understand the life history and ecology of Swainson's hawks across portions of their range with a low human footprint. The species' strong association with certain agricultural landscapes has shaped much of our knowledge regarding their natural history. Native landscapes with low human disturbance are diminishing rapidly throughout the western United States, and YNP may now represent a unique place to advance our knowledge of Swainson's hawk breeding ecology and habitat.

THE FOREST HAWKS

Within the family of birds that include hawks and eagles, a group of species with distinct morphological similarities, adapted for maneuverability, characterizes the genus *Accipiter*. It is diversely represented around the world; only three species occur in North America, and all breed within YNP. These include the sharp-shinned hawk, the Cooper's hawk, and the northern goshawk (Fig. 14.2), ranging in size from smallest to largest, respectively. The natural history of these small to medium-sized hawks, with their short, yet broad wings and long tails, includes a strong association with forested communities. This association, coupled with their ability to move rapidly through these environments, make observations less common and often brief. For this reason, locating nest sites and monitoring reproduction for all three species is very difficult.

Much of what we rely on regarding the population status of sharp-shinned hawks and Cooper's hawks is estimated from spring and autumn migration counts across North America. Consequently, these estimates lack a fine-scale understanding of where these birds are from and how their reproductive efforts vary in relation, limiting our understanding of their breeding ecology. However, unlike the most difficult to monitor sharp-shinned hawk, Cooper's hawks have been undergoing a range expansion since the mid to late 1970s (Farmer et al. 2008), adapting to a wide range of non-forested and urban environments. These adaptations have afforded more opportunities to study Cooper's hawk breeding ecology in portions of its range (Rosenfield et al. 2020). Northern goshawks exhibit a less migratory and more sedentary life history, and population estimates for the species are generally weak at a broad scale (Squires et al. 2020). At a finer scale, northern goshawk occupancy, abundance, and breeding ecology are studied more extensively, greatly driven by forestry practices known to degrade habitat.

With extensive forest communities found throughout the park, we acknowledged early on that YNP has the potential to support a sizable accipiter population. These communities provide the appropriate nesting substrate and a prey base of small birds and mammals to promote occupancy, survival, and reproduction. Additionally, we recognized that data are limited regarding the reproductive demographics of these species across their ranges. However, any one of these species would require a challenging independent effort to locate occupied territories and monitor reproduction. This requirement did not align with one primary objective of the YRI, which was to establish data on multiple species previously unmonitored in YNP. Therefore, we painstakingly made the decision to exclude accipiters from the YRI.

Despite the exclusion of these forest specialists from our monitoring efforts, we commonly document encounters, recording the time and location of each. Most of our observations occur while individuals soar above their forest environments or hunt the margin between forested and non-forested communities. Given the elusive nature of these hawks, these encounters are typically met with excitement among the Bird Program crew and the courtesy of alerting each other that one is in view. Occasionally, some Cooper's hawks and northern goshawks have deliberately alerted us to their presence with distinct vocalizations (Cooper's and goshawk) and sometimes aggressive encounters (goshawk), a definite sign that a nest site with young is nearby. These encounters are most typical of the northern goshawk, the largest of the three species and notoriously aggressive near their nest site; they are even known to injure unwelcome visitors to their territory. Regardless of these specific encounters, we rarely, if ever, locate a nest, even when we are certain one is near.

Conservation concerns surrounding these species commonly include their response to human-caused and naturally occurring forest disturbance. Differences in breeding habitat between species is difficult to discern, given their overlap in habitat requirements (Rosenfield et al. 2020). However, research has shown that body size is positively correlated with the size of tree each species selects for their nest (Siders and Kennedy 1996). For example, on average, a sharp-shinned hawk will select a smaller tree to build a nest than that of a Cooper's hawk or northern goshawk. Additionally, differences in forest structure appear to influence the presence of each species (Rosenfield et al. 2020). Both tree size and forest structure are attributes of forest age, which undoubtedly are influenced by forest disturbance. Depending on the frequency and magnitude of these disturbances, habitat

OPPOSITE PAGE:

FIGURE 14.2 Uncommonly observed, as it lives in forested areas, an adult northern goshawk perches in Yellowstone National Park. PHOTO BY HOWARD WEINBERG

may be threatened or limited, particularly for the northern goshawk, which is commonly associated with mature forests that require long periods to regenerate (Squires et al. 2020).

YNP may be the ideal laboratory for research regarding the response of these species to natural disturbance. The extensive forest communities, and the multitude of wildfire scars at varying stages of succession, offer a unique opportunity to apply principles of experimental design to an ecological study. With increasing rates of forest fires throughout the western United States, in part attributed to a changing climate (Higuera et al. 2021), there is a strong need to better understand the population response of these species to this disturbance of forest communities.

Each fall, most sharp-shinned hawks and Cooper's hawks migrate out of YNP to wintering areas as far south as Central America (Bildstein et al. 2020, Rosenfield et al. 2020). Both species will regularly use rural and urban environments during these times of the year, and they are known to hunt songbirds at backyard bird feeders. For northern goshawks at Yellowstone's latitude, the way they spend winter is less predictable; some likely reside in and around the park, while others may exhibit a relatively short migration. These goshawk movements may occur across elevational gradients or resemble a more typical north–south route. For all species, the decision to migrate is partially driven by the availability of prey, which is greatly influenced by the harsh winter conditions within YNP.

Though the local accipiter population is poorly understood, YNP provides excellent opportunities to encounter and observe these forest residents during the breeding season. These encounters may come in the form of a brief blur, followed by the question "What was that?," or an in-your-face and terrifying experience of a northern goshawk fiercely screaming and diving as you run for safety. Ultimately, questions regarding population size, reproductive demographics, habitat preferences, and migratory strategy will be left to the uncertain future of accipiter research in YNP.

PRAIRIE FALCON

While trying to relocate a peregrine falcon territory in June 2011, relying on records with no coordinates or details of territory location and only the name Upper Black Canyon, we stopped to scan a cliff band along an incredible section of the Yellowstone River corridor. This was not our first attempt to locate this territory, but when in search of nesting raptors, one must often try and try again. Not long after our arrival, we caught sight of a falcon moving fast across the cliff face. Elated by the possibility that the search for the peregrine territory was over, we diligently followed the falcon with our binoculars to a perch. However, instead of seeing the dark gray color on the wings' upperside and the broad black facial stripe that forms the iconic masked appearance of the peregrine, we observed a falcon with pale-brown upperparts and a narrow black facial stripe. This was a prairie falcon, a species similar in size and shape to the peregrine. The falcon promptly flew from its perch and entered a small cavity in the cliff. Unable to see into the cavity, we were certain that this was the falcon's nest site and that peregrines must not be present. Fortunately, as prairie falcons are a species of interest, we carefully documented this territory and planned our return to determine if the falcon pair would be successful in fledging young. Later that day, still on the search for peregrines, and approximately 5 miles (8 km) up-river, we located a second prairie falcon nest site on a small, yet prominent cliff emerging from a forested slope. We again documented our findings, enthusiastic that we had located two previously unknown territories. These discoveries were more than enough to ease any discouragement that would normally come with another failed attempt to locate the Upper Black Canyon peregrines.

DAVID B. HAINES

A LITTLE ABOUT PRAIRIE FALCONS

The prairie falcon is typically found in open arid environments across western North America, with a breeding range that reaches from southern Canada down to Mexico (Fig. 14.3; Steenhof 2020). Exhibiting a less specialized diet than the peregrine, a species that subsists almost exclusively on avian prey, the prairie falcon is known to exploit small mammals, birds, and reptiles. As a particularly vocal species during the breeding season, their presence is often heard more than seen. With patience and keen observation, this vocal nature, common among many falcon species, can greatly aid in one's ability to witness their beauty and behavior. Like the peregrine, prairie falcons most commonly select ledges and cavities on cliffs for breeding. They will, however, uncommonly use trees and man-made structures. Unlike many raptors, falcons do not build nests; the prairie falcon simply scrapes the substrate's surface and lays its eggs directly on the ground or in unoccupied stick nests made by ravens or other raptor species (Steenhof 2020).

At a broad geographic scale, the prairie falcon is one of the more enigmatic raptor species; little is known about overall population size and status. As a species, it does not exhibit, on average, typical migration patterns and is resident across portions of its range (Steenhof 2020), and estimates of continent-wide population size and trends are difficult to interpret (Farmer et al. 2008). Therefore, our understanding of the species is limited to regional studies throughout their range. These studies have provided invaluable knowledge regarding locally specific prairie falcon ecology but are often unable to account for how the species responds to resources and conditions found elsewhere. For this reason, we initially proposed that the prairie falcon be included in the monitoring efforts of the YRI.

THE TRIALS OF COMPETITION

The objective of locating and monitoring prairie falcons in YNP was quickly met with challenges that again limited us to monitoring only a few territories and documenting opportunistic encounters. One challenge was that the places where we located nest sites were not often consistent year to year. Often, this inconsistency appeared to be the result of both peregrine falcons and golden eagles displacing the prairie falcon from our documented nest sites. The territories within the Black Canyon of the Yellowstone, located in 2011, provide a great example. The first territory located on that fruitful day of discovery has been occupied by peregrines since 2012 and is now identified as the Upper Black Canyon territory. The second territory has varied over the years. Since 2012, we have

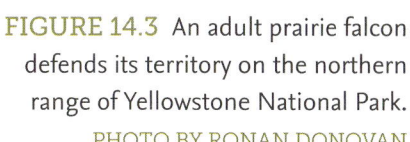

FIGURE 14.3 An adult prairie falcon defends its territory on the northern range of Yellowstone National Park.
PHOTO BY RONAN DONOVAN

observed a local pair of golden eagles commonly perch on the cliff face, and in 2013, eagles initiated nest building, but no eggs were laid. In 2018, the cliff was occupied by a pair of peregrine falcons that successfully fledged three young but have not been observed since that time. At both Black Canyon territories, we still observe prairie falcons in the area, suggesting that they have not been fully displaced but are more likely using less-prominent cliffs along tributaries of the Yellowstone River.

How the prairie falcon was distributed through YNP historically or, more importantly, how the species was distributed in the absence of the peregrine falcon (1970–1983) is virtually unrecorded. What little is available comes from peregrine falcon survey notes, conducted by Bob Oakleaf, a now retired Wyoming Game and Fish biologist. In the late 1970s and early 1980s, in the interim of peregrine falcon occupation (i.e., post-extirpation and pre-reintroduction; *see chapter 10*), Oakleaf documented prairie falcons at cliff sites in the northern range of YNP. Today, peregrines and golden eagles occupy two of these sites, and consistent with what we observed in the Black Canyon, prairie falcons are seen in nearby areas, but nest sites are unknown. Large prominent cliffs in the northern range of YNP are abundant, but are, nonetheless, a fixed resource, meaning they are not renewable or increasing. Therefore, given the relatively dense population of the golden eagle (*see chapter 11*) and the dramatic increase of peregrine falcons since the mid-1980s (*see chapter 10*), the proposed competitive dynamic between the prairie falcon and these two species is at least partially supported.

Competition for nest sites may further be supported by an observation made at another well-known peregrine territory. It is unclear if prairie falcons migrate from YNP during winter; however, they are in the park and on territory before peregrine falcons return, a species we know migrates. While conducting early-season occupancy surveys, we observed prairie falcons exhibiting quintessential pair-bond behavior. In addition to the pair soaring and vocalizing together, we observed copulations and courtship displays at the exact ledge peregrine falcons breed in most years. We were left un-

certain about what this meant for the territory, and our following visit was met with a peregrine falcon pair and no sign of the prairie falcons. Though we must speculate, it appeared that the peregrine falcon had outcompeted the prairie falcon upon their return to YNP that spring.

WHERE TO BREED?

All documented nest sites have been located on cliffs, but at one location, where adults and fledglings have been observed in multiple years, the nest site is unknown, and no known cliffs are immediately available. Given our knowledge of golden eagle and peregrine falcon occupation in neighboring areas with available cliffs, we suspect the prairie falcon pair may be using a tree cavity or an unoccupied tree nest. If true, the use of an uncommon nest site could perhaps be driven by what appears to be a disproportionately high abundance of ground squirrels in this area. Although interesting anecdotally, we do not believe the use of trees is prevalent, based on most of our observations and what we know about the species.

The northern range of YNP, with an environment that best reflects ideal prairie falcon habitat in the park, is where most of our observations have been made during the breeding season. Across the northern range, we have documented 11 areas where either prairie falcons are observed regularly or physical nest sites have been identified. The number of prairie falcons on the northern range is a minimum count, and with no formal searches, it is likely there are more. Nonetheless, we suspect that density is relatively low, given the infrequent number of sightings during extensive survey and monitoring efforts for other raptor species. Further, despite changes that may have influenced where prairie falcons nest on the northern range, it may be reasonable to assume that abundance has remained constant, given the locations of past and present observations.

Unlike the northern range, the forested environment that dominates much of the YNP's interior does not likely support many territorial prairie falcons. In this area, we rarely encounter the species during the breeding season, and on those rare occasions, encounters are in open areas like Hayden Valley. Alternatively, the high-elevation alpine environment of the Absaroka Range, along the park's eastern boundary, has been an area with an unexpected number of encounters during the breeding season and even strong evidence of successful breeding. While surveying above 9,000 feet (2,740 m) through the Hoodoo Basin in 2016, we observed copious amounts of whitewash (fecal matter) staining an area of rocks. There were obvious ledges that were suitable for nesting, so we surmised that this was the sign of a falcon, but on this day, none were observed. However, two days later, while on neighboring Parker Peak, in view of the area where whitewash was observed, an adult prairie falcon and two incredibly vocal fledglings flew below our perch on the peak. Given the timing of this encounter (mid-July), the nest site had to have been near since young could have only recently fledged. There is already a narrow window of time during which conditions are suitable for breeding at the lower elevations of YNP, so successful breeding at these high elevations is an ecological wonder and a tribute to the resourceful nature of the prairie falcon.

With little known regarding the population status of the prairie falcon in many parts of its range, we will continue to document our encounters with this species. With what appears to be relatively low density, research regarding the demography and habitat of the prairie falcon in YNP could be a difficult undertaking. However, maintaining records of where we know they occur today may help to serve as an index for how the distribution of this species in YNP will respond to future environmental factors, such as climate-induced habitat change.

AMERICAN KESTREL

In the early 2000s, a previously unimaginable conservation concern was proposed to the widespread community of raptor ecologists: North America's most abundant raptor species, the American kestrel, could be in decline

(Bird 2009). By the end of the decade, this concern had become an apparent reality, evidenced by occupancy data from multiple nest-box programs and the analysis of Breeding Bird Survey (BBS) data and raptor migration data across North America (Farmer and Smith 2009, Smallwood et al. 2009). In response to these findings, hypotheses related to the cause of the decline included the negative effects of West Nile virus (Nemeth et al. 2006), interspecific predation (particularly by Cooper's hawk; Farmer et al. 2008), climate change (Steenhof and Peterson 2009), environmental contaminants (Smallwood et al. 2009, Rattner et al. 2015), and the loss of habitat (Farmer et al. 2008). However, over a decade later, no empirical evidence has substantiated any one of these hypotheses as a single cause for the decline (McClure et al. 2017). What can best be concluded is that the decline of the American kestrel has been occurring over many decades (preceding 1966; McClure et al. 2017, McClure and Schulwitz 2022), and the cause is more likely a piecemeal of factors across disparate populations. Importantly, though these declines have been detected over a broad geographic area, some populations appear to be stable or even increasing (Bird 2009, Smallwood et al. 2009, Steenhof and Peterson 2009).

The American kestrel is the most widespread falcon in North America, so it is not surprising that the species occurs commonly throughout YNP. This uniquely patterned, strikingly colorful, and smallest North American falcon is best associated with open to semi-open environments across a variety of native and non-native communities, including urban areas (Fig. 14.4). In these areas, kestrels most commonly capture and consume insects (e.g., grasshoppers and beetles) and small rodents (e.g., voles and mice), but ultimately their diet reflects availability, and other prey includes songbirds and small reptiles. As a secondary cavity nester (one that does not excavate its own cavity), breeding areas are best characterized by the availability of natural (e.g., tree) or man-made cavities interspersed within small to large areas of low-growing vegetation (Fig. 14.5). Importantly, this is a species that most commonly searches for prey while perched, so the availability of perch sites is an important habitat component (Smallwood and Bird 2020).

Relative to many raptors, kestrels are easily identifiable by their unique plumage, and unlike many raptors, plumage is a key characteristic that differentiates males from females. With similar black, rufous, and slate-blue coloring on the head, males have slate-blue across the dorsal surface of their wings and a tail that is mostly rufous in color, with a black sub-terminal band, as opposed to the rufous wings and barred rufous tail of the female. The common nature of this species, the use of prominent perch sites, and their undeniable beauty often make them a favorite for many birders and non-birders alike.

Like many of the species discussed in this chapter, we initially proposed to include the American kestrel in the YRI. This was driven in part by the conservation concerns mentioned above, which include an apparent steady decline of kestrels throughout the northern and southern Rockies since at least the late 1960s (McClure et al. 2017). Further, the majority of work monitoring reproduction has been strongly weighted toward the use of nest boxes, a management strategy not employed in YNP. Therefore, we recognized the opportunity to gain insight into the status of a population that is restricted mainly to natural cavities. Unfortunately, like other species discussed in this chapter, our ambition was greater than our resources, limiting our ability to maintain occupancy and breeding data, despite a concern for kestrel habitat inside YNP.

FIGURE 14.4 Male American kestrel perched on sign at the edge of the Lamar Valley. PHOTO BY TOM MURPHY

CAUSE FOR CONCERN?

The availability of nest cavities is an apparent factor limiting American kestrel population growth, a constraint that is at least partially supported by nest-box programs. For example, in areas where natural cavities are absent or uncommon but forage quality is good, nest boxes have been responsible for successful breeding in formerly unoccupied areas (Smallwood and Bird 2020). In the mountains of the western United States, aspen plays a significant role for cavity-nesting birds (Dobkin et al. 1995). Large live and dead aspen commonly provide a relatively high abundance of cavities and are generally favored by both primary and secondary cavity nesters (Dobkin et al. 1995).

In northern YNP, American kestrels are commonly associated with large-diameter aspen snags (Hollenbeck and Ripple 2008). However, because of a previously abundant elk population through the early 2000s, aspen stands are generally degraded in Yellowstone. It is presumed that the extirpation of large carnivores (e.g., wolves and cougars) and other management strategies influenced elk densities and foraging behavior, which led to increased browsing

FIGURE 14.5 Male American kestrel, a cavity nester, delivers insect prey to young in the nest.

PHOTO BY TOM MURPHY

of young aspen (Ripple and Beschta 2007). The abundant elk population has since been reduced by the recovery of large carnivores and other management measures outside of YNP. However, the recruitment of young aspens in northern YNP was historically minimal, and consequently, it is assumed that large aspens will decline as these mature trees die without sufficient replacement (Hollenbeck and Ripple 2008). Hollenbeck and Ripple (2008) estimate that large-diameter aspens will continue to decline in YNP through the year 2045, and recovery to historic levels

may not occur until the year 2150, if ever. As the availability of large aspens declines, it may be reasonable to presume that kestrel abundance could reflect this loss of habitat. In the future, if aspens cannot maintain their current levels, YNP may not have the habitat to support its current kestrel population. This is important since kestrel abundance at the northern reaches of Yellowstone is significantly greater at aspen stands inside the park (Hollenbeck and Ripple 2008), a finding that perhaps has implications for kestrels beyond the boundaries of YNP.

RAPTOR DIVERSITY IN YELLOWSTONE NATIONAL PARK

THE TRIALS OF MONITORING THE AMERICAN KESTREL IN YELLOWSTONE

Whether roadside or in a remote corner of YNP, American kestrel encounters are common. Between 2011 and 2015, we opportunistically recorded nearly 100 locations where individuals, pairs, nest sites, or recent fledglings were observed. These locations included, but were not limited to, much of the northern range, open valleys of the interior, thermal basins, montane meadows, and river corridors. Nest sites were observed at aspen stands, but the use of cavities in live and dead pine trees was not uncommon. Upon our initial attempts to monitor breeding kestrels, we quickly recognized some daunting challenges, like how to identify the age and number of nestlings in the cavity of a deteriorating snag that is 30 feet (9 m) above the ground. Effective measures to meet such challenges required equipment and time not available to us; thus, we relinquished our efforts to monitor reproduction. This realization made clear why most work surrounding American kestrel breeding activity involves nest boxes, where one can simply lift the top of the box and peek inside.

To estimate kestrel abundance in the northern range of YNP, we conducted a roadside point-count survey twice each breeding season from 2012 to 2015. A primary objective of this survey was to establish abundance estimates for multiple raptor species in this region of the park. With an abundance estimate and a repeatable survey method, our hope was to establish a baseline from which population change could be measured. However, despite their common nature, our detections of kestrels along the survey routes were too low to adequately inform our models, and abundance was strongly underestimated. With the identified challenges of monitoring reproduction and a sampling method that was not adequate for reliable abundance estimates, our attempt to better understand the status of the American kestrel in YNP was largely unsuccessful.

As mentioned above, encounters with the American kestrel in YNP are common, a trend that reaches well beyond this landscape. The common nature of the species across North America is at least partially responsible for why a negative trend in population growth went unnoticed for so long, and why it is difficult to identify what factors have influenced the decline. As concern for the species continues to mount, we recognize the importance of better understanding the population dynamics of kestrels in YNP. With focused objectives and the appropriate equipment, estimating abundance and monitoring reproduction is still achievable and a management need.

HELP FROM THE PUBLIC

For nearly the last 20 years, annual visitation to YNP has exceeded 3 million and in 2021, there were nearly 4.9 million visits. Feeling safe to assume that a portion of these visitors are aspiring, casual, or avid bird-watchers, we initiated a raptor-sightings program in 2010 during a pilot program for the YRI. The intention was to simply quantify when and where species were observed and hopefully raise awareness for these remarkable animals. Unlike many of the charismatic species found in YNP, raptors can be encountered in our everyday life, even in some of our largest cities. The power of connecting our experiences in wild places like YNP to our own backyards is an important factor in understanding the need for species conservation.

With just over 100 sightings reported in 2010, participation in the raptor-sightings program increased rapidly, with over 700 in 2012. However, submissions declined in the following years, and by 2015, we had received a total of 1,650 sightings. We anticipated that submissions would continue to increase as the program gained traction, but promotion was not likely sufficient to maintain high participation rates. Nonetheless, with one confirmed report of a burrowing owl (a new raptor for YNP's bird checklist!), 25 species were represented in this reporting system (Table 14.1). An added benefit was that some reports assisted us in locating territories and nest sites for species of particular interest to the long-term Bird Program and the YRI.

TABLE 14.1 RAPTOR SIGHTINGS REPORTED TO THE YELLOWSTONE BIRD PROGRAM, 2010–2015

SPECIES	SIGHTINGS (%)
American kestrel	109 (7)
Bald eagle	256 (16)
Boreal owl	11 (1)
Broad-winged hawk	1 (0)
Burrowing owl	1 (0)
Cooper's hawk	21 (1)
Ferruginous hawk	16 (1)
Golden eagle	103 (6)
Great gray owl	47 (3)
Great horned owl	20 (1)
Long-eared owl	5 (0)
Merlin	10 (1)
Northern goshawk	43 (3)
Northern harrier	76 (5)
Northern pygmy-owl	9 (1)
Northern saw-whet owl	5 (0)
Osprey	264 (16)
Peregrine falcon	71 (4)
Prairie falcon	34 (2)
Red-tailed hawk	349 (21)
Rough-legged hawk	9 (1)
Sharp-shinned hawk	16 (1)
Short-eared owl	3 (0)
Swainson's hawk	92 (6)
Turkey vulture	63 (4)
Unknown raptor	16 (1)
TOTAL	**1650**

Beyond the YRI and our own raptor-sightings program, there are several other available avenues of citizen-contributed data to inform us of what, when, and where species are found in YNP, and we encourage those who are interested to participate. For example, eBird (Sullivan et al. 2009), with reporting worldwide and a platform that coordinates well with modern-day technology, has the potential to help inform management through citizen-based science. The raptor-sightings program was tailored to our interests and allowed us a stronger connection to the visitors' experience and, if necessary, easy communication. Unfortunately, without continual promotion, reporting programs in the park cannot sustain participation, and perhaps broad-reaching platforms like eBird are more reliable for visitor-sightings data. Importantly, the practical and scientific value of reporting species sightings comes with some level of consistent participation. From there, the potential benefits include, but are not limited to, the identification of species previously unobserved in YNP, species previously unknown to breed in YNP, or even population trends over time. With no sign that visitation to YNP will decline, the public remains a viable and valuable contributor to our collection of data on species presence.

NOTABLE MENTIONS

Given the detail we have provided thus far for some of the raptors we know breed in YNP, it is important to at least acknowledge some other species we know or suspect are breeding and others that migrate through or spend winter in and around the park.

(1) The **northern harrier**, with its slender body and tail, white rump patch, unique, owl-like facial disk, and distinct low coursing and buoyant flight, breeds in some of YNP's open wetland and shrub-steppe communities (Fig. 14.6). Like the American kestrel, adult male and female harriers can be identified by their unique plumage. The adult male, sometimes referred to

FIGURE 14.6 Two immature northern harriers flush a small songbird (a common prey) from the grass as they slowly fly low over the ground during late fall in Yellowstone National Park.

PHOTO BY TOM MURPHY

as the "gray ghost," is gray on its upperside and mostly white on the underside, with contrasting black wingtips, as opposed to the female, which has a brown upperside and a buffy underside with brown streaks. The juveniles, however, can only be differentiated by size; the sexes share the same dark brown upperside and deep rufous underside. Because it is a ground nester, usually in dense vegetation, nest sites are uncommonly observed; however, observations are common, as they search for prey while on the wing. Encounters in YNP are infrequent but broadly distributed; we have observed harriers from the shrub-steppe and wetlands of the northern range south through the remote reaches of the Yellowstone River drainage. No data exist to substantiate prey selection in YNP, but based on what we know about the species, harriers likely feed on small mammals (e.g., voles and mice), songbirds, and amphibians. The species is generally considered to be nomadic (Smith et al. 2020b), and few territories have been opportunistically located in only some years. We have no data regarding the breeding population in YNP and how it may fluctuate given their nomadic nature, but based on encounters, harrier abundance appears relatively low.

(2) In recent years, encounters with **turkey vultures** have become increasingly common in YNP. The range of the turkey vulture has been expanding northward since the early to mid-20th century (Kirk and Mossman 2020). The cause for this expansion remains unidentified, but hypotheses include the effects of carrion availability and a changing climate. This long-winged, blackish-brown vulture, with a mostly featherless head, can be easily observed while in flight. As it soars above, rarely flapping, what best identifies the turkey vulture is the contrast between the silvery appearance of the flight feathers' (wing and tail) underside and the black body, combined with a distinct teetering and seemingly unsteady flight. The unsteady flight, identifiable from great distances, is a result of what we refer to as low wing loading, in which the ratio of the bird's mass (low) to the surface area of their wings (high) slows flight

speed and decreases stability. As birds that are only scavengers, turkey vultures opportunistically feed on a wide array of small to large carrion, providing what we refer to as an ecosystem service or, in this case, a clean-up crew (Grilli et al. 2019). No nest sites have been located in YNP, but with pairs often seen soaring together across the park, breeding is strongly suspected. Areas where we suspect breeding range from relatively large open cliffs in the northern range to diminutive cliffs tucked away into the forest near Yellowstone Lake. Although it is notoriously difficult to find nest sites, it is only a matter of time before we can confirm the breeding status of this species in YNP.

(3) The **merlin** is a small, fast, and robust falcon, only slightly larger than the American kestrel, but easily differentiated by appearance and life-history strategy. Three of the nine subspecies found in the world breed in North America, and all exhibit their own unique plumage characteristics (Warkentin et al. 2020). However, for the sake of brevity, we will not provide detail between subspecies, but will quickly touch on key characteristics of the species. Adult males are most distinct, with plumage color on the upperside ranging from a pale blue-gray to a blackish gray, opposed to the adult females and juveniles, which resemble each other and are brown on the upperside. All are streaked on the underside of the body; the most defining characteristic is the brown and white to black and white banded tail (Fig. 14.7). The species can be found across a variety of open to semi-open environments and, like the other falcons, does not build a nest, but instead uses old corvid (e.g., crow, raven, magpie) and raptor nests (Warkentin et al. 2020). With a diet that is strongly weighted toward avian prey, merlins primarily forage on the wing and commonly capture prey in midair. The merlin has not been documented as a breeder in YNP, and sightings are very uncommon during the breeding period. The bulk of our encounters occur during migratory passage in spring and autumn. Nonetheless, two subspecies are known to breed within the region and can even be found nesting in developed areas like Bozeman, Montana.

FIGURE 14.7 A female merlin (prairie subspecies) perched in a snag near Stephen's Creek at the northern edge of Yellowstone National Park during fall 2019.
PHOTO BY GREG ALBRECHTSEN

Therefore, it is reasonable to suspect that the species may breed in YNP, even though we are certain abundance would have to be low. Only time will allow us to either confirm or deny this suspicion.

(4) The **ferruginous hawk** is probably best recognized for being the largest hawk in North America. There are two distinct color morphs (light and dark); the light morph is significantly more abundant and strikingly beautiful, sometimes appearing almost entirely white from the underside. The flight feathers (wings and tail) of both color morphs are generally white on the underside, a characteristic that is useful for identification when a hawk is seen soaring. Associated with open grassland and shrub communities, the species uses trees and shrubs, cliffs and rock outcroppings, man-made structures, and the ground for nesting. Prey varies with availability, but ferruginous hawks are best associated with rabbits, ground squirrels, and prairie dogs (Ng et al. 2020). Found breeding in the grasslands and shrub-steppe of the surrounding states, the ferruginous hawk does not breed in YNP, despite the appearance of suitable habitat in the northern range. Encounters are most common during spring and autumn migration, and these hawks are regularly observed foraging throughout Hayden Valley in autumn.

(5) A winter visitor from the far north, the **rough-legged hawk** can be seen passing through YNP in both spring and autumn as the species moves between breeding and wintering grounds. Not all just pass by, however; some spend their winter in and around YNP, enduring conditions that many other species move south to avoid. With breeding grounds at the subarctic and arctic latitudes, the rough-legged hawk is well adapted to cold, harsh conditions, and perhaps winter in YNP is just short of a simple walk in the park. At their northern breeding grounds, rough-legged hawks primarily nest on cliffs and forage for small mammals, primarily lemmings and voles (Bechard et al. 2020b). In YNP, small mammals such as voles and mice remain active through winter but generally gain refuge under the snow, so in high-snow years, prey detection by rough-leggeds is likely difficult, and scavenging of carrion may help to compensate. Two color morphs (light and dark) are generally recognized for the species; however, high variation within the morphs is prevalent (Fig. 14.8). As with the ferruginous hawk, one of the best times and places to see rough-legged hawks in YNP is during late autumn in Hayden Valley. In this open landscape, they are commonly seen perched or soaring as they forage for the remaining small mammals that have yet to gain refuge under the forthcoming snow. By late spring, rough-legged hawks are absent from YNP and well on their way back to the northern reaches of the continent and beyond.

▶ **FIGURE 14.8** Light-morph rough-legged hawk in flight during the winter of 2022. PHOTO BY SCOTT HEPPEL

▼ **FIGURE 14.9** Adult Cooper's hawk perched at the northern edge of the Greater Yellowstone Ecosystem near Bozeman, Montana, in early spring of 2020. PHOTO BY KYLE MOON

CONCLUSION

The content of this chapter provides some of what we know about the diverse community of raptors in YNP but, even more so, highlights how much we don't know. Ultimately, many raptors use Yellowstone to breed or as part of their seasonal passage (Fig 14.9). In some of our recent work, presented in other chapters of this book, we describe how we have begun to scratch the surface of raptor ecology in this unique landscape. As we continue to build upon our knowledge of raptors in YNP, we inch forward toward opportunities to study their community ecology—that is, how the various species, and the resources required by each, interact to drive population dynamics. The ongoing work of the Yellowstone Bird Program and the focused research that comes from other interested parties are the forces that drive these opportunities to address more complex ecological questions. Nonetheless, for those who visit YNP, the chance to encounter a species of hawk, falcon, eagle, or owl are widespread, constrained only by the distraction of those magnificent ungulates and large carnivores that garner so much of the public attention.

15 FALL RAPTOR MIGRATION

LAUREN E. WALKER, KATHARINE E. DUFFY, DAVID B. HAINES, and DOUGLAS W. SMITH

Fall raptor-migration counts in Yellowstone National Park (Yellowstone; YNP) are not for the faint of heart. They often entail sitting for six or more hours in below-freezing temperatures, with winds upward of 12.5 miles per hour (20 km per hour). Sometimes you're sitting on snow. Sometimes you get snowed on. And, of course, as with any field work in Yellowstone, wildlife encounters are common; on several occasions, grizzly bears were spotted near the count site. For park biologists, and our dedicated volunteers, the efforts are worth the rewards.

E VERY FALL, NUMEROUS raptors migrate south through YNP along the Rocky Mountain or "central" flyway. Although local populations of raptors have been a topic of study in the park for decades, fall raptor migration was largely ignored until recently, with the initiation of the Yellowstone Raptor Initiative (YRI) in 2011. In anticipation of that broad effort, park biologists conducted a pilot study in 2010 and observed numerous Swainson's and red-tailed hawks migrating through Hayden Valley in central YNP. This prompted the placement of our first migration count site just off the main loop road in Hayden Valley, where raptors were counted in September and October, from 2011 through 2015 and again in 2017 (Fig. 15.1, Table 15.1). During the YRI, from 2011 through 2015, park biologists and volunteers conducted counts five times each week, including occasional simultaneous

TABLE 15.1 RAPTOR SPECIES TALLIED DURING FALL MIGRATION COUNTS IN HAYDEN VALLEY, YELLOWSTONE NATIONAL PARK, 2011–2017

SPECIES	2011	2012	2013	2014	2015	2017	TOTAL (%)
American kestrel	73	62	64	155	104	16	474 (7)
Bald eagle	93	68	26	95	60	19	361 (5)
Broad-winged hawk	0	7	1	35	9	0	52 (1)
Cooper's hawk	31	32	28	75	85	46	297 (4)
Ferruginous hawk	34	20	32	29	10	3	128 (2)
Golden eagle	241	134	35	187	105	18	720 (11)
Merlin	13	11	6	12	7	3	52 (1)
Northern goshawk	10	7	9	14	3	3	46 (1)
Northern harrier	55	30	27	119	131	18	380 (6)
Osprey	12	14	18	22	11	8	85 (1)
Peregrine falcon	10	8	8	7	5	1	39 (1)
Prairie falcon	4	11	5	11	3	3	37 (1)
Red-tailed hawk	571	235	177	382	402	131	1898 (28)
Rough-legged hawk	70	130	23	108	61	1	393 (6)
Sharp-shinned hawk	65	72	68	109	80	15	409 (6)
Swainson's hawk	357	46	171	208	68	5	855 (13)
Turkey vulture	9	22	0	29	2	2	64 (1)
Unknown raptor	198	44	19	80	102	19	462 (7)
TOTAL	1846	953	717	1677	1248	311	6752

OPPOSITE PAGE:
FIGURE 15.1 Volunteers with the Yellowstone Bird Program spent hundreds of hours surveying migrating raptors from Hayden Valley.
PHOTO BY HOWARD WEINBERG AND KATHARINE DUFFY

raptor counts from Observation Peak. In 2017, volunteer biologists conducted counts two days each week.

Red-tailed hawks, Swainson's hawks, and golden eagles accounted for 51% of the total individuals (3,473 of 6,753) observed migrating through Hayden Valley between 2011 and 2017 (Table 15.1; Baril et al. 2017a, 2017b). Red-tailed hawks were, by far, the most observed raptor and accounted for 28% of all detections. While the bulk of Swainson's hawk and American kestrel migration occurred in September, many species continued to migrate well into October (Baril et al. 2017a, 2017b). For golden eagles and rough-legged hawks (Fig. 15.2), migration

FIGURE 15.2 A light-morph rough-legged hawk takes flight. These hawks are commonly observed during migration counts in October as they pass through or begin their winter stay in the Greater Yellowstone Ecosystem. Note the feathers that reach down the legs to the feet, the dark belly band, and the distinct dark carpal patches in the wings.
PHOTO BY SCOTT HEPPEL

through Yellowstone did not even begin until October and likely extended into November, beyond our typical survey window (Baril et al. 2017a, 2017b).

Despite our success and interesting findings, we ultimately found Hayden Valley to be limited in its utility as a raptor migration site. Poor weather and visibility frequently plagued the site, and winter road closures in Yellowstone's interior necessitated an early end to the migration observation season. Thus, in 2018 and 2019, we initiated fall raptor-migration surveys in the Rescue Creek area of northern Yellowstone, on days when volunteer biologists were available. Although our data from Rescue Creek are relatively limited (fewer observation days, conducted in only two survey years), the new perspective is interesting, and preliminary data reinforce a few of our assumptions from the Hayden Valley results. As with previous surveys, red-tailed hawks were frequently observed from Rescue Creek and made up 22% of all our observations in 2018 and 2019. Contrary to observations in Hayden Valley, however, golden eagles were the most common migrating raptor detected at Rescue Creek and comprised 31% of raptor observations. On October 8, 2018, we counted 54 golden eagles migrating south into Yellowstone National Park, with a total of 96 golden eagles over 9 days of observation in 2018. During 2019, we detected a total of 102 golden eagles over 15 days, during 66.4 hours of observation. The abundance of migrating golden eagles indicates that the Rescue Creek area is a favorable location for future, more focused raptor migration counts, especially during October and possibly into November, depending upon the availability of experienced raptor counters.

To educate visitors about raptor migration in Yellowstone National Park, coauthor Katy Duffy conducted free public raptor programs once each September from 2010 through 2019. Programs consisted of a raptor ecology and identification presentation, followed by the Hayden Valley Hawkwatch, a public raptor count. Each year, these popular programs attracted 40 or more visitors for the raptor ecology talk and as many as 100 for the outdoor raptor count. Duffy also taught a raptor migration class for the Yellowstone Institute each September from 2010 through 2019, engaging additional volunteers and visitors.

In total, over eight survey seasons, volunteers, staff, and visitors have documented 7,388 raptors belonging to at least 17 species, providing Yellowstone's Bird Program with baseline information on the abundance, diversity, and timing of migration through the park. Ultimately, this information will enable park biologists to detect any broad-scale changes in raptor migration patterns and develop more effective strategies for protecting migrating raptors and their migration habitat (Fig. 15.3).

FIGURE 15.3 Sharp-shinned hawk in flight. Migration count sites provide many opportunities to observe this species, a forest resident during the breeding season, in wide-open country as they make their way to latitudes south of Yellowstone National Park. PHOTO BY SCOTT HEPPEL

16 LISTENING FOR OWLS IN YELLOWSTONE

KATHARINE E. DUFFY and HOWARD J. WEINBERG

Toot-too-too-too-toot-it, bell-like notes ring out clearly in the frigid, still night. "Boreal owl, north side of the road," softly but excitedly exclaimed one warmly dressed volunteer. "Got it!" replied the other similarly bundled observer as she recorded data. And another owl survey begins.

KATY DUFFY

SURVEYING FOR OWLS at night offers an auditory glimpse into the lives of owls found in Yellowstone National Park (Yellowstone; YNP). Without ever seeing an owl, what we hear explains what individual owls of several species are doing. Roadside surveys, coupled with knowledge of owl nesting ecology and behavior, allow us to document sites where owls of selected species advertise for mates or give other vocalizations. While these surveys do not tell us if the owls are nesting successfully, they fulfill the goal: determining an index of owl presence during the nesting season.

Started in 2013 as part of the Yellowstone Raptor Initiative (YRI), the study continues. Although Yellowstone has records documenting breeding for seven owl species, the park had never conducted systematic owl surveys, necessary for owl conservation (Takats et al. 2001). The flammulated owl and the long-eared owl are listed as species of concern at the continental level, and in the Northern Rockies bird-conservation region, the region that encompasses Yellowstone, by the United States Fish and Wildlife Service (USFWS 2021). The boreal owl, flammulated owl, and northern pygmy-owl are listed as Bird Species of Greatest Conservation Need by the Wyoming Game and Fish Department (WGFD 2017). Boreal owls, northern saw-whet owls, and northern pygmy-owls comprise the primary target species for surveys. Secondary target species include long-eared owls and flammulated owls, and we also record observations of both great horned and great gray owls.

Surveys occur primarily in the northern tier of Yellowstone, from Mammoth to the Northeast Entrance, and follow a set protocol at stops located at least a half-mile (0.8 km) apart. Twelve to 20 roadside surveys, conducted each year through 2021, have detected six species: boreal owl, northern saw-whet owl, northern pygmy-owl, long-eared owl, great gray owl, and great horned owl. Surveys have revealed patterns of owl breeding activity over the nine years of the study.

HOW TO WATCH AN OWL

What a privilege it is to see and watch an owl, especially when the owl is not aware of your presence. Witnessing owl behavior that has not been influenced or impacted by people can provide a remarkable window into how owls live. To observe owls this way, watch from a safe distance (Fig. 16.1). When observers intentionally approach too closely, they startle and disturb the owl. When they pursue an owl that flees because they approached too closely, they are harassing the owl. When they are part of a crowd that surrounds an owl, they could prevent an owl from listening and looking for prey it needs to survive. A crowd of people in pursuit of an owl also damages prey habitat and chases small rodents, which are essential prey for owls.

All wild owls can seem tame, but they are not. Owls active in daytime, such as great gray owls and northern pygmy-owls, might appear tolerant of human presence, especially while the owls are focused on hunting. Owls discovered while roosting often freeze to avoid detection by potential predators and so appear to tolerate people. Owls (and all birds) are sensitive to human disturbance during nesting. Indications that an observer is far too close to a nest include an agitated owl that makes alarm calls or circles the observer. That's when it's time to move away and watch from a distance—75 feet (23 m) is the park regulation, and that might even be too close. Respectful photographers use long lenses so that an owl's behavior is not changed.

On the other hand, observing without disturbing an owl provides satisfaction by allowing one to see firsthand how owls use their myriad adaptations, from superb directional hearing to silent flight to spectacular courtship and hunting displays. These are breathtaking experiences!

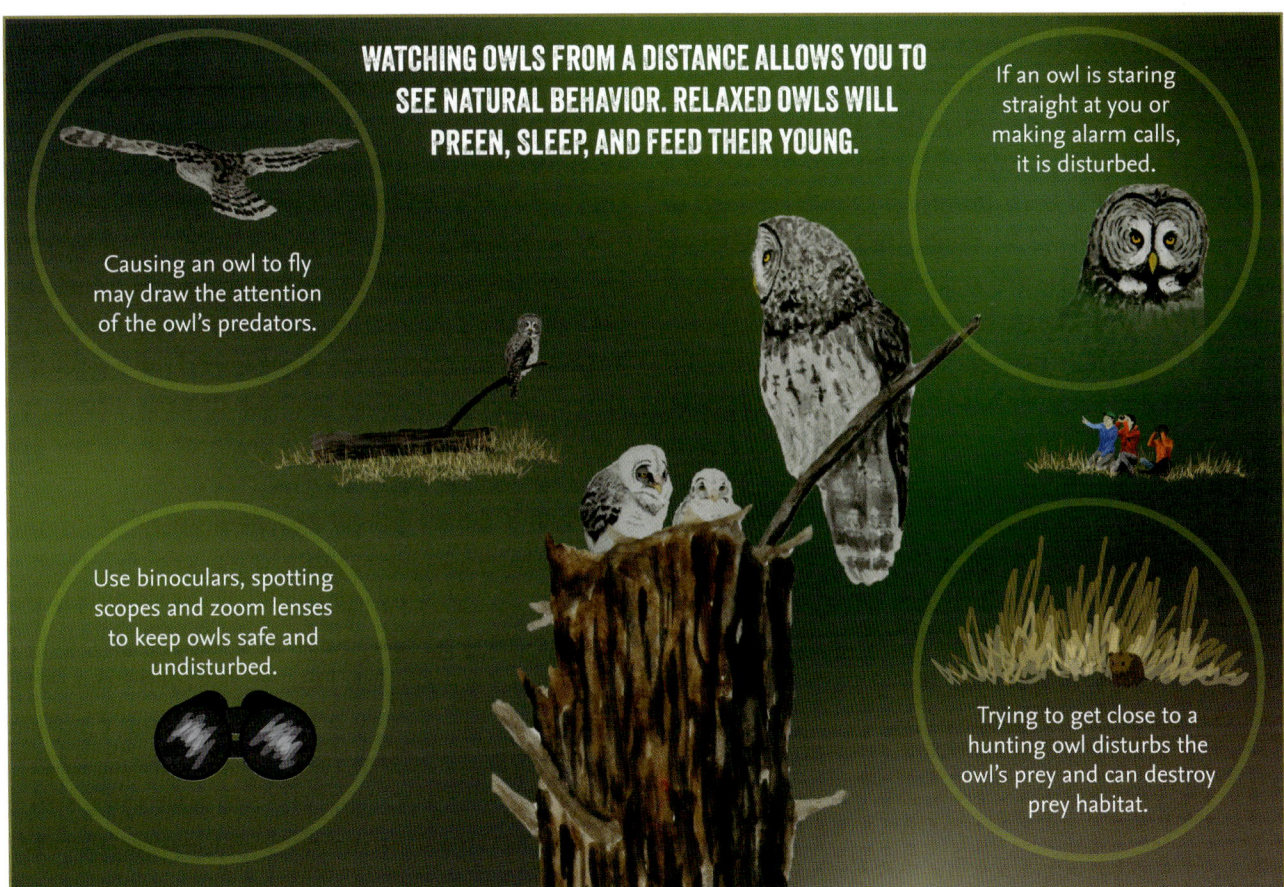

WATCHING OWLS FROM A DISTANCE ALLOWS YOU TO SEE NATURAL BEHAVIOR. RELAXED OWLS WILL PREEN, SLEEP, AND FEED THEIR YOUNG.

If an owl is staring straight at you or making alarm calls, it is disturbed.

Causing an owl to fly may draw the attention of the owl's predators.

Use binoculars, spotting scopes and zoom lenses to keep owls safe and undisturbed.

Trying to get close to a hunting owl disturbs the owl's prey and can destroy prey habitat.

OWL BASICS

Every owl chooses a location that will provide sufficient prey to enable it to nest successfully. Most of the owl species we study prey on small rodents; while northern pygmy-owls also eat birds, great horned owls choose from a wide array of prey, and flammulated owls are strictly insectivorous. Each species can hunt from a perch, with boreal owls, northern saw-whet owls, and northern pygmy-owls specializing in "perch and pounce" hunting, in which the owl listens and looks for prey close to its perch and drops down to capture what it heard or saw. Long-eared owls and great horned owls also fly over the landscape as they seek to detect prey. Flammulated owls seek insect prey, such as noctuid moths flying around treetops. While many owl species exhibit only nocturnal activity, great gray owls and northern-pygmy owls are active during the day.

Owls do not make their own nests. Small owls appropriate a cavity, typically one excavated by a woodpecker, often in an aspen, as a nest site. Long-eared owls, great horned owls, and great gray owls often use an abandoned stick nest, a large cavity, a tree with a broken-off top, or a witch's broom, which is an abnormal growth of twigs in a tree sometimes caused by fungus or other pathogens. Female owls of all species incubate the eggs without help from the male; the male owl's role is to feed the female and the young nestlings until they can maintain their body temperature without brooding by the female. At that point, the female joins the male in capturing prey to feed their fast-growing nestlings.

Although there are many similarities among the owl species we studied, numerous differences exist, including the specific habitats each species prefers, its favored prey, and the timing of its nesting. An introduction to the individual traits of each species, as demonstrated by their behavior in Yellowstone, follows.

BOREAL OWL

Dusk settled into the dense forest of tall subalpine fir and Engelmann spruce, each tree draped with dark horsehair lichen. Snow blanketed the ground. Out of the darkness of the February evening, several series of whistled toots told us a male boreal owl was advertising for a mate. Then he switched to singing non-stop, and the sound was muffled. "What just happened?" asked my friend, and I happily explained, "He's attracted a female! Now he's in his chosen nest cavity, singing away, hoping to entice her to stay and mate with him! It's exactly what we want to hear!"

KATY DUFFY

In Yellowstone, as in other parts of the Greater Yellowstone Ecosystem, boreal owls inhabit dense forests of mature conifers because one of their main prey, the red-backed vole, flourishes here. Grassy meadows interspersed with scattered clumps of aspen complete the landscape of their breeding habitat. Other prey such, as deer mice, live in the same forests and meadows too (Fig. 16.2). Strongly asymmetric ears allow boreals to pinpoint prey with extraordinary accuracy.

A male boreal owl will give a series of whistled *toots*, his primary song or "staccato song," to proclaim his availability as a mate to any females within earshot (Hayward and Hayward 2020). Male boreal owls typically advertise their availability starting in early February and until each male has found a mate or until the end of April, whichever happens first.

When a female boreal owl demonstrates interest by approaching a singing male, she gives frequent soft calls, eliciting an instant response from the male. He moves into the cavity he has chosen as a nest, located very near where he's been singing. Once in the cavity, the male boreal's song changes to a prolonged song ("prolonged staccato"), and it sounds softer because the cavity muffles the sound. The prolonged staccato seems to go on endlessly. If the female boreal chooses the male, she will stay and nest with him.

FIGURE 16.2 Boreal owl with a flying squirrel it caught. To keep prey from freezing on cold winter days and nights, or to thaw stashed prey that has frozen, boreal and other small northern forest owls hold their prey close to the warmth of their bellies.
PHOTO BY HOWARD WEINBERG

Year-to-year variation in the number of boreal owls detected has occurred (Fig. 16.5): there were no boreals in 2017, but there were nine in 2016 and 2019, when several male boreal owls advertised from suboptimal habitat, forests without mature Engelmann spruce and subalpine fir trees. Although the annual variation could not be conclusively explained, we suspect prey availability is probably involved. Since boreal owls are somewhat nomadic throughout their range, perhaps some males roam until they encounter places with abundant prey.

Based on the number of years of occupancy, two sites in the northeastern part of Yellowstone appear to contain prime breeding habitat, as boreals occupied one of these sites for seven years out of a possible nine and the other one for six years. Two sites hosted advertising male boreals for five years, one site for four years, two sites for three years, and four sites for two years. Due to the nature of current owl studies in the park, no boreal owls

are marked or transmittered, so we cannot determine if the same adults use the sites year after year. Most of these sites consist of dense, mature mixed forest of Engelmann spruce and subalpine fir, along with scattered stands of aspen—preferred boreal owl habitat. These mature conifer forests, dominated by Engelmann spruce and subalpine fir, tend to be at least 200 years old, based on tree diameter (R. Renkin, pers. comm.).

Advertising male boreals also chose eight sites where we detected advertisement in just one year, with most of these sites occurring in suboptimal habitat. In addition to roadside surveys at locations in prime habitat, backcountry surveys in the Slough Creek drainage, where optimal habitat occurs, during February or March of six winters, yielded advertising male boreal owls at two sites. Similarly suitable habitat for boreal owls also exists in many other parts of Yellowstone, including Craig Pass and the Lake–Bridge Bay area. Again, boreal owls demonstrate dependence on forests of mature mixed conifers, primarily Engelmann spruce and subalpine fir. What will happen as climate change causes these usually moist forests of mature trees to dry out? Both Engelmann spruce and subalpine fir retain their lower branches, providing ladder fuel to bring fire from the ground to the crown of trees, rendering these forests extremely vulnerable when lightning strikes. Where their essential nesting habitat burns in Yellowstone, boreal owls are unlikely to nest for hundreds of years, until spruce and fir trees mature.

NORTHERN
SAW-WHET OWL

December in Yellowstone. Chickadees and nuthatches calling in defiant protest in a wooded grove. We stopped skiing to listen. Moments like these warrant investigations, as long as you're not disturbing anything and bears are not present. So I left the trail and skied down into knee-deep snow. My skiing skill at the time could be generously described as "experienced beginner" when on a trail and, well, a lot worse off trail. Managing to stay upright, I ski-crunched through the snow until I was stopped a short distance later by a tangle of willows and saplings. The protest continued as birds called and swooped at something in a Douglas-fir. A blockish form on a low branch took shape. One more turn of the binocular focus and a beautiful northern saw-whet owl came into view! I immediately turned back to Katy and quietly called "Saw-whet." Katy had stayed up on the trail, as I had attempted many such explorations in the past, and often enough came up empty. She came down in a flash.

HOWARD WEINBERG

Saw-whets can remain all year in Yellowstone, although other studies have found that many of these tiny owls (males weigh only 2.5 oz/70 g, females about 4 oz/113 g)

migrate south for winter, with females typically moving farther south than males. By early February, males begin to advertise for mates (Rasmussen et al. 2020). A male saw-whet announces his availability to females by singing and singing and singing whistle-like sounds. If the sounds weren't so beautiful, they would seem monotonous. If he has not found a female, his whistle-like calls will continue into May.

Our surveys demonstrate that saw-whets in Yellowstone prefer the same habitat as boreal owls, but they also choose forests comprised mainly of Douglas-fir and forests where lodgepole pine dominates. Detection of more advertising male saw-whets in some years, such as 2019 (Fig. 16.5) could perhaps be due to increased prey availability. In certain years, such as 2021, we did not hear saw-whets advertising at a few sites until May.

One site we surveyed had an advertising male saw-whet for six years, another site five years, and one other site four years. Saw-whets also picked six sites for three years and six other sites for two years. In contrast, 10 sites hosted advertising males in just one year. As we witnessed with boreal owls, occupancy likely reflects the quality of the habitat (nesting substrate and prey availability) for northern saw-whet owls (Fig. 16.3).

Photos taken of fledgling saw-whets display the variability in nesting chronology that can occur. For example, a photo of a fledged saw-whet taken in early June 2015 indicated successful nesting had already occurred near Tower (D. Renkin, pers. comm.). Another encounter, a young saw-whet, observed on its daytime roost because mountain chickadees were loudly scolding it, was still in juvenile plumage in October (K. Duffy, pers. obs.). The small rodent on the branch next to this saw-whet probably was a gift from a parent that was still providing it with food. The presence of the second young saw-whet could have resulted from a nesting effort that began much later in the spring than that for the owlet seen near Tower. When the owlets leave the cavity about one month after hatching, they have a dark facial disk, their upper breasts are dark brown, and their lower breasts and bellies are buffy. This plumage is usually replaced

FIGURE 16.3 Northern saw-whet owls choose old woodpecker cavities or other natural cavities in trees for their nests. PHOTO BY RONAN DONOVAN

within two to four months with a paler facial disk and a white breast and belly streaked with brown, which is identical to adult plumage.

NORTHERN PYGMY-OWL

One winter, I was lucky enough to observe several northern pygmy-owls in Yellowstone, and more than once, Clark's nutcracker calls led me to the owl. This time, I was watching a perched pygmy-owl from a distance when it flew behind a stand of trees. Then, a nutcracker began to call from this area, so I moved toward the sound. Using my binoculars I located the nutcracker on the ground among the snow, fallen branches, and dried grass. I realized a moment later that a pygmy-owl was on the ground too, and it had a rodent in its talons! The two birds were in the midst of an intense standoff. The nutcracker was nearly twice as large and armed with a formidable bill. Showing its fierceness, the tiny owl took a mantling position, with wings held up and out to the side. This was not a bird to be taken lightly! The nutcracker moved and shifted side to side. The owl turned its body to match the shifts. There they postured. Then perhaps a subtle sign passed between the two, or the nutcracker realized that this meal was not to be had easily, and the battle was over. The owl flew away with its rodent prize.

HOWARD WEINBERG

Despite its tiny size, just 2–3 ounces (57–85 g), pygmy-owls demonstrate how feisty they are by often pursuing—and killing—prey considerably larger than they are (Holt and Petersen 2020). And they can retain their prey even when challenged by another bird more than twice its size (Fig. 16.4). Spots on the backs of their heads resemble eyes and perhaps aid these tiny owls to deter predators or competitors approaching from the rear.

Because northern pygmy-owls are active during the day, especially morning and early evening, they are sometimes observed in winter when they perch at the top of a live tree or in a snag in the Mammoth area, in mixed conifer forests across the northern range, and in other parts of the park. This owl appears to be about the size of a closed fist of a small adult, with a blocky, rounded head and a tail held at an angle. Its characteristic posture while perching allows it to be distinguished from other birds, such as Clark's nutcrackers and Townsend's solitaires, that perch in similar conspicuous places.

Males advertise for a mate during March and April mornings and early evenings by singing their single-note

FIGURE 16.4 Diminutive in size but extremely feisty, a northern pygmy-owl rises to the challenge of a Clark's nutcracker that seeks to steal the owl's prey. The pygmy-owl won this standoff and flew away with its deer mouse. PHOTO BY HOWARD WEINBERG

song, sometimes from a visible perch. When a male has attracted a female, softer calls, presumably made by the female, can be heard. Based on surveys in Yellowstone, pygmy-owls choose a variety of forest types, from mature mixed conifers to Douglas-fir, if suitable nest cavities exist. Because surveys targeting saw-whets and boreals start at dusk and end several hours later, diurnal northern pygmy-owls may not be detected (Fig. 16.5) except at stops prior to darkness and during surveys specifically targeting them, surveys we conduct before dusk. One site near Mammoth hosted advertising northern pygmy-owls for four years and two other sites for two years each.

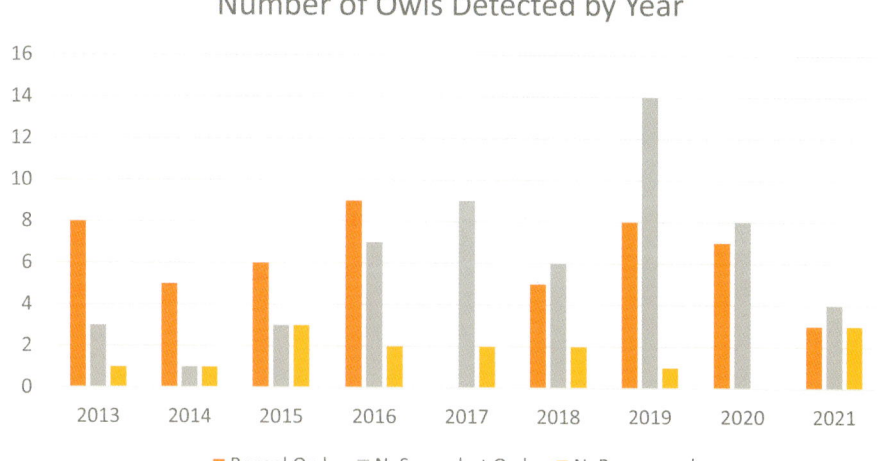

Number of Owls Detected by Year

Boreal Owl ■ N. Saw-whet Owl ■ N. Pygmy-owl ■

FIGURE 16.5 Number of boreal owls, northern saw-whet owls, and northern pygmy-owls detected each year by roadside surveys across northern Yellowstone.

NESTING HABITAT OVERLAP FOR SMALL FOREST OWLS

Boreal owls, northern saw-whet owls, and northern pygmy-owls can claim similar habitat, even the same general area, in Yellowstone. For example, an area west of Tower hosted an advertising male saw-whet in 2013, 2017, 2018, 2019, and 2021; an advertising male northern pygmy-owl in 2015; and an advertising male boreal in 2016 and 2020. Owls were not detected at this site in 2014.

Some sites had two species advertising at different times within one year: at a stop just east of Mammoth, a northern pygmy-owl advertised on March 30, 2015, while a northern saw-whet owl advertised there on April 29, 2015. At the same site in 2016, a northern pygmy-owl and a northern saw-whet owl advertised simultaneously from either side of the road. A northern pygmy-owl claimed this site in 2017 and 2019, while a saw-whet occupied the site in 2021.

What's up with this? Apparently, there is sufficient overlap in the habitat requirements of these three small forest owls that they can be attracted to the same locations in northern Yellowstone. Each of these species preys largely on small rodents, although northern pygmy-owls frequently prey on birds too.

LONG-EARED OWL

One summer evening while I lived and worked at Old Faithful, I took a break from planning for a major event and did an easy hike up to Observation Point near Geyser Hill. But it was already past sunset, and hikers on their way down the hill reminded me that it was going to get dark soon. I answered that I had my headlamp, knew the trail as I walked it frequently, and was making noise to alert any bears. Well, it did get dark, and that's when I heard eeeyip calls, vocalizations that reminded me of the food-begging calls of juvenile great horned owls, but much weaker. I thought the maker of the sounds was either the wimpiest great horned owl fledgling or. . . The sounds came from the direction I was going, so when I was close to the source, I flashed my headlamp toward the calls and lit up two fuzzy long-eared owlets staring down at me from their perch. After an ecstatic few moments of admiring the owlets, I hurried on down the trail, practically floating along from the sheer delight of stumbling on youngsters of my favorite owl species.

KATY DUFFY

Why were these long-eared owl fledglings near the Observation Point trail? They were only capable of weak flight, which means they were at least five weeks old, but not much older, and still close to their nest. Long-eared owlets jump from their nests when they are about three weeks old, but they are not capable of flight till they are five weeks old. Their parents probably appropriated an abandoned stick nest, originally made perhaps by common ravens or by Cooper's hawks that nested in the area. The mixed conifer forest where the owlets were begging would have deer mice and perhaps red-backed voles, while moist grassy meadows nearby probably harbored montane or meadow voles, often a favorite prey of long-eared owls.

Long-eared owls typically advertise in March and April in Yellowstone; their call is a deep, single-note call. Once mated, the female responds to the male with a much softer call from the nest (Marks et al. 2020). Long-eareds choose a variety of conifer forests for nesting in the Greater Yellowstone Ecosystem, including forests comprised primarily of lodgepole pine as well as forests of mature mixed conifers.

Despite nesting in forests, they usually hunt in open areas, so observations tend to occur in parts of Yellowstone where moist meadows of dense grasses (montane vole and meadow vole habitat) occur in proximity to forests with abandoned stick nests. Two years in a row, long-eareds nested in a campground in the northern part of the park. The first year, the owlets parachuted down out of a flimsy old nest as campers watched in surprise. The second year, the old stick nest the long-eared owls appropriated was tucked tightly into a dense Engelmann spruce merely a few feet from a nearby campsite's tent pad. Thick matted grasses, perfect vole habitat, lined creeks adjacent to this campground. In addition to these known nesting efforts, surveys have detected long-eareds at just two sites.

GREAT GRAY OWL

Early one evening during the third week of May, seven of us left the trailhead before sunset and skied to Shoshone Lake. A full moon rose in the east when we reached the lake. The sky turned pink as the sun set in the west, and we skied back across DeLacy Meadows,

the moon illuminating our way. We stopped to gather quietly for a last look at the meadow, peaceful in the moonlight, before skiing uphill through the trees for the last half-mile (0.8 km) to Craig Pass. While we paused, we heard a distant male great gray hoot in response to a female's food-solicitation call from the other side of the meadow. Her call told us she was already incubating eggs and expected the male to deliver prey to her as she sat tight on the nest. A magical ending to a perfect moonlight ski!

KATY DUFFY

For nests in Yellowstone, great gray owls choose an old stick nest, maybe one made by a common raven or northern goshawk, perhaps a witch's broom in a large spruce, or a broken-top conifer. Great gray owls begin nesting in Yellowstone by early May, even where snow still covers the ground. Nestlings jump from the nest when they are three to four weeks old, but they are not capable of flight for another one to two weeks. Although fledglings climb angled trees during their flightless period, they are vulnerable to predators such as northern goshawks and American martens. Fledglings beg loudly for food from the adults; the male has been reported to feed them for a few months (Bull and Duncan 2020).

Great gray owls, arguably the most charismatic owl found in the Greater Yellowstone Ecosystem (Fig. 16.6)

OPPOSITE PAGE:

FIGURE 16.6 Viewing owls from a respectful distance allows them to hunt successfully by listening for their preferred prey. PHOTO BY TOM MURPHY

FIGURE 16.7 (A) Great gray owls carry prey, such as this vole and other small rodents, in their bills. (B) Great grays remain in Yellowstone until snow gets too deep, hunting by listening for prey under the snow, then flying to the location of the sound and plunging through the snow feet-first to catch prey they heard.
PHOTOS: (A) BY RONAN DONOVAN;
(B) NPS PHOTO/KATHARINE DUFFY

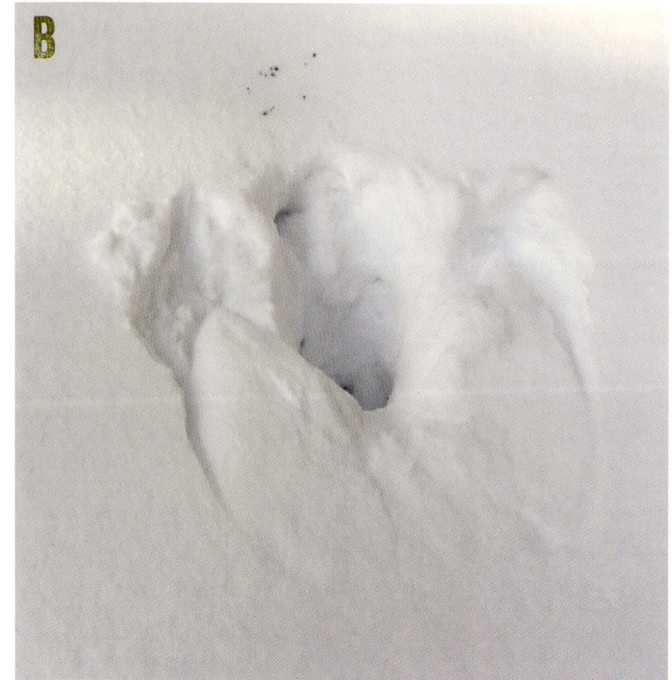

and perhaps all North America, stand more than 2 feet (0.6 m) tall, with a wingspan greater than 4 feet (1.2 m). But these attractive owls are all fluff—females can weigh 3 pounds (1.4 kg), but males weigh less than 2 pounds (0.907 kg). These somewhat diurnal owls prefer open forests, often of mature mixed conifers (Engelmann spruce, lodgepole pine, Douglas-fir, and subalpine fir) near meadows (Franklin 1987). Great grays hunt from a perch at the edge of a meadow for pocket gophers and voles, their main prey items in Yellowstone.

Surveys have detected great grays vocalizing at four locations in the park's northeast corner. Several backcountry surveys in the southeast corner of the park found great grays too. Great gray owls also frequent forests on the northern range, in the northwest corner of the park, in the Canyon area, and adjacent to the park's South Entrance Road.

Great grays can remain through winter (Fig. 16.7), providing snow does not get too deep or crusty, based on plunge holes found in the park and other areas of the Greater Yellowstone Ecosystem (K. Duffy, pers. obs.). They listen for sounds of pocket gophers or voles scurrying under the snow and dive through the snow, feetfirst, to catch what they hear. Signs of a hunting great gray: a plunge-hole in the snow, bracketed by wing prints, with marks where its feet dragged as the owl took off. A drop or two of blood in the plunge-hole indicates a successful hunt.

GREAT HORNED OWL

Three fuzzy owlets screeched loudly as the adult female great horned owl flew to the gutter on the roof of the administration building and pulled out a dead Uinta ground squirrel she had stashed there earlier. As cameras clicked away, the female flew straight back to the nest and tore the squirrel into bite-sized pieces that she presented to each owlet. This owl nest is in a witch's broom in an old Engelmann spruce. The nest tree grows between houses on Officers' Row at

Mammoth, an area always busy with visitors. Owls use this nest nearly every year, so it is undoubtedly the best known and most photographed owl nest in Yellowstone.

KATY DUFFY

Great horned owls, found from Alaska to Florida, occur widely in Yellowstone, too. Although not typically considered diurnal owls, powerful great horneds readily hunt abundant Uinta ground squirrels in daylight and take other prey at night. The heaviest owl in the park, great horned females weigh from 3 to nearly 4 pounds (1.4–1.8 kg), while males weigh 2.5 to 3 pounds (1.1–1.4 kg).

At about six weeks of age, young great horned owlets become "branchers" and start moving out of the nest onto nearby branches. They become capable of weak flight a week or so later. When the Mammoth nest is active, the antics of the youngsters before and after they leave the nest, as they learn to fly, land, and become independent, are entertaining and often downright hilarious.

In Yellowstone, as in other parts of their range, great horned owls maintain their territories year-round and frequently appropriate old red-tailed hawk nests by late winter, so their habitat can overlap that of red-tailed hawks: open areas, dry or wet meadows near trees supporting stick nests or witch's-brooms, big cavities, sometimes even cliffs with old stick nests. Great horneds nest in or near most developed areas in the park, including campgrounds, as well as being widespread in appropriate habitat everywhere else.

Surveys have detected great horneds at two to 11 locations each year and nearly 30 specific places over all the years. Pairs frequently duet, giving their recognizable hoots as they prepare to nest, and respond to vocalizations of other owls. Surveys have also detected female great horneds giving distinctive solicitation calls, somewhat reminiscent of owlets' food-begging calls, to entice their male partners to feed them. One of the only owls seen during nocturnal surveys, great horneds sometimes perch on snow poles, primarily in Lamar Valley, on stumps, and on bridge railings.

FLAMMULATED OWL

Although we conducted several surveys for flammulated owls in May and June of 2015 and 2019 and June of 2021, none has yet been detected. These strictly insectivorous owls prefer ponderosa pine, a tree species that does not grow in the park, but they also could inhabit forests of large Douglas-fir with aspen stands nearby. Although Yellowstone has very little specific flammulated owl habitat near roadsides in the northern part of the park, where we conduct most surveys, we participate in statewide surveys for this diminutive owl because it is considered a Bird Species of Greatest Conservation Need in Wyoming. Plus, surveys constitute the most likely way to find another owl species in Yellowstone.

CONCLUSION: WHY MONITOR OWLS?

Our monitoring efforts present an annual index of sites where owls of several species have advertised for and courted mates in the northern part of the park. These data have improved our knowledge of owls' occurrence and use of the park. This is key for park managers because it highlights both knowledge gained of these seldom-seen birds and knowledge gaps (e.g., causes of year-to-year variation in owls detected and the possible variation in small mammal populations) as well as management needs like protection of certain habitat types. For example, our preliminary data on boreal owls have pinpointed the required breeding habitat for this species: extensive forests of mature Engelmann spruce and subalpine fir, the same habitat that boreal owls inhabit in Grand Teton National Park (R. Wallen and K. Duffy, unpublished data).

Acquiring baseline data will enable biologists and managers to estimate change to or loss of essential habitat for owls. Potential effects of climate change include a new fire regime of greater frequency, intensity, and size of wildland fires. For nesting, boreal owls require forests that take hundreds of years to reach maturity. Widespread burns in mature spruce–fir forests could ultimately result in long-term eradication of boreal owls within Yellowstone. When their habitat is gone, boreal owls will be gone.

Long-term monitoring will continue, perhaps coupled with installation of autonomous recording units (ARU) in a variety of habitats to allow us to analyze owl vocalizations at many locations on a nightly basis during the breeding season. Studies of small mammal populations where roadside owl surveys occur could also provide data to explain annual variation in owls detected. Owls symbolize all species that seem invisible due to their mostly nocturnal lifestyles. Understanding their requirements allows for better protection of them.

These scientific endeavors, conducted solely by unpaid volunteers, constitute a labor of love: labor because of the challenges of conducting surveys on cold winter nights, love because of a deep appreciation for and amazement at the many adaptations owls possess. Because we still have a lot to learn about Yellowstone owls, we will continue to gather data and listen intently for owl sounds on winter and spring nights, celebrating their indomitable presence. And there's a special reward for those who endure frigid nights surveying for owls: hearing an owl call emanate from the darkness is an ethereal experience, every single time.

Visit the Yellowstone's Birds website (https://press.princeton.edu/resources/yellowstones-birds-video-collection/v16) to watch an interview with Katy Duffy.

4 WATERBIRDS

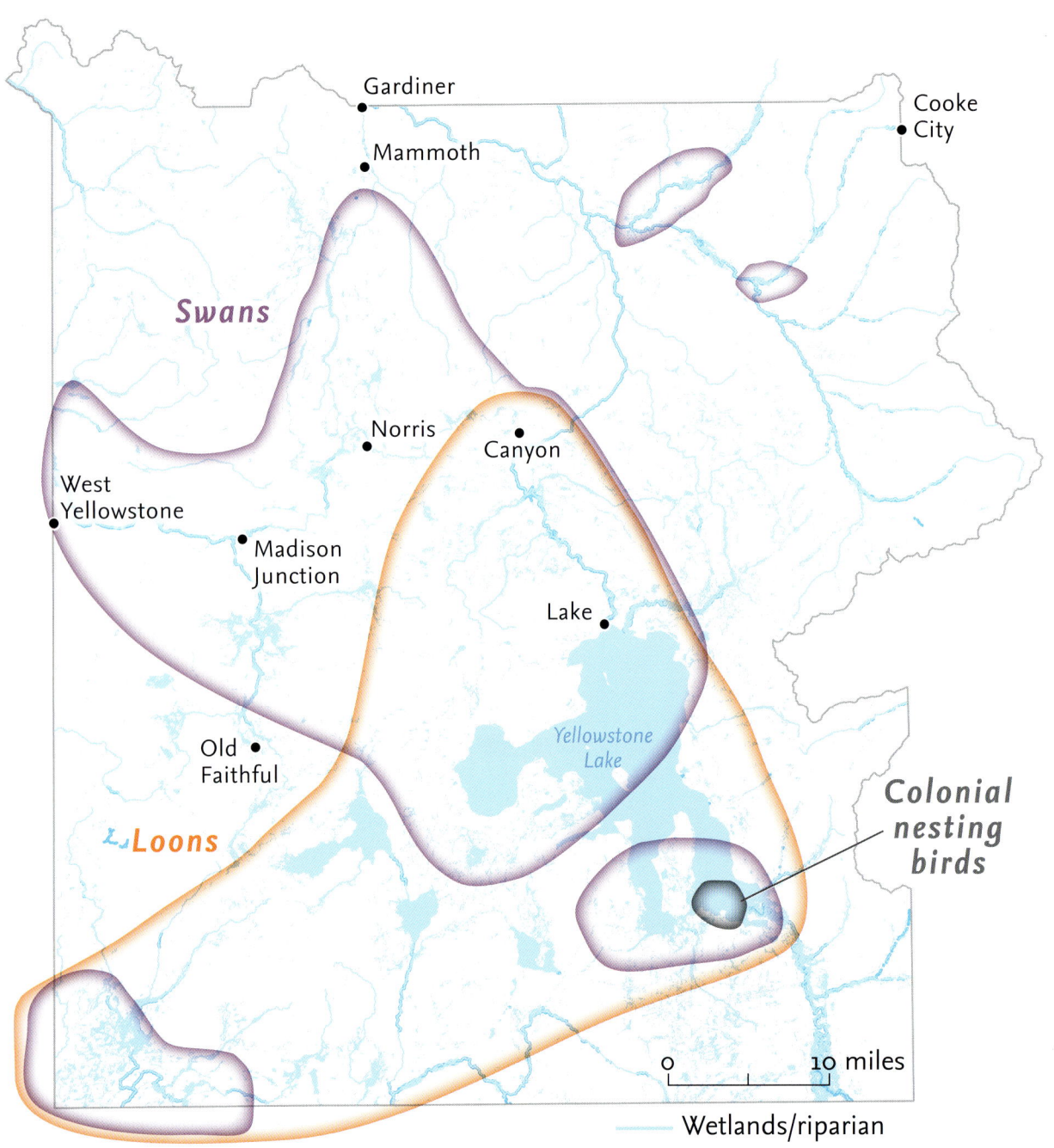

Gardiner

Cooke
City

Mammoth

Swans

Norris

Canyon

West
Yellowstone

Madison
Junction

Lake

Yellowstone
Lake

Old
Faithful

**Colonial
nesting
birds**

Loons

0   10 miles

Wetlands/riparian

YELLOWSTONE'S ICONIC BIRD
THE TRUMPETER SWAN

17

DOUGLAS W. SMITH, EVAN M. SHIELDS, WILLIAM LONG, RUTH SHEA, LAUREN E. WALKER, and JAY ROTELLA

She looked confused. Instantly, a cygnet had appeared when moments before there were just eggs. This wasn't usually how it worked, and her mate was equally perplexed. He was paddling in the water below the floating nest, looking at the cygnet that had plopped out, then back at the nest to see what would come next: nest, cygnet, nest, cygnet. A lesser scaup coasted by, and the day-old swan swam behind momentarily following it. "It's all over," we thought. Known for imprinting on the first thing they see after hatching (just like Konrad Lorenz's infamous greylag geese imprinting on and following him, back in the 1930s), we worried the young bird would swim off with the scaup. Then, just as fast, the tiny young trumpeter swan turned back to focus on the big, white, confused male—to us humans, the dad. For a moment, it looked like the introduction would fail. Both parents watched closely and slowly caught on. They helped the other cygnets from the nest, guiding them down the matted grass, then began swimming with them, caressing with their beaks, watchful, ever watchful—all easy to see and understand if you have ever studied a swan parent its young. It all happened so fast. We were grafting cygnets, a technique perfected elsewhere to help restore swans to former habitat after population declines and range retractions. Yellowstone was now one of these restoration areas.

DOUGLAS W. SMITH

THIS FIRST REINTRODUCTION occurred on Grebe Lake in 2012, one of only two trumpeter swan nesting territories in Yellowstone National Park (Yellowstone; YNP) that year (Fig. 17.1). This territory had not produced young since 1952, possibly because a fish hatchery was constructed there. Although the hatchery was gone, the swans were still not reproducing. The fall before, we had put in a floating nest platform to bob with fluctuating water levels to eliminate the potential of shoreline nest flooding, terrestrial predator attack, and human intrusion—all documented threats to nesting swans.

The swans had used their new nest site, and now we had taken the eggs, hatched them, and brought back the young. When we took the eggs, we replaced them with identical

FIGURE 17.1 Park staff release 110-day-old cygnets along the Yellowstone River in Hayden Valley. Swan releases commenced in 2012 as part of a restoration plan after a scientific meeting was unable to determine the cause of the decline and settle on a management strategy. Since releases began, the swan population has increased. PHOTO BY RONAN DONOVAN

wooden eggs that fooled the incubating female (she does most of the incubating and only occasionally gets breaks from the male to feed away from the nest, or she covers the eggs and leaves them unattended). We kept the swan eggs in an incubator and, after about 35 days, they hatched out in captivity. We then immediately rushed the precocial young birds back to the park, hiking them in the 3 miles (4.8 km) to the lake in a soft-sided cooler with heat packs. We also carried an artificial head that looked identical to an adult swan's, which we used to periodically poke at the cygnets with muffled honks, our attempt to mimic a swan vocalizing to its young—all this to imprint the young birds on a swan until we could get them back to their parents.

Today was the day of their return, and it all had been a little confusing to the parents. Known to be among the best parents in all the animal world, it took them only a few minutes before they accepted their young. The ancient, instinctual tug of parenthood was too strong. They had a family. After the reunion, the group swam off into the lake to begin the summer-long task of rearing and parenting. Grebe Lake once again had swan young—cygnets. Maybe the future was bright.

SWAN HISTORY IN YELLOWSTONE

The past had been anything but bright. In the 1800s, the future of Yellowstone swans appeared promising, but it didn't turn out that way. Swan fate in the park was tied to their fate elsewhere on the continent. Trumpeter swans, once abundant across North America, were nearly wiped out by the early 20th century due to overhunting and habitat destruction. The killing of swans for their skins and feathers is well documented from fur-trading times during the mid-1700s and is frequently mentioned in journals, although trumpeters were not distinguished from tundra swans (the other native North American swan species) until after 1831 (Mitchell and Eichholz 2020). They were easy to kill because they were a big white bird that lumbered across the sky with haunting calls that sounded like a trumpet. Their feathers were used for writing quills and to adorn hats and clothing, while skins were used for women's powder puffs. Records of killing continued through the 1800s and into the 20th century, indicating over two centuries of swan exploitation (Banko 1960). John James Audubon, the noted early ornithologist, preferred a trumpeter swan quill to the best steel pen of the day: "so hard, and yet so elastic the best steel pen of the day might have blushed, if it could, to be compared with the trumpeter swan quill" (quoted from Banko 1960). This exploitation was so early and over such a large area that the historic range of trumpeters is uncertain, but they probably occurred over a much wider area than originally presumed, and their range most likely included most of the continental United States, Canada, and Alaska (Mitchell and Eichholz 2020).

The full extent of the killing is unknown, but by 1933 the continental United States breeding population of trumpeter swans consisted of only about 70 individuals, mostly found in the Greater Yellowstone area (Hansen 1973, Pacific Flyway Council 2017). By 1968, the tally of the North American population had only increased to about 3,719 swans (Gale et al. 1987). Most of these (around 2,600) were in recently discovered populations in Alaska, and around 300 were in Canada. The Yellowstone region (YNP, Grand Teton National Park, nearby Red Rock Lakes National Wildlife Refuge, and the Island Park area) had about 600 birds and was the last trumpeter swan stronghold in the continental United States, saved by its remoteness and the additional protections provided by the parks and refuge. But, even within the confines of YNP, they barely survived.

The first records of trumpeter swans in Yellowstone came from the Washburn-Langford-Doane expedition in

FIGURE 17.2 Wing marks in the snow made by a swan trying to escape coyote predation, which ultimately the bird was not able to do. PHOTO BY SCOTT HEPPEL

1870. Later, in the 1880s, statements in journals and from the collection of specimens describe and document summering swans as abundant on Yellowstone Lake. Observations declined greatly in the 1890s as the species neared extinction due to continued killing outside the park. M. P. Skinner (1927) recorded swans in the park in 1915 and found a nest on Lewis Lake in 1919 (swans haven't nested there since the late 1930s and have not been documented on Lewis at all since 1961). That same summer, a fisheries worker reported six cygnets at Delusion Lake, which incidentally was a reintroduction site for four swans nearly a century later in 2014. In 1920, between 20 and 30 swans were reported wintering in open water at the outlet of Yellowstone Lake, a location where swans still overwinter today (Fig. 17.2).

THE EARLY YEARS

The next four decades, and specifically the park's situation, were best described by Banko (1960). Compared to current numbers, the 1930s through the 1950s could be called the heyday for cygnets, and probably for the entire population in Yellowstone (Fig. 17.3). The peak population

hit 87 in 1954. At the beginning of this period, swans were garnering enough attention, due to their possible extinction, that the young, brilliant, and first wildlife biologist of the National Park Service (NPS), George Melendez Wright, came to the park and sought out trumpeters to study. He and Ben Thompson took notes on swans from 1929 through 1932 at several locations in and near YNP. They gathered the most information about swans at Trumpeter Lake in northern Yellowstone, and in the interior, at the remote Tern, Fern, and White lakes near Pelican Valley, where they observed a nest and 11–13 swans between the three lakes (none are there today). At East Tern Lake, they saw attempted raven and otter predation on the eggs; to find out what happened, the two waded out to the nest. "We determined then to visit the nest and taking off our clothes waded out in the shallow lake.... We found five eggs unharmed by the otter" (Wright 1930–31). Elsewhere in his notes, Wright wrote that he suspected the issue was the well-meaning biologists and recommended the area be designated as a natural area void of trails so the swans could nest there undisturbed. Today a trail does go through the area, although it receives little use and is closed until early July because of heavy grizzly

FIGURE 17.3 Population trends of trumpeter swans and cygnets 1931–2019, Yellowstone National Park. The mid-20th century was the swans' heyday, and since that time, they have declined for unknown reasons; the decline is likely due to multiple causes.

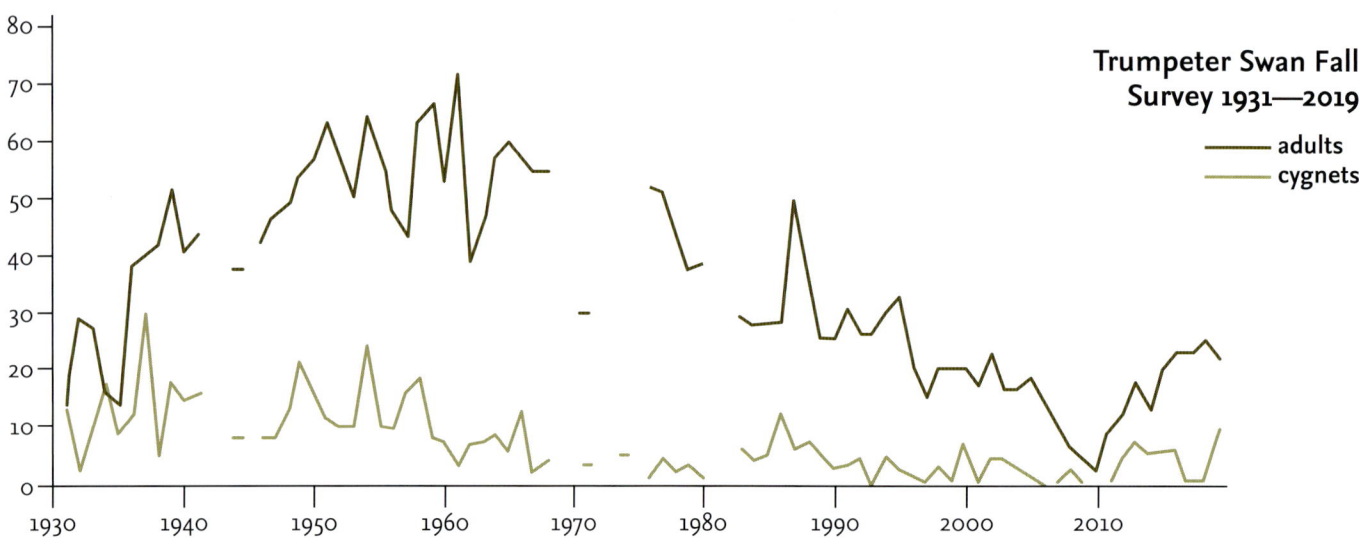

bear presence in the area. The East Tern Lake swan territory was active until 2000, but it has since gone quiet for unknown reasons. To jump-start recovery, and based on this historic occupation, three birds were released there in 2013, but they didn't stay. Over the following years, occasionally one would show up, but no pairs established there, despite the marsh still looking suitable for nesting.

Not long after these Yellowstone trips, Wright, with Park Superintendent Roger Toll and others, were concerned about saving swans in the region and helped establish Red Rock Lakes as a National Wildlife Refuge in 1935. To a swan, this refuge is just over the hill from Yellowstone, and the two areas may be linked (more on that later). Together, with Henry's Lake, which lies between, they represented the last stand of the once-abundant swans in the continental United States. This NPS-led effort cemented Yellowstone as famous for swans—hard to do in a land known for geysers and mammalian megafauna like bears, bison, and elk (wolves would come later).

From about 1930 through 1950, the population (adults and cygnets) fluctuated between 30 and 80, and cygnet production averaged about 12 per year (Banko 1960). What was important to realize, during this time, was that not all lakes or wetlands were good for swans. Although YNP has about 210 lakes, very few provided the suitable nest sites and abundant food swans required to fledge cygnets. Even during this "heyday" of good cygnet production, only 11 lakes fledged any cygnets. Some referred to these lakes as "engines," as they were spread across the park and, for each local area, may have sustained other marginal sites around it.

THE DECLINE

The Yellowstone swan population was relatively stable in the 1950 and 1960s but began to decline precipitously after this period (Proffitt et al. 2009). Numbers of adults fluctuated from 44 to 87 from 1951 through 1968, but then crashed by 50% in 1971, slightly rebounded in the mid-1970s, then dropped again to a 50-year low of 24 adults by 1986 (Gale et al. 1987). Alarmed, and unsure why numbers

were falling, the park installed artificial nest platforms that helped some pairs nest in 1986. Compared with 18 nests less than a decade earlier in 1978, only 7 swan pairs nested in 1986 (Gale et al. 1987). Additionally, park-wide cygnet production was low, much lower than the mid-century average of 12 fledged per year (Fig. 17.3).

Despite concern and management intervention, productivity and population size did not improve. Only 57 cygnets were fledged from 1987 to 2007, and the probability of fledging at least one cygnet per nest was only 15%. On average, only three cygnets were produced park-wide per year (Proffitt et al. 2010). The number of breeding pairs also stayed low. As a result of this period of low reproduction, the population reached its historic all-time low of four total birds in 2009 and two in 2010 (although, during the annual survey in 2010, fog obscured Riddle Lake, where two birds had been seen all summer). Trumpeter swans had bottomed out in YNP, and talk of extirpation was whispered by some.

What happened? Swan declines are hard to spot and diagnose because they are such a long-lived bird, with a 20- to 30-year life span (Mitchell and Eichholz 2020). Numbers appear stable, then the bottom falls out as adults start dying due to lack of recruitment to replace them. Declines creep up on you. Importantly, production in some of the key areas—the "engines"—had fallen off. Why this occurred continues to be debated, but some believe human disturbance from growing park visitation was a factor. Many other sites that were left largely undisturbed did not have the habitat to produce enough cygnets to support the population. Secondly, through much of the supposed heyday, the YNP population was thought to be propped up in part by Red Rock Lakes. Swan numbers and reproduction there were high; a modest banding program occurred between 1945 and 1984, and a few swans with bands were spotted in YNP—a pair near Lamar Valley and another at Seven-Mile Bridge near West Yellowstone (McEneaney 1986, McEneaney and Sjostrom 1986). This immigration could have masked falling production of YNP cygnets. Was Yellowstone propped up by Red Rocks National Wildlife Refuge?

All of this leads to an important observation: Yellowstone's swans are iconic because they were, with the population at Red Rocks Lakes, the last surviving trumpeter swans in the continental United States. This fact presumed these areas to be good swan habitat—yet more likely, it was their last stand, due to remoteness preventing human killing. Since then, all have wondered why the swans have faltered, struggling to retain their former glory (what glory?), and what to do. Maybe they cannot be restored. Yellowstone swans are truly wild—winter and summer—and most trumpeters in North America today thrive only where they have human assistance on wintering grounds in the form of agricultural crops. These are not Yellowstone swans.

THE SWAN WORKSHOP: THE SEARCH FOR A SOLUTION AND WHAT TO DO

Regardless of the cause, by 2010, YNP swans were clearly in trouble. Thus, in 2011, the park organized a workshop to guide management action and help save the dwindling population. By then, many causes for the decline had been suggested, and trumpeter swan and wetlands experts from all over North America were invited to determine two things: Why did Yellowstone swans decline, and what should be done about it? Neither goal was achieved (Smith and Chambers 2011). Workshop participants could not agree on a cause for the decline, probably because the decline was multi-causal and human nature tends to favor our own, single-cause theory. Participants also couldn't agree on a course of action. Some recommended immediate action despite not knowing the cause because the birds were in peril and at risk of extirpation. Others wanted more research before taking action, arguing that immediate action was akin to blindly flailing about, hoping for the best.

Although few conclusions were drawn, the reasons discussed for the decline were: (1) habitat reduction due to a warmer and drier climate over the last 40 years and/or issues related to climate change, (2) less immigration from Red Rocks Lakes, (3) increased predation, (4) human

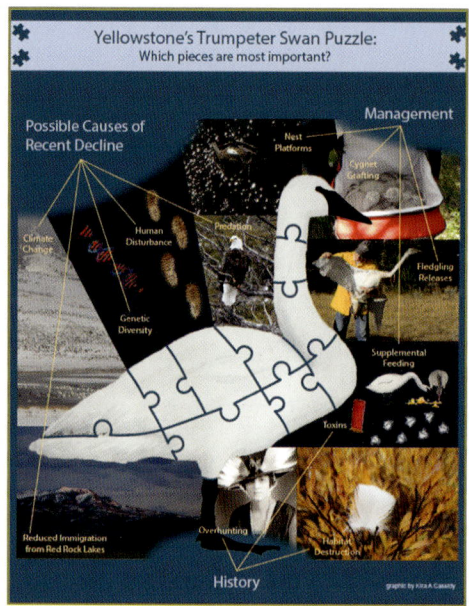

FIGURE 17.4 A conference was convened in 2011 to help determine the cause of and a course of action to halt swan declines in Yellowstone. Neither objective was achieved, likely because the decline is due to multiple causes, and, as is common, many participants favored their one favored cause, preventing a consensus.
ARTWORK BY KIRA CASSIDY

disturbance, (5) genetic depression caused by recent population bottlenecks, and (6) all the above (Fig. 17.4; Smith and Chambers 2011).

WHY DID SWANS DECLINE IN YELLOWSTONE?

Certainly, weather and climate have changed since Yellowstone's swan heyday, but what had it done to wetland complexes used by trumpeter swans? Previous research had found that clutch sizes and fledging success were higher on larger wetlands (Proffitt et al. 2010), but had wetlands shrunk? Some suggested that Yellowstone, due to its high elevation and brief summer breeding season, was always only marginal swan habitat and the climate had tipped just enough to nudge out swans. Trumpeter swans, although in some ways tough and resilient, are in other ways fragile, especially when raising young takes an inordinately long time each season. In fact, with 140–160 days from egg laying to fledging, trumpeter swans have the longest breeding period of any bird in North America (Mitchell and Eichholz 2020). Any change, on either end, that might cut the season short could prevent swans from being successful.

Although there are not many computer models geared to trying to understand why trumpeter populations fluctuate, the few attempted offer disparate results. One commonality, however, is the impact of cool, wet springs: they're bad for swans. In fact, swans will wait it out, forgoing breeding until conditions are more favorable. When you live for 20 to 30 years, why try and nest when conditions are poor? Wait until next year. Evolutionarily, swans only need to replace themselves, and given their long lives, they have lots of time to do it. For a songbird that lives only four or five years, this is unheard of. They must try to breed every year, no matter what. In Yellowstone, three recent springs (2017, 2018, and 2019) have been cool and wet; of the two active breeding pairs, one pair (Riddle Lake) did not attempt to nest any year; the other (Grebe Lake)

FIGURE 17.5 Swan Lake was last occupied by trumpeter swans in 1967. It was reoccupied in 2019 due to park restoration efforts, and young successfully fledged there in 2021. PHOTO BY LISA CULPEPPER

skipped 2017 and was unsuccessful in 2018. Two new pairs, however, nested in 2019, one at Swan Lake and the other in the southwest corner of YNP in the Bechler region, possibly due to the lower elevation and a more favorable micro-climate (Fig. 17.5).

The cool, wet spring argument has found recent support from graduate student Laura Cockrell from Eastern Kentucky University. She also found support for the hypothesis that large wetlands increase site occupancy and production. Her conclusions relied on remote sensing data, however, which, although useful, were not conclusive, leaving the issue unresolved (Cockrell 2014).

We have already discussed the Red Rock Lakes immigration hypothesis, which brings us to the next hypothesis: increased predation. Today known for its predators, Yellowstone has not always been this way. In the early 20th century, wolves and cougars were eradicated and other predators suppressed (Smith et al. 2020a). Wolves and, it was thought, especially coyotes (Childs 1934, Barrows 1936) probably occasionally preyed on swans and eggs. But besides wolves (there were none then), coyote numbers are hard to assess based on the available literature. Bears were always present, a famous exception, and in the late 20th century increased significantly (White et al. 2017). Fewer carnivores probably also meant fewer scavengers like ravens, crows, and magpies; known for their scavenging at mammalian carnivore kills, these three species are also known to be nest predators of many birds, including swans. This is why George Wright and Ben Thompson waded out to the swan nest at East Tern Lake—a raven had been there, and they suspected egg predation. Bald eagles, another waterfowl predator (especially on cygnets), also declined last century because of DDT (dichloro-diphenyl-trichloroethane) pollution. During the mid-20th century, all predators were probably low.

Then everything came back: wolves were reintroduced in the 1990s, cougars restored themselves through natural dispersal, and grizzly bears, always here, increased (Smith et al. 2020a). Coyote numbers are uncertain and could have been high in the mid-20th century due to control actions ending in the mid-1930s (Murie 1940), but they may have increased in the 1970s due to the banning of poison around the park (Schullery and Whittlesey 1999). Bald eagles weathered the DDT storm but suffered through the loss of a major food source—cutthroat trout on

FIGURE 17.6 Swan Lake in mid-October 2021. Early freeze-up when cygnets are not fledged can make them particularly vulnerable to predation—by bald eagles, in this case. PHOTO BY LISA CULPEPPER

Yellowstone Lake, which they survived by switching their diet to waterfowl (Baril et al. 2013). Swans are waterfowl. Adults are too big to prey upon, but during the summer, the cygnets are not. The other predators, except cougars, all prey on swans. Wolves, coyotes, grizzlies, eagles, and ravens do for sure (Fig. 17.6). Grizzly bears have been documented destroying nests, and a bobcat was observed eating a swan (McEneaney 2006). George Wright wrote about coyotes hunting swans at Yellowstone Lake in the winter of 1932 and ravens that "rapped" (tapped with the

beak) on eggs at Tern Lake, and coyotes killed three cygnets at Swan Lake in 2019 and the incubating female on her nest in 2020 (Fig. 17.7). Coauthor Shea, however, did not see much evidence of predation in the 1970s (Shea 1979). In fact, she located dead cygnets that hadn't even been scavenged, and she only remembers one predator kill—likely a mink, since the head had been chewed off. Bald eagle predation was never mentioned.

In sum, what predation did to swans during their mid-century "heyday" was not fully documented, and

FIGURE 17.7 Coyotes commonly prey on swans in Yellowstone. PHOTO BY LISA CULPEPPER

although there are anecdotal reports, it was likely low. Within the last few decades, however, Yellowstone has become predator rich. Has this impacted swans? Even with similarly low levels of predation, the formerly larger population would have been more resilient. Thus, while predation may not have been the cause of the population decline, predation could be playing a role in maintaining current low population numbers.

Our next hypothesis, human disturbance, is controversial. Of course, if true, it's a hard problem to solve. Yellowstone is a national park, created in part for human enjoyment, and this is where the debate begins. Early on during the swan heyday, the park had relatively few visitors, and nesting swans were likely left more alone. Then the park stocked some lakes, advertised fishing, and encouraged

use, leading, hypothetically, to an increased disturbance regime. Proffitt et al. (2010) found no relation between distance from the breeding territory to a road and swan production, nor a visitation effect as measured by the number of visitors per year in the park, but some feel that this measure does not adequately describe the human disturbance swans might experience.

Another key factor—the "engine" issue—comes into play: some swan territories are better than others. Proffitt et al. (2010) found about 80% of cygnet production comes from 40% of the territories. Shields (2021) found data on habitat quality to be generally sparse but, based on a few environmental factors potentially linked to habitat quality, always found "good" territories to be far fewer than "poor" ones. In short, and these two independent analyses

FIGURE 17.8 (A and B) With the loss of cutthroat trout in Yellowstone Lake as a food item for bald eagles, eagles have adapted by aggressively preying on swan cygnets, especially at Riddle Lake, where swans are not accustomed to people. Swans flee the shore when people approach, leaving their cygnets vulnerable to eagle predation in the open water. NPS PHOTOS/DOUGLAS SMITH

agree, there are a handful of productive territories, which if disturbed for any reason, could lead to population declines. For example, Trumpeter Lake in northern Yellowstone near the road, rated by Shields (2021) as one of the "good" territories, had 33 cygnets from 1931 to 1957, three cygnets from 1958 to 1995, and none since then, and now the swans are gone. Trumpeter Lake has also experienced significant drying. Grebe Lake is now a popular hiking route and was formerly productive. East Tern Lake, also rated highly by Shields (2021), is not close to the road but has a trail next to it (which Wright was against). Riddle Lake, arguably the single best territory park-wide (Shields 2021), is closed until mid-July due to grizzly bear management. But once the lake opens to the public, cygnet mortality ensues.

When cutthroat trout began to decline in Yellowstone Lake in the mid-1990s, bald eagles had to find something else to eat (*see chapters 12 and 21*). Waterfowl were abundant, especially Canada geese, which have dramatically increased in the park in recent decades (*see chapter 20*; USFWS 2020, Walker et al. 2020), and goslings are especially easy prey. Cygnets are too.

Our research trips to Riddle Lake after it opened to human visitation found bald eagles patrolling the lake every time we went. Predictably, the cygnets would disappear each year shortly after the lake opened on July 15. On several instances we saw bald eagle attacks. On one occasion, during an airplane monitoring flight, Smith saw a cygnet floating dead next to the adults amid feathers in the middle of the lake; nearby an eagle was perched in a tree drying its wings (Fig. 17.8). We didn't circle long enough to watch the eagle retrieve the dead cygnet from the water. Another time, Lisa Baril, a Bird Program technician, saw an eagle swimming to shore, using its wings as oars, after an unsuccessful attack. Despite cygnets hatching most years, none survived.

While the solution is obvious now in retrospect, this was a puzzle that took us several years to figure out (Fig. 17.4). The key piece was when Smith saw the adults with the dead cygnet in the middle of the lake shortly after July 15. Why would they go out into the open water, away from the thick vegetation along the lake shoreline that offered them protection? People. Due to long-term protection until mid-summer (i.e., the lake is closed until July 15 as part of the bear management protocol), these birds were unused to people and became hyper-sensitive. Enter the patrolling bald eagle. Pushed out to the middle of the lake by human activity on the shoreline, the young were exposed in the open water, and it was just a matter of time until the eagle picked off all the cygnets.

Our response was to close the lake until September 1 (not popular with the public) to allow the swans to continue to hide in the thick, shoreline vegetation. We chose September 1 because we thought by then the Canada-goose sized cygnets would be immune from predation. They were not. We now keep the lake closed until September 15 when the swans have young (we open the lake if they do not reproduce, as is often the case during cold, wet springs), and even then, they can usually only raise one cygnet.

The detrimental effects of human disturbance are reinforced by swan behavior. Parenting differences between Riddle and Grebe lakes is stark (Fig. 17.9). At Grebe, the parents are relaxed and let the cygnets wander meters from them. At Riddle, no such thing happens: the cygnets are always close to the adults, and the parents look like nervous wrecks. Yet guarding like this, they still lose most of them. At Grebe, people walk up to feeding swans in weed beds and the swans do nothing, except maybe pose for photographs; at Riddle, they beeline it to open water. The bane of swans—a big white bird on black water that is never able to hide—is that people always walk up to them. At Grebe, this has produced habituated swans. At Riddle, this has produced intolerance to people at a lake patrolled by a deadly predator. These behaviors are the outcomes of having campsites and trails next to one lake versus a lake closed during the nesting period with no camping areas. Was this the case at East Tern Lake as well? But what is the problem, predation or human disturbance? This issue has been known for some time (Condon 1941), yet we still struggle with it.

In addition, Canada geese may compete with swans. Both species are very fond of the same aquatic vegetation, *Potamogeton* spp., commonly called pondweed, and this may be just part of the overlap between the two species. Their surging numbers in the last 50 years coincide with the swan decline, but geese are confounded with visitation, time, and bears, so teasing out an effect was not possible (Shields 2021). Certainly, this is deserving of more study. The last hypothesis proposed by the workshop has since been disproven. Genetic testing has revealed that Rocky Mountain swans, although not possessing a lot of genetic diversity due to North America–wide population reductions, are no less diverse than other swan populations (Oyler-McCance et al. 2007). This is not likely the reason Yellowstone swans declined.

FIGURE 17.9 Swan parenting styles: (A) At Riddle Lake, where bald eagles prey on cygnets, nervous parents rarely let the cygnets go far from them; (B) At Grebe Lake, with no bald eagle predation, parents are much more relaxed. NPS PHOTOS/DOUGLAS SMITH

THE FUTURE

So, where are we? We keep finding new pieces to this puzzle, and some would argue that this is true of all nature study: it is forever withholding, then slowly revealing, trickling out clues, always unevenly and unexpectedly, then it all changes. This has certainly been the Yellowstone swan story. We have to consider that the swan decline likely occurred because of a combination of many things, including changing wetland conditions brought on by a changing climate, combined with human disturbance and predation and the interaction between them, as well as possibly less connectivity to Red Rock Lakes National Wildlife Refuge.

Coauthor Evan Shields recently completed his master's degree at Montana State University studying the swan decline in YNP (Shields 2021). Shields investigated the availability and utility of existing datasets that could be used within statistical models to explore all the main hypotheses for swan decline discussed here. There were difficulties inherent in working with such a long dataset—90 years (1931–2020)—and many environmental variables were unknown through time. Most significant among his findings was the presence of swans outside of Yellowstone; when numbers were high outside the park, it was indicative of a better year inside the park, suggesting a connection to swan populations outside the park. Many variables he tested were correlated with one another and trended through time, making separating one of them out a challenge. For example, grizzly bears and increases in park visitation naggingly, but not strongly, hung around in his analyses as possible factors, loosely supporting the predation and human disturbance hypotheses. Again, this supports a multi-causal decline.

Other unstudied factors could also play a role. What about beavers (muskrats have never been very common in YNP)? Swans prefer nesting on islands in the middle of lakes possibly because of the protection that affords. There are few such islands in the lakes of Yellowstone, but beaver lodges provide ideal island nest sites. Swans in Alberta, for example, commonly nest on old beaver lodges (W. Long, pers. comm.). Beavers declined throughout Yellowstone during the 20th century (Smith and Tyers 2012), likely forcing swans to nest along lake shorelines, where they are more vulnerable to people walking up to them, fluctuating water levels, and terrestrial predators, resulting in nest failure. Like eagle predation at Riddle Lake, there could be an interaction between a natural phenomenon and human disturbance.

Of course, the ultimate question is even if we figure out the answer to why swans declined, is there anything we can do? National parks are not meant to prop up species that are on their way out for "natural" reasons (Smith and Peterson 2021). In cases of human-caused declines or extirpation (think wolf eradication in Yellowstone or lake trout introduction), intervention is sometimes approved. Where does the swan decline fall?

While we actively continue to try to answer these questions, in the meantime, our grafting and fall swan releases show signs of working. New pairs are popping up at Swan Lake (Fig. 17.5) and in the Bechler region, and possibly on Yellowstone Lake as well. Our fall count is consistently in the mid-20s, up 10-fold from a decade ago. Trumpeters from the north, from Canada and possibly Alaska, continue to pour in by the hundreds most winters (Fig. 17.10). This regal bird, ancient in our memories, almost destroyed by us, has been nurtured back across the continent, aided by school groups, landowners, and societies that love them. Their fidelity to each other in a lifelong pair bond, replaceable only by death, and their caring and doting parenting of their young serve as an example to all who care to watch. Yellowstone has always had them. How do we keep them? Should we keep them if it requires perpetual management? What would the park be without them?

Perhaps this is beyond science and policy and in the realm of poets and writers. William Butler Yeats wrote a poem about swans, "The Wild Swans at Coole": "But now they drift on the still water, Mysterious, beautiful; Among what rushes will they build" (Yeats 1919). And E. B. White fashioned a novel about them: "And if you looked up, you would have seen, high overhead, two great white birds. . . .

YELLOWSTONE'S ICONIC BIRD

Ko-hoh, ko-hoh, ko-hoh! A thrilling noise in the sky, the trumpeting of swans" (White 1970). Imagine Grebe Lake without them drifting still and Hayden without their thrilling trumpeting to the sky. The trumpeter swan's last stronghold, Yellowstone, would not be the same without Yellowstone's iconic bird.

FIGURE 17.10 Many migrant trumpeter swans from Canada overwinter in Yellowstone National Park, swelling their numbers considerably, sometimes to several hundred, and intermingling with the small resident population (20–25 individuals). These trumpeters on the west side of the park look for small open patches of water near the Madison River. PHOTO BY SCOTT HEPPEL

Visit the Yellowstone's Birds website (https://press.princeton.edu/resources/yellowstones-birds-video-collection/v17) to watch an interview with Douglas W. Smith.

COMMON LOONS OF YELLOWSTONE AND THE GREATER YELLOWSTONE ECOSYSTEM

VINCENT SPAGNUOLO, JEFF FAIR, ARCATA LEAVITT, DOUGLAS W. SMITH, DAVID EVERS, WALTER G. WEHTJE, and LAUREN E. WALKER

Crouching silently along the shore of a remote lake, deep in the vast wilderness of Yellowstone National Park, I hear the voices of chorus frogs and sandhill cranes, and when I'm lucky, I catch the occasional distant howls of a wolf pack piercing the wind. The only signs of human life are a single-engine plane in the distance, the quiet chatter between rangers and the communications center on my park radio, and the sounds of my own breathing. Suddenly, the resolute silence of the Yellowstone wilderness is split open like lightning by an inexplicably loud and haunting sound that echoes across the lake: the calls of a pair of common loons. They exchange wails before breaking out in duetting tremolos, a sign of disturbance, followed by the male's pugnacious yodel. Loon pairs are highly territorial; they own their lake and let everyone know it. Here in Yellowstone, their calls remain uncommon, one of the rarest and most unforgettable sounds you'll hear in this wild Wyoming landscape.

VINCENT SPAGNUOLO

LOONS ON AN ISLAND OF WILDERNESS

If you stand along Yellowstone Lake in spring and scan the waterbirds that forage in the shallow waters beyond the lakeshore, you may see a few common loons in breeding plumage (Fig. 18.1). While most are migrants, stopping over to rest and feed before continuing north on their way to Montana and Canada, some summer in the park, where they form the nucleus of the southernmost breeding population of this species in the world. One of five extant species of loons in North America, common loons are most abundant in Ontario and Quebec, where over half of the breeding population resides (Evers and Taylor 2014). In the western United States, loon populations

COMMON LOONS OF YELLOWSTONE AND THE GREATER YELLOWSTONE ECOSYSTEM

FIGURE 18.1 Two loons chicks enjoy a ride from their parent in southern Yellowstone National Park. PHOTO BY CHARLIE HAMILTON JAMES

are well-studied in Montana and Washington, but not in Wyoming. The isolated Wyoming breeding population has persisted in the high-elevation and rugged environment of the Greater Yellowstone Ecosystem (GYE) and has only recently become the focus of researchers and landscape managers.

From the mid-1800s through the late 1970s, common loons across North America experienced a range-wide northward contraction due to a variety of human impacts. Habitat degradation, lake-level management that flooded or stranded nests, and human disturbance were all important factors on the breeding grounds. Bycatch in the gillnets of marine and Great Lakes fisheries occurred on migration and on the wintering grounds, while shooting for sport and direct persecution and lead poisoning from ingested fishing tackle occurred throughout the year. By the 1970s, local extirpations and declines were so dramatic and widespread that concerned groups initiated aggressive conservation efforts in many parts of the range, particularly New England, the Midwest, and Ontario. These efforts produced considerable advance-

ments in our understanding of loon natural history, their demographics, and human-driven threats, as well as strategies and tools to study and conserve these charismatic birds. Conservation actions taken by federal and state governments, local conservation groups, and private citizens have allowed loons to rebound and recover significant portions of their range from coast to coast.

When Yellowstone National Park (Yellowstone; YNP) was established in 1872 to protect this unique landscape, it also facilitated the recoveries of declining or extirpated species such as bison, grizzlies, and trumpeter swans. Although we know that loons experienced a significant range contraction, we have little information about the Yellowstone loon population before 1989. Today, the individuals breeding in the park comprise the majority of a small and isolated GYE loon population. Disjunct from the nearest breeding population to the north by over 220 miles (350 km), these loons occupy a small island of suitable habitat on the Rocky Mountain front and are hemmed in by drier mountainous areas to the south and the steppe and plains habitat to the east. To further complicate our

FIGURE 18.2

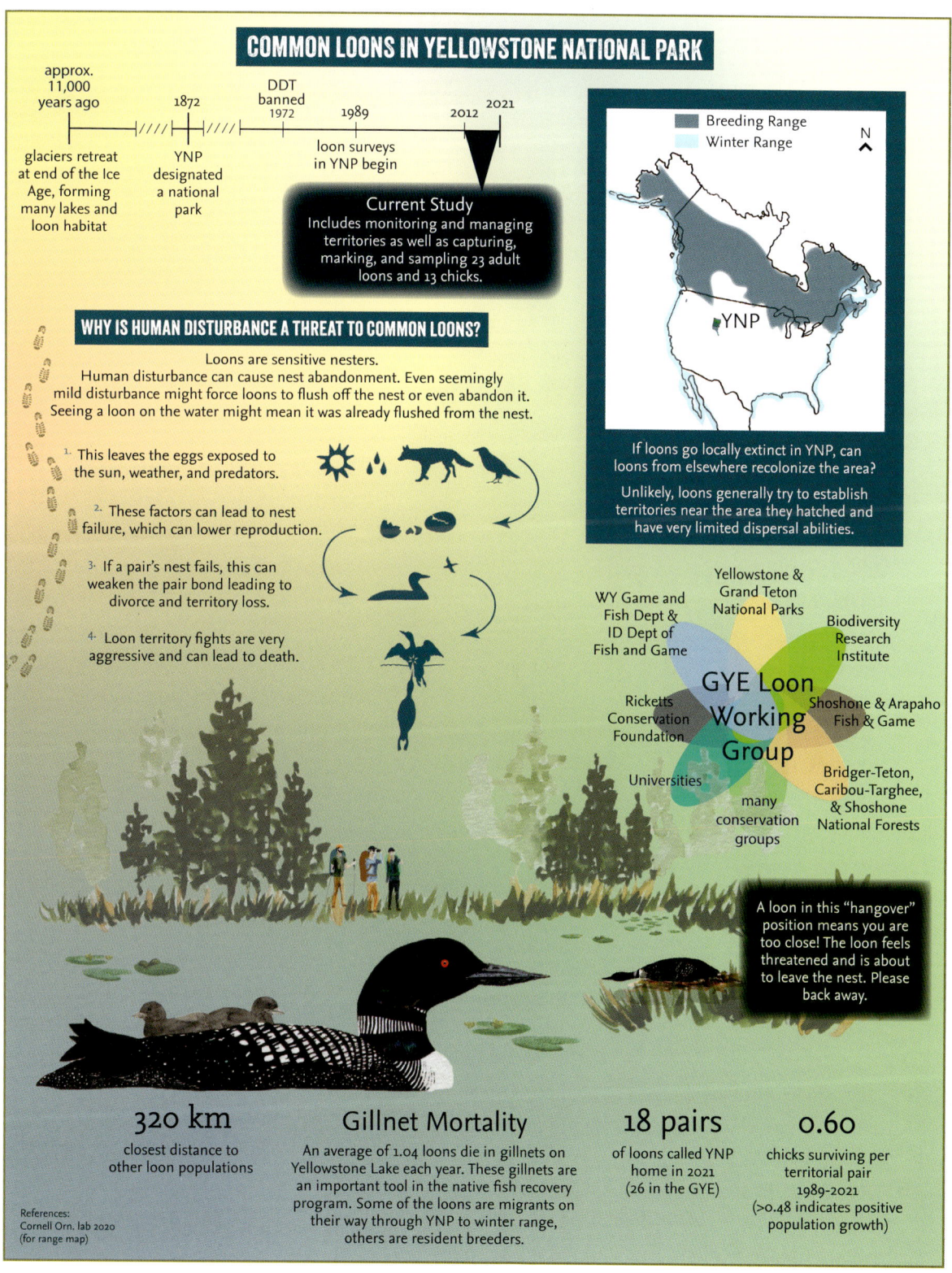

COMMON LOONS IN YELLOWSTONE NATIONAL PARK

approx. 11,000 years ago — glaciers retreat at end of the Ice Age, forming many lakes and loon habitat

1872 — YNP designated a national park

DDT banned 1972

1989 — loon surveys in YNP begin

2012

2021

Current Study
Includes monitoring and managing territories as well as capturing, marking, and sampling 23 adult loons and 13 chicks.

Breeding Range
Winter Range

N

YNP

If loons go locally extinct in YNP, can loons from elsewhere recolonize the area?

Unlikely, loons generally try to establish territories near the area they hatched and have very limited dispersal abilities.

WHY IS HUMAN DISTURBANCE A THREAT TO COMMON LOONS?

Loons are sensitive nesters.
Human disturbance can cause nest abandonment. Even seemingly mild disturbance might force loons to flush off the nest or even abandon it. Seeing a loon on the water might mean it was already flushed from the nest.

1. This leaves the eggs exposed to the sun, weather, and predators.

2. These factors can lead to nest failure, which can lower reproduction.

3. If a pair's nest fails, this can weaken the pair bond leading to divorce and territory loss.

4. Loon territory fights are very aggressive and can lead to death.

GYE Loon Working Group

Yellowstone & Grand Teton National Parks

WY Game and Fish Dept & ID Dept of Fish and Game

Biodiversity Research Institute

Ricketts Conservation Foundation

Shoshone & Arapaho Fish & Game

Universities

many conservation groups

Bridger-Teton, Caribou-Targhee, & Shoshone National Forests

A loon in this "hangover" position means you are too close! The loon feels threatened and is about to leave the nest. Please back away.

320 km
closest distance to other loon populations

Gillnet Mortality
An average of 1.04 loons die in gillnets on Yellowstone Lake each year. These gillnets are an important tool in the native fish recovery program. Some of the loons are migrants on their way through YNP to winter range, others are resident breeders.

18 pairs
of loons called YNP home in 2021 (26 in the GYE)

0.60
chicks surviving per territorial pair 1989-2021 (>0.48 indicates positive population growth)

References:
Cornell Orn. lab 2020 (for range map)

ARTWORK BY KIRA CASSIDY

COMMON LOONS OF YELLOWSTONE AND THE GREATER YELLOWSTONE ECOSYSTEM

understanding of the history of this population, recent genetic evidence reveals that Yellowstone's loons are more closely related to those that breed in Saskatchewan and other Canadian prairie provinces, more than 500 miles (800 km) from the GYE, than to those found relatively nearby in Montana and Washington. Although their present breeding habitat today in Wyoming remains largely protected, Yellowstone loons remain at risk due to the impacts of human activity, the population's isolation, and climate change. Furthermore, annual surveys over the past 10 years have found an average of only 20 territorial pairs in the GYE, making common loons the rarest breeding bird in Wyoming. While every pair across this landscape is of critical importance, the heart of this breeding population is the territories found within the borders of Yellowstone National Park (Fig. 18.2).

CRITICAL ASPECTS OF COMMON LOON NATURAL HISTORY

- Common loons are long-lived; they can live to 35 years of age or longer.
- They exhibit limited dispersal proclivity and are slow to recolonize a former range. On average, they disperse 8 miles (13 km) from their natal lake; established breeding adults rarely disperse more than 5 miles (8 km).
- Their diet primarily consists of fish but also includes aquatic invertebrates and amphibians.
- Pairs are highly territorial and aggressively defend breeding territory, even occasionally fighting to the death.
- They establish nests on shorelines, islands, marsh, floating bog mats, etc.
- They lay one or two eggs, and rear only one brood per year; a pair may renest after an early nest failure.
- Loons winter on near-shore marine waters, or sometimes on rivers and lakes that remain ice-free.
- Juveniles mature on near-shore marine waters for one or two years before returning to the breeding grounds.
- On average, they are six years old at first breeding.
- They exhibit high fidelity (about 80%) to their territory, often to a specific nest site.
- The annual survival rate for adults is high (about 92%).
- Critical threats include human disturbance during nesting, shoreline development, environmental mercury loads, lead poisoning, fluctuating water levels, oil spills, fishery bycatch, and cyanobacteria outbreaks.

LOON LANGUAGE

- Their wail is like a wolf howl and is often used between mates, presumably meaning *Where-arrrrrre-yooooou?*
- Tremolo: loon laughter (vibrato) is often in response to stress or surprise, and the only call given both in flight and on the water.
- The yodel begins with a wail-like crescendo unique to the individual and usually ends with several repeated phrases; it is a sound of wild insanity, given only by males as a territorial warning.
- The hoot is a softer, usually interfamilial, syllable of recognition.
- A chick's call is a quiet, short whistle repeated several times, perhaps to get parental attention; it is easy to mimic and has been used by biologists to lure in the adults for banding.

DESCRIPTION
AND LIFE HISTORY

Common loons are long-lived birds, reaching up to 35 years of age (Paruk et al. 2021). In breeding plumage, they show a black bill and head, bright red eyes, vertical striping on their neck that's overlaid by a black collar, a white chest, black flanks, and a black-and-white checkered pattern on their back. In contrast, wintering birds are relatively drab. Adults molt into a plumage in which the throat and chin are white, the bill turns bluish-gray, and the body is a dull gray-brown color. Juveniles resemble wintering adults but are more boldly patterned and paler. All loons are supremely aquatic, with a streamlined body; their legs are placed so far back that they are nearly incapable of walking on land. In another adaptation for efficient swimming, the loon's tarsus (the bone that connects the foot to the body) is flattened to reduce turbulence when moving underwater. Combined, these traits mean loons can only push themselves onto land, which limits their nest sites to shorelines and other near-shore sites.

After wintering along the Pacific Coast and in the Gulf of California, GYE loons arrive at regional lakes as soon as the ice has melted. Many of their breeding lakes are at higher elevations, and loons can be found congregating at the lower-elevation Jackson Lake and other open water in April and early May, while they wait for other nearby lakes to thaw. Once the ice melts on their breeding lakes (in Yellowstone, this generally occurs in late May or early June), the breeding pairs arrive to claim their territories. Rather than produce many offspring at once, loons invest significant energy into protecting and raising only one or two chicks per year over a very long breeding life. Therefore, controlling a productive nesting territory is of primary importance. Arriving early allows established pairs to defend their territories from intruding birds. Common loon territorial disputes can be violent; adults and their chicks can be injured, sometimes even killed. A favored tactic by offensive loons is to dive underwater and spear the belly of a bird swimming on the surface, punching a

hole through the victim's breastbone or abdomen. Dissections of dead loons have shown that loons may suffer and survive many attacks like this throughout their lives, with up to eight partially healed punctures recorded on both male and female loons. Swimming loons often peer under the water to look at what's under them. During the breeding season, they're looking for prey but also checking to see that they're not being attacked by another loon.

Within the GYE, common loons breed on lakes of a variety of sizes, ranging from 87,680-acre (35,483-ha) Yellowstone Lake to water bodies only 20 acres (8 ha) in extent. Habitat criteria include a safe nest site and enough space for the adults to be able to gain flight from the lake's surface. A loon's weight, high wing-loading, and leg placement mean they must run and flap their way across nearly 100 yards (90 m) of open water before they are able to become safely airborne. Interestingly, although common loons are considered primarily piscivores, or fish eaters, some of the most productive breeding territories within the GYE are on fishless lakes. Loons nesting on these lakes may substitute a mostly fish diet with one of amphibians and invertebrates, such as dragonfly larvae and leeches, especially when feeding chicks. Adults may fly back and forth to nearby lakes that have fish, perhaps to both supplement their diet and confirm the presence of sufficient local prey biomass to fledge chicks.

Given the inability of common loons to walk on land, their nest sites are limited to lake shorelines, marshes, floating vegetation mats, and small islets. Their need to place the nest near the lake edge makes the nests vulnerable to changes in lake level, wave action, boat wakes, terrestrial predators, and direct human disturbance by people fishing or hiking along the shoreline. Once a nest site is selected (generally by the male), the pair creates a small depression or builds a shallow nest bowl in which the female lays one or two eggs. The eggs are olive-colored with darker blotches, almost twice the size of a chicken egg, and look remarkably like gull eggs. This led early ornithologists to place them in the same taxonomic order as gulls and shorebirds. More recent genetic studies, however, place loons in their own order, Gaviiformes, and suggest their closest modern relatives are Procellariiformes (albatrosses, petrels, and shearwaters). Like these species, they have a conservative breeding strategy. If their nest fails early in the breeding season (before mid-June), they may start over and build a new nest. If the nest is lost later during the next cycle, however, the pair is unlikely to renest—they are resigned to trying again the following summer.

In successful nests, loon eggs hatch after an average of 28 days of incubation. After briefly drying off in the nest, the one or two chicks are ready to follow their parents onto the water. Although they are good swimmers from the get-go, the chicks spend much of their first few weeks of life on their parents' backs. This keeps them warm as well as safe from predators. During this time, one parent will carry the chicks on its back while the other hunts and delivers food. After about two weeks, the chicks spend more and more time on the water and gradually learn to feed themselves. Finally, by late summer or early fall, the parents leave their young to learn to fly and migrate south on their own. In the GYE, many adults and some juveniles congregate on Yellowstone Lake at the end of the breeding season before heading south. Based on limited data, we believe that GYE loons migrate to winter off the Pacific coast of Mexico and perhaps in the Gulf of California. Juveniles find their way to the wintering grounds on their own and stay there for one to two years before venturing back north to their natal grounds to begin the process of establishing a breeding territory and finding a mate (Paruk et al. 2021). Individuals must compete for limited suitable breeding sites, and it often takes several years to secure a territory and a compatible mate; on average, common loons breed at six years of age, but may begin as early as four years old.

LOON CONSERVATION IN YELLOWSTONE AND THE GYE

In the era of modern loon conservation (1975–present), Yellowstone National Park and the surrounding GYE were among the last population areas in the United States where loon researchers initiated a comprehensive conservation effort. While some bird species garnered attention early in Yellowstone National Park's existence (e.g., trumpeter swans; *see chapter 17*), loons were initially not given much attention as a species of concern or interest. Starting in the late 1980s, however, biologists from the park as well as the Wyoming Game and Fish Department (WGFD) surveyed the region for loon pairs and chicks on a near annual basis. These counts focused solely on documenting the presence of adults and chicks, with no proposed management goals or accompanying research. Monitoring efforts varied from opportunistic observations and reports from non-biologist park staff to limited systematic surveys of lakes known to host loon pairs by staff with little to no formal loon training. Despite some sharing of data among the park and other federal and state agencies, no coordinated GYE effort existed. Within Yellowstone, efforts were taken to minimize publicizing the location and status of loon territories. Yellowstone's stance through the early 2000s was that the best management for loons was to leave them alone—a popular misconception regarding loon conservation.

Between 1989 and the early 2000s, the GYE population appeared relatively stable, fluctuating around 20 observed territorial loon pairs, mostly on consistently occupied territories in or near the park (Fig. 18.3). The full extent of annual survey efforts during this period remains unclear, however, and likely varied somewhat between years. In 2011, agency biologists were startled to observe only a single loon chick and vacancies in several historically occupied territories in Yellowstone. Fearing they were facing a decline in the loon population and lacking the necessary knowledge or resources to accurately assess the population status or propose effective management action, biologists from Yellowstone and WGFD raised the alarm. Acting on their concerns, the agencies contacted expert loon biologists at the Biodiversity Research Institute (BRI) and requested assistance in monitoring and understanding the status of the GYE loon population (GYE Committee 2022).

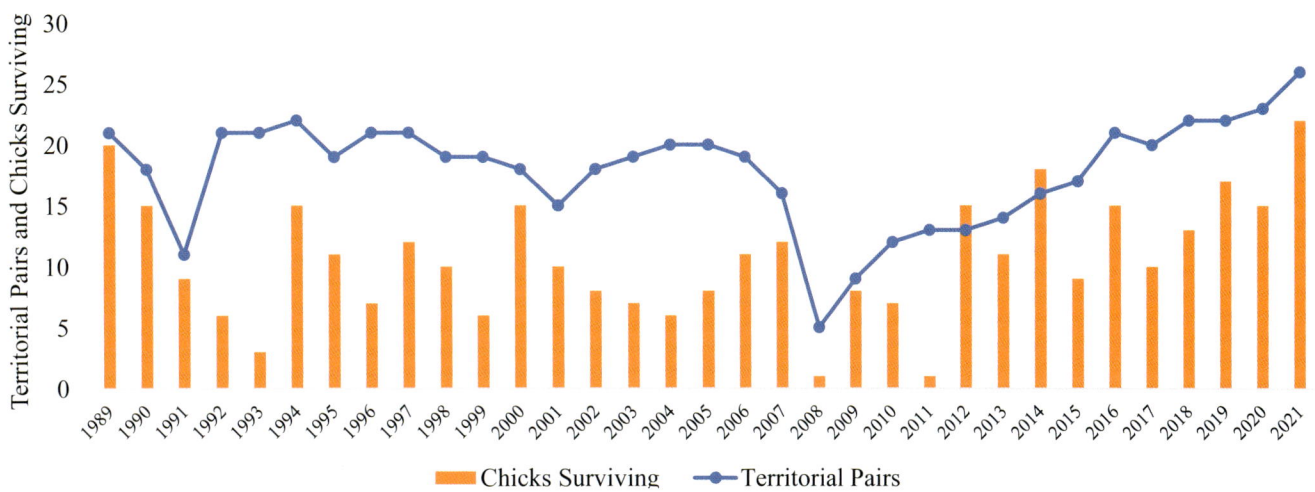

FIGURE 18.3 Territorial loon pairs and number of chicks surviving per year in Yellowstone National Park and the Greater Yellowstone Ecosystem, 1989–2021.

In 2012, a pilot study assessed the existing information on the GYE loon population, quantified breeding territories, identified knowledge gaps, and informed the agencies of loon conservation approaches and strategies. Yellowstone, WGFD, BRI, and other agencies within the GYE developed a project spanning the full 2013 breeding season, with funding from Yellowstone and WGFD. Supplemented by significant additional funding from the Ricketts Conservation Foundation (RCF) to BRI, the study of GYE loons was quickly extended and expanded with more intense survey and research efforts through 2018 as part of a larger effort to strengthen and restore loon populations across the southern extent of the species' historic range in the United States. In 2019, RCF assumed lead monitoring and conservation roles, with BRI focusing on research needs, namely loon capture.

Following the example of conservation efforts in other loon breeding population areas, the GYE Loon Working Group was formed in 2013 to coordinate efforts and share data and information among the various GYE agencies and loon researchers (GYE Committee 2022). The working group now serves to facilitate more effec-

tive and efficient loon research and conservation efforts in the GYE. Combined, increased communication and coordination, expert involvement, secured funding, and retrospective analyses have laid the groundwork for a long-term effort to ensure the persistence of breeding loons in Yellowstone and the wider ecosystem.

DEVELOPMENT OF THE YELLOWSTONE LOON STUDY

Our knowledge and understanding of the common loon population in Yellowstone have grown steadily over the past several decades. Beginning in 2013, biologists conducted annual surveys of the loon population in Yellowstone and throughout the GYE and monitored for the presence of territorial pairs, unpaired adults, hatched chicks, and chicks that survived to six weeks of age (an age at which their mortality rate drops considerably and we consider them likely to fledge from their natal lake). Over time, we've expanded our survey area and monitored more lakes, enabling the detection of newly formed and previously undiscovered pairs; improved tracking of unpaired adults; and evaluated potential, as yet unused, breeding habitat. A critical part of the work has also been to assess the natural and anthropogenic threats to loon survival and reproduction in key habitat areas, including current and historically active territories as well as habitats utilized by unpaired adults. To increase our understanding of Yellowstone loon demography, dispersal, conservation issues, and migration, loons both inside and outside the park have been captured and marked with unique combinations of a metal federal leg band and colored plastic leg bands to enable individual identification (GYE Committee 2022). Additional samples (described in more detail below) were collected from captured loons that will further help in our understanding of the status of the GYE loon population.

Monitoring loons in the GYE is not easy, and efforts to collect comprehensive data on the region's loons are challenged by several factors. The rugged, high-elevation (6,000–8,000+ feet/1,828–2,438+ m) terrain and the distances to be covered (on and off-road, as well as off-trail, and sometimes requiring overnight backcountry camping) has impeded surveys compared to loon population studies in the east and Midwest. In the GYE, a few dozen loons are scattered across tens of thousands of square miles, a study area nearly twice the size of Massachusetts.

Aerial surveys from fixed-wing aircraft provide only brief snapshots of loon activity on a lake but can quickly cover large swaths of habitat and have proved invaluable in supplementing and guiding the ground-survey efforts. Through aerial and ground-survey efforts, we found territorial loon pairs in some unexpected locations, including several small lakes, some nearly covered with emergent vegetation and some fishless. Loons are often considered to be dependent on fish to survive and raise chicks, but loons in the GYE often breed successfully on fishless lakes. This discovery ultimately added many lakes to the project's coverage that might have been written off as too small, too shallow, or too vegetated in other parts of the species' range.

In many other US loon population areas, residents and volunteers are happy to report on "their" loons. Not so in Yellowstone. Most of the park's visitors are here for a short time and are not savvy about the local wildlife. Additionally, in contrast to many "loon" lakes in the Northeast and Midwest, few observers live on Yellowstone's lakes or stay for the entirety of the loon nesting season. The exceptions are the park staff, particularly rangers, who are regularly in the backcountry and play a major role in the current Yellowstone loon project. Rangers often have intimate knowledge of the loon activity within

SMALL AND FISHLESS LAKE ECOLOGICAL CONSIDERATIONS

- Small lakes covered extensively with emergent vegetation, often fishless, are generally deemed unsuitable loon habitat outside the GYE.
- Loons on fishless lakes in the GYE are thought to feed on aquatic invertebrates and fill their need for calcium by consuming the leeches and amphibians that are more numerous due to a lack of predatory fish.
- Outside the GYE, loons nesting on fishless lakes are often less productive, but loons in the GYE are often highly productive on small, fishless lakes.

- Loons in the center of the continent are the smallest and lightest in general (likely due to longer migration distances to marine waters for wintering), needing less prey biomass, and requiring smaller open water areas for takeoff.
- A small lake with limited open water may inhibit disruption by intruding loons.
- Fishless lakes and vegetation-covered lakes are less likely to suffer human disturbance by anglers and other recreationists.

DON'T BOTHER THE LOONS—JEFF FAIR

When I entered the realm of loons as a young field biologist back in New Hampshire in 1978, I carried a deep respect and reverence for their wildness and spirit-voices. I had no intention to interrupt their ancient behaviors by capturing them and holding them in my human hands, nor to out them from their individual anonymity with colored bands on their legs. I simply wanted to observe them and learn from what their behaviors told me. And I did learn. But soon there came along a proven safe technique to capture and mark them individually, so that we could learn even more, much more, from them as individuals in their own societies. I worried about violating their primeval peace but recognized the ecological need to understand them better to mitigate human-made changes to their world and save them

from disappearance. At first, I distrusted the handling of them, but I later attended the early captures to judge the impacts on them. I was encouraged that proper handling had little effect on their lives, and eventually capitulated my concerns and held these loons and their chicks and their eggs in my own hands to help dissect their precious secrets and lay bare their ecologies and needs. I wanted only to help provide the great power of scientific understanding of their basic realities and some of their magic to avoid their demise. This is what brought me, in my elected field of wildlife research, to the sacred wild waters of Yellowstone National Park in 2013 as part of this team to discern how to safeguard the songs of the loons, like the choruses of wolves, as voices of the wild here in this rare and tiny population.

their districts, aid in implementing management actions in the field, facilitate biologists' access to their study areas, and regularly engage with members of the public around the loon nesting sites. Park staff, interns, and volunteers from all backgrounds make meaningful contributions to loon conservation by reporting loon sightings and sharing intimate knowledge of backcountry sites. Many of the region's federal and state agency personnel have joined the working group and helped cement a coordinated regional goal of common loon conservation.

DANCING WITH LOONS: CAPTURE AND BANDING, A RESEARCH TOOL

Loons have their own dances. One is called the penguin dance, performed by territorial (or would-be) pair members when confronted by trespassing loons. It consists of a vertical stance while treading water. And then there is the circle dance, when two to five (or maybe more) loons gather on a pair's territory, apparently accepted but maybe distrusted, swim in a circle, and dive simultaneously.

There is only one dance that loons do with humans, and the humans always lead. It's called capture (Fig. 18.4).

To assess the status of the Yellowstone loon population, researchers and managers must go further than counting birds. Loons are long-lived and have an interesting social structure, which is apparent in their interactions with each other and their environment. Researchers and managers invest substantial time to better understand each territory, pair, and individual—an understanding made possible by leg-banding individuals with unique color-band combinations. Marking loons in such a way offers insight into an individual's movements, age, health, and survivorship, as well as into the dynamics of territorial pairs, their territory fidelity, divorces, movements, mortality, and territorial takeovers. Further, it allows us to track the nesting success of individuals and pairs, which is important for understanding localized threats and developing management priorities. For example, in other well-studied populations, approximately 20% of the pairs produce 50% of the young, and these "superior" individuals (or individual territories) may have characteristics

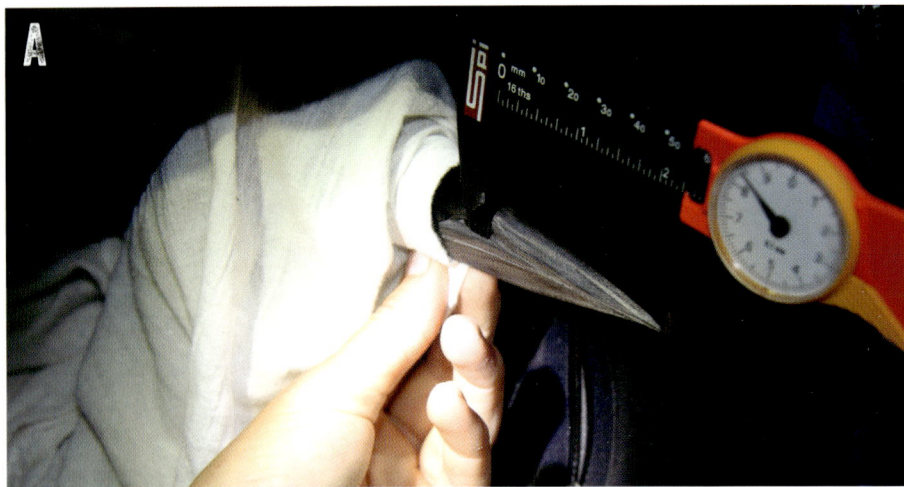

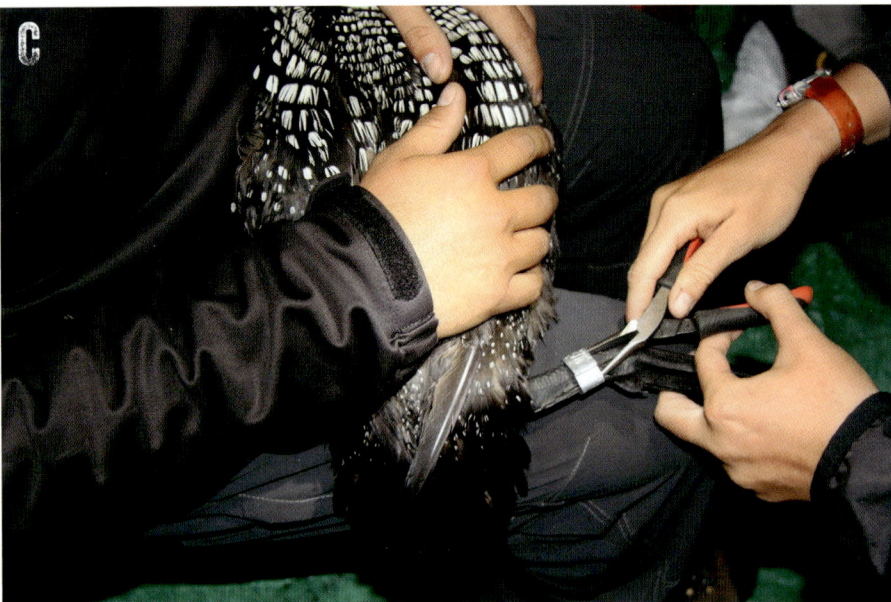

FIGURE 18.4 Loon capture panel: (A) Measuring a bill. (B) Measuring a wing. (C) Putting on a band. NPS PHOTOS

FIGURE 18.5 Loon day capture with "Leroy" (front), the decoy. NPS PHOTO

that could be more important for future conservation goals (Paruk et al. 2021).

But to put bands on a bird, you've got to catch it first. Capturing and handling loons is the most intimate part of our monitoring activities, and we operate with great care and proven methods and under rigorous regulatory guidance (Evers 1993, 1994). There are only two viable methods to capture loons: diurnal and nocturnal (Fig. 18.5). (It is nearly impossible to get near a loon free on the water in daylight; many early loon biologists tried and failed.)

The diurnal method of loon capture was developed by United States Geological Service (USGS) biologists and BRI in Arctic Alaska as they studied yellow-billed loons at latitudes where there is no darkness under the midnight sun all summer. In these attempts, we use a common loon decoy, a floating mist net, and loon-call recordings to try to bring the loon over the net. This technique has been used with limited success in Yellowstone, while the nocturnal method remains substantially more efficient.

Nocturnal capture is more commonly used with common loons and, in the wilds of Yellowstone, usually involves a minimum of three or four people, a canoe, a powerful spotlight, a long-handled dip net, recorded loon calls, and paddles, all of which is lugged down a rugged trail up to several miles long to the lake (Fig. 18.6). The technique works best on loon pairs with young chicks, as those adults are more likely to stay at the water's surface to defend their chicks. The dance begins by locating the loons in daylight.

COMMON LOONS OF YELLOWSTONE AND THE GREATER YELLOWSTONE ECOSYSTEM

After dark, the canoe is launched with the spotlighter crouched in the bow, the netter on or behind the bow seat, and one or two paddlers aft. The loons are (usually) eventually found in the spotlight, which also blinds them to the canoe's approach. Sometimes loon calls are played, or the netter mimics the calls of a chick to bring the adults in closer; if a loon comes within reach of the netter, it is caught with the dip net. The loon is then brought directly into the canoe and its head covered with a thin linen towel to calm it; it is held to avoid injury until being delivered to shore for banding. Once ashore, we band the loon with two bands on each leg: one numbered aluminum USGS band and three variously colored plastic bands in a unique combination. As often as not, however, the loon evades the net, and the dance continues.

ASSESSING LOON HEALTH: INDIVIDUALS AND POPULATION-WIDE

Capture and handling of wildlife is stressful for the animal, so we aim to gain as much information as we can while we have the bird in hand. After captured loons are banded, we also take measurements of the loon (e.g., weight, wing length, and bill length) and collect samples, including a small volume of blood, along with a few feathers (clipped, not pulled), and samples of fecal matter and oral and cloacal swabs. Blood is later analyzed for complete blood counts, blood chemistry, genetics, parasites, viruses, mercury and other heavy metals, and other contaminants, including biotoxins from cyanobacteria (colloquially known as blue-green algae). Feathers inform stable isotope analyses and provide additional information on mercury that, combined with the blood sample, help form a year-round picture of a loon's exposure. As loons finish their molt on the wintering waters, feathers collected during the breeding season carry contaminants from their winter diet. Blood sampling, in contrast, reflects their recent exposure to contaminants on the breeding grounds. Combined, these samples and analyses are part of efforts to build baseline health and mercury profiles for Yellowstone loons, which will ultimately contribute to an understanding of the species across its range (Evers et al. 1998, Ackerman et al. 2016, Kneeland et al. 2020).

Data from the samples collected from loons help us identify and assess current and future threats to the population. Blood samples from Yellowstone loons have

FIGURE 18.6 Senior author Vince Spagnuolo getting ready for nighttime capture. NPS PHOTO

identified individuals with avian malaria, "low to moderate risk" exposure to mercury (based on well-established effect levels; Evers 2018), and unexpectedly high levels of BMAA (a neurotoxin produced by naturally occurring cyanobacteria; β-methylamino-L-alanine; K. Low, pers. comm.). In general, blood sampling has also revealed relatively high red blood cell counts compared to loons sampled in other populations, likely due to Yellowstone's high elevation (M. Kneeland, pers. comm.).

Although it was previously suspected that Yellowstone's loons were most closely related to the geographically close populations in Washington and northwestern Montana, the results of DNA (deoxyribonucleic acid) testing indicate that Yellowstone loons are, in fact, more closely related to loons in the Canadian prairie provinces. This revelation supports the hypothesis that Yellowstone's loon population is a relic left behind by glacial retreat (Larison et al. 2021). Knowing their closest genetic kin will allow biologists to select the best infusion of genetic relatives if the local population ever exhibits low genetic diversity or pending extirpation.

YELLOWSTONE PARK LOON DEMOGRAPHICS

The best measure of the status of a healthy loon population is a combination of two factors: the number of territorial (and thus potential nesting) loon pairs and the number of chicks produced per pair that survive to fledging age. Further, as loons are long-lived, these demographic measures must be tracked over decades of equal-effort surveys to fairly assess loon population trends.

Although the GYE loon population has been surveyed annually since 1989, monitoring efforts through as late as 2012 were incomplete and inconsistent, hindering efforts to reliably estimate population size and trends over time. Although we observed what appear to be significant declines during the period between 2007 and 2012, we are uncertain if they actually occurred or if the apparent trends are a result of monitoring inconsistencies. Some territories occupied prior to 2007 were abandoned, and while a few have since been recolonized, others remain vacant today.

In 2013, the modern era of loon monitoring in Yellowstone began. It took a few years for loon biologists in the GYE to standardize survey techniques and cover the immense study area efficiently, so we may have missed some pairs on territories for which there was little to no historic data. The data for these years may therefore underestimate the true loon population at the time (Table 18.1). While it is difficult to compare population estimates across years with varying survey effort and expanse, the consistent annual monitoring of a subset of historic territories provides a rough benchmark to the population status through time (GYE Committee 2022). From 2017 through 2021, when the study was well-established, the portion of the loon population found in Yellowstone National Park appeared to hover around 15–18 territorial pairs, comparable to reported historic pair counts going back to 1989. In the GYE, recent seasonal tallies of 20–25 territorial pairs are similar to estimates from 1989 through 2006. From 2013 through 2021, 67% of the GYE's territorial loons summered in Yellowstone. The park also supported 66% of the region's nesting pairs and 62% of chicks surviving to six weeks of age, cementing Yellowstone National Park as the heart of the Wyoming breeding loon population.

Obtaining an absolute count of loons in the GYE is a significant challenge, and comparison to years with fewer data is difficult. The recent tallies, even when combined with historic records, do not permit us to suggest a quantitative population trend for either the park or the GYE. They do, however, allow for a qualitative assessment of the population status and provide a solid basis to compare with the next decade of survey results. After years of close study, it is heartening to discover new loon pairs and observe the recolonization of previously vacated historic territories.

While annual population tallies can be compared to loosely assess demographic and distribution trends, future stability is guided by the survival and productivity

COMMON LOONS OF YELLOWSTONE AND THE GREATER YELLOWSTONE ECOSYSTEM

TABLE 18.1	OBSERVED LOON DEMOGRAPHICS IN YELLOWSTONE NATIONAL PARK (YNP) AND THE GREATER YELLOWSTONE ECOSYSTEM (GYE), 2013–2021								
		2013	2014	2015	2016	2017	2018	2019	2020	2021
YNP	Territorial pairs	8	10	11	13	15	16	15	16	18
	Nesting pairs	6	10	11	9	8	11	12	13	13
	Chicks hatching	8	16	6	9	8	12	11	7	14
	Chicks surviving	8	14	4	9	8	9	9	7	13
GYE	Territorial pairs	14	16	17	21	20	22	22	23	26*
	Nesting pairs	11	14	15	14	12	17	18	18	21
	Chicks hatching	13	20	11	15	10	17	19	15	23
	Chicks surviving	11	18	9	15	10	13	17	15	22

* Includes newly discovered territorial pairs in the Wind River Range ~87 miles (140 km) southeast of Yellowstone National Park

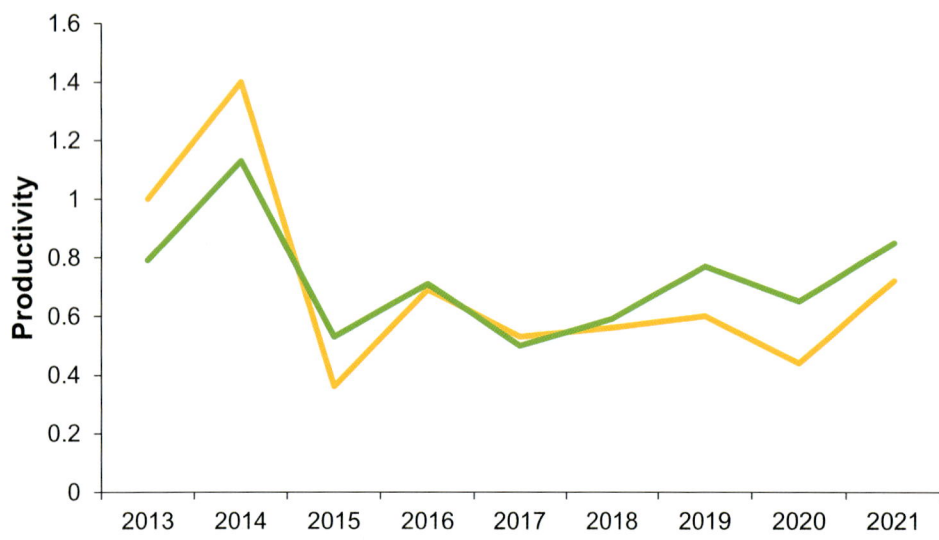

FIGURE 18.7 Observed loon productivity in Yellowstone National Park (yellow) and the Greater Yellowstone Ecosystem (green) from 1989 through 2021. Productivity is calculated from the number of chicks surviving to fledging age, divided by the total number of territorial pairs.

of the individuals currently in the population. Recent observations suggest the local loon population produces an average of 0.7 surviving chicks per territorial pair each year (Fig. 18.7). This average well exceeds the 0.48 chicks surviving per territorial pair benchmark of productivity to maintain a steady population, as estimated from research on other populations (Evers 2007, Paruk et al. 2021). Productivity measures suggest the Yellowstone National Park and GYE loon population may be growing, although the applicability of the 0.48 benchmark is also dependent on adult and juvenile survival rates. Once sufficient data can be collected on color-marked individuals, the survival rates of loons in the GYE can be compared to other regional standards (Mitro et al. 2008) and inform a better assessment of the GYE's population stability.

Despite these positive findings, the Yellowstone loon population is tiny, entirely isolated from the rest of the range, and thus vulnerable to extirpation. In Wyoming, the common loon is a Species of Greatest Conservation Need (WGFD 2017). Because loons are slow to colonize new territories, natural recolonization would be unlikely if this population were to be extirpated. While we continue to track these loons and their individual stories, it is imperative to understand and mitigate the human-caused threats they face.

HUMAN-CAUSED THREATS TO YELLOWSTONE'S LOONS

Human disturbance of nesting loons has long been suspected as a widespread and serious issue affecting Yellowstone loon productivity and population stability. Loons nest on the water's edge, preferring small islands, peninsulas, open marshes, or bog mats, which can render their nests and eggs both conspicuous and easily accessible to humans (Fig. 18.8). If a potential threat, human or otherwise, approaches, the incubating adult may leave the nest as a defense mechanism for itself and to avoid revealing the nest location. However, if the threat lingers long enough, keeping the incubating adults away from the nest, the nest may fail due to abandonment, exposure, or depredation. A single person fishing along the shore of a small lake for a few hours may be enough to cause a loon pair to abandon their nest and fail to reproduce that year. Beyond reducing the productivity of a single pair in a single year, nest failures can also contribute to territorial pair instability, leading to mate switching or territory takeover by intruding loons. These secondary effects may further reduce nesting and productivity across multiple breeding seasons as well as at neighboring territories.

Within Yellowstone, most loon territories are at some level of risk of disturbance by visitors. Although it is often difficult to identify the direct cause of nest failure, many failures occur at lakes with high human use. While only one human-caused nest failure has been observed by staff through 2021, human disturbance was a suspected cause or factor in most nest failures from 2013 to 2021. Compared to many breeding loons in areas with larger human populations and heavier recreational pressure, loons in Yellowstone National Park appear to exhibit a heightened sensitivity to human presence, rendering them particularly vulnerable to human-caused nest failures. Nest disturbance by park visitors has been documented during in-person surveys and from trail cameras, although these events have not necessarily resulted in nest failures. In efforts to reduce disturbance of all shoreline-nesting waterbirds in Yellowstone, biologists across the park have recently increased coordination of field efforts to minimize the potential negative impacts of our research while enabling the implementation of necessary monitoring and management actions.

While lake shorelines in the park are undeveloped and motorboat intrusions are only of concern on Yellowstone and Lewis lakes, human disturbance remains a growing threat. Park visitation has increased steadily over the past several decades, nearing 4.9 million visits in 2021, and recreational experiences are shifting toward the backcountry as more visitors seek out less-crowded areas. Lakes are often target destinations for hikers, campers, and anglers, and despite Yellowstone's protections, many lakes experience a relatively high volume of human traffic. Shoreline trails and campsites create intermittent disturbance and may present an ecological trap for nesting loons. Loons typically begin nesting before peak summer visitation; thus, what may appear to be a peaceful shoreline for a loon to nest on in early June may become a popular human hiking destination by early July. Once visitors begin disturbing loons—often unintentionally—it is too late for the loons to select a more remote site for their nest. These disturbances become more disruptive and widespread as more visitors seek to explore lakes and see or photograph wildlife. Some areas in the park are particularly popular with anglers, which can lead to locally

FIGURE 18.8 Photo of a loon nest with two eggs. The simple nest and proximity to water make loon nests vulnerable to flooding, especially with more unpredictable precipitation and snowmelt due to the effects of climate change.
PHOTO BY NICK FERRAUOLO

high disturbance due to repeated and prolonged visitation. However, even at territories that rarely or infrequently experience human visitation, a short-term disturbance could have a significant impact and cause nest failure. Observations of loon territories in Yellowstone National Park suggest that pairs nesting in areas with less recreational pressure are disturbed more easily (e.g., from a greater distance), and disturbance events result in longer periods of time off-nest than pairs in territories with more regular recreation and human exposure.

To help mitigate human disturbance of nesting loons, biologists in Yellowstone National Park have relied primarily on targeted area closures. These closures may include shorelines, coves, trails, campsites, or entire lakes, and are generally designed to be brief, ideally paralleling the nesting period. Loons are most sensitive early in incubation, so many closures are deployed before nesting begins, sometimes even before ice-out, to ensure protection throughout the nesting cycle. In 2019, the neighboring Caribou-Targhee National Forest enacted first-time closures on five active loon breeding lakes adjacent to Yellowstone's southern border, resulting in triple the number of loon chicks surviving on those lakes compared to years before the closures. This is evidence of the potential efficacy of closures and of the likelihood that human disturbance may be a primary threat to the success of nesting loons throughout the GYE.

In areas popular with Yellowstone's visitors, where closures would be difficult or impossible to implement, loon researchers have also utilized artificial floating nest platforms or "loon rafts." Loons will usually opt for a small-island nesting site when one is available, and an anchored raft with proper construction and nesting materials resembling those on a small island can encourage them to nest offshore and away from more vulnerable locations. The installation of a nest raft is often accompanied by seasonal closures of the area to the park's visitors, and while these rafts were initially developed to mitigate the effects of flooding or the stranding of loon nests by

water-level fluctuation and to reduce shoreline predation, they have been used on three loon territories in the park to lure the nesting pairs away from sites of human disturbance. Once used by a territorial pair, the rafts can be moved incrementally farther away from disturbed sites each nesting season. The key danger is that, if the raft is evident and the visitors uninformed, it may attract curious humans and negate its purpose. Like many parts of wildlife management, it's a delicate balance that can be tipped in a positive direction by human education.

GILLNET BYCATCH

The restoration of native cutthroat trout to Yellowstone Lake, and the removal of invasive lake trout, is a conservation project of top priority to the park. Cutthroat trout are prime prey for loons, other piscivorous birds, and many mammal species, and this effort is deemed critical to the proper function of the local ecosystem (Koel et al. 2019b, Williams et al. 2022). Yellowstone Fisheries and Aquatic Sciences (YFAS) has demonstrated success with setting gillnets to remove lake trout at a large scale, and their efforts have resulted in both a decline in lake trout and an increase in the lake's cutthroat trout population. However, a consequence of setting thousands of miles of gillnet in Yellowstone Lake each year is the infrequent but regular bycatch mortality of loons and other diving birds.

Because the GYE supports so few loons, the loss of even a few breeding adults can have significant consequences for the population (Paruk et al. 2021). Since the first records of avian bycatch in 1998, more than two dozen adult common loons have been drowned in gillnets (an average of a little more than one loon per year). Much of the loon bycatch in Yellowstone Lake occurs in late summer, during the loon's fall migration. Thus, most loon deaths may be individuals beginning their migration from more northern breeding populations. However, two drowned individuals were banded breeding loons from the GYE population, and biologists suspect at least one other was a locally breeding adult, based on the time of year the bird was caught.

Despite the threat to Yellowstone loons, mitigation of gillnet bycatch is complicated. A restored cutthroat trout population will benefit a range of species, including loons, osprey, eagles, terns, bears, and otters, by stabilizing food-web dynamics. Loons will benefit because cutthroat trout are relatively small and spend more time near the surface, and thus are more suitable prey than the significantly larger lake trout that inhabit deeper areas of the lake. In response to bycatch mortality, YFAS has been working closely with the Yellowstone Bird Program to implement depth and area restrictions for gillnet sets and to adjust the timing and duration of sets in loon territories. Ensuring adult loon survival and lake trout removal is imperative to a healthy ecosystem, but the coexistence of the two creates a challenging management scenario for Yellowstone. Collaborative efforts are necessary to ensure that both loon and cutthroat trout populations flourish. The ongoing cooperation between YFAS and the Bird Program serves to minimize loon bycatch to a level that we hope will not significantly impact long-term loon population stability.

MIGRATION AND WINTERING GROUNDS

While Yellowstone loons currently benefit from the park's protection on the breeding grounds, they face unknown challenges the remainder of the year. Because common loons spend their first two plus years on the inshore marine waters that constitute their wintering grounds, and later approximately half of every ensuing year wintering there, the components of winter habitat are just as important to loon health as those on the breeding grounds. To address challenges and threats on the wintering grounds, however, we must know where they are. Until recently, this was one of the greatest unsolved mysteries of the Yellowstone loon population.

To track the migration route and determine the wintering area of Yellowstone's loons, over the past decade, we attached tracking devices called geolocators to a handful of GYE loons. Geolocators self-record the timing and duration of daylight, offering a crude set of latitude and longitude points. Though far less accurate than GPS (global positioning system) or satellite trackers, geolocators are less expensive, less invasive to attach, and are small enough to be attached to a loon's colored leg band. Unfortunately, although small and externally affixed to a loon's leg, geolocators do not transmit data on their own and need to be recovered to collect the recorded data.

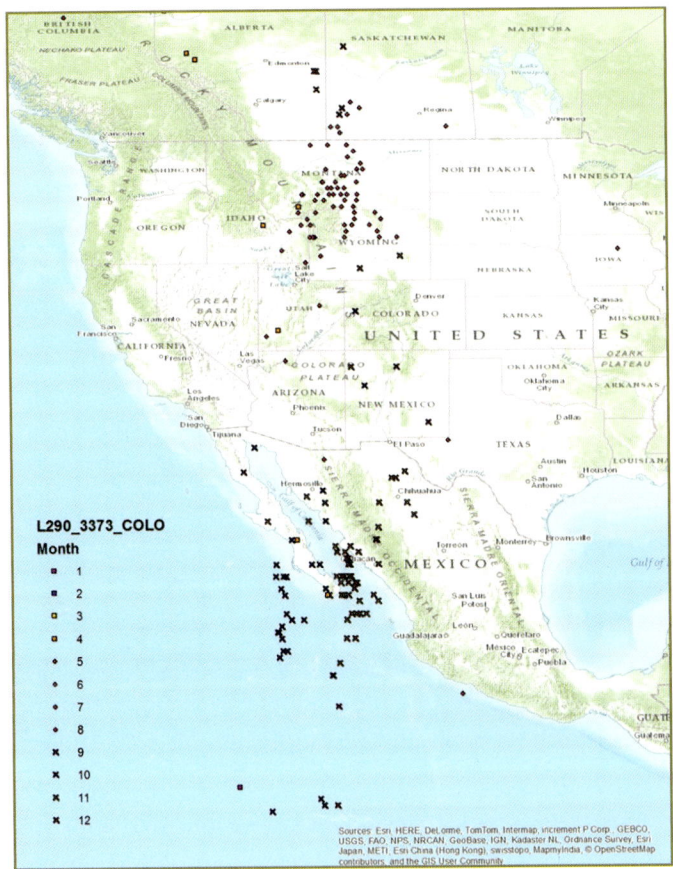

FIGURE 18.9 Where do Yellowstone loons go in the winter? Locations of a migrating loon, captured at Wolf Lake in Yellowstone National Park and tracked to the Gulf of California.

While our efforts to date have shed some light on Yellowstone loons' migration and wintering habits, many aspects remain a mystery. Of the eight geolocators we deployed between 2013 and 2014, four have been recovered. Due to equipment failure, only one carried data. Data from the single successful transmitter illustrated an overland migration route south and west to the waters of the southern Gulf of California in Mexico (Fig. 18.9). We plan to deploy more geolocators to provide a more informative dataset.

LOOKING AHEAD

Back at the lake, the mournful wails of the loons fade into the distance as the eagle that had disturbed them flew out of sight. That fading call brought with it a

moment of calm excitement, as two freshly hatched chicks were revealed on an adult's back. This is the first time we've observed this recently recolonized Yellowstone territory hatch chicks, and I couldn't be more elated. This loon pair is part of what may be the most precarious population across the entire species' range, and they have no idea how easily the scales could tip toward extirpation. Perhaps ignorance is bliss? Regardless, these loons fight with all they have in order to raise these chicks, hold their territory, and simply survive, completely unaware of the awe they inspire in those of us who study them. It's moments like this one—the milestone findings that we speculate on, yearn for, and talk of late into the night and on long drives across the ecosystem—that truly make our annual effort worth it. The push to ensure the persistence of this population demands so much both mentally and physically, and seeing the loons' perseverance fuels our own ambition to fight the good fight for them.

VINCENT SPAGNUOLO

The previous 10 years of monitoring have provided invaluable insights into the factors that affect common loon survival and breeding success within the GYE. Over the next 10 years, continued and more in-depth and expansive surveys must be carried out to determine the drivers (and threats) of population growth, as well as patterns of productivity and recolonization both in the park and across the GYE. To summarize the goals of the GYE Loon Working Group and support thoughtful and informed loon management and research in Yellowstone, biologists are developing a conservation plan for the GYE breeding population. The Greater Yellowstone Common Loon

COMMON LOONS OF YELLOWSTONE AND THE GREATER YELLOWSTONE ECOSYSTEM

Conservation Plan (Plan; GYE Committee 2022) will better organize the goals and objectives among the involved federal and state agencies, further increasing collaboration among the working group members. The Plan focuses on management and research priorities, addressing the threats, data needs, and other issues facing this population. The Plan also proposes potential improvements to the study, such as the increased use of trail cameras and the deployment of autonomous recording units (ARU) to record loon sounds. The use of remote monitoring technology, along with the increased use of aerial survey methods, may reduce the need for on-the-ground visits to known and consistently occupied territories, enabling biologists to expand their survey area and explore new potential habitats. Finally, while until recently little information about Yellowstone's loons was shared with the public, largely due to concerns about increased disturbance, the Plan supports and incorporates public education as a management tool to reduce human disturbance going forward.

Beyond the known threat of human disturbance, the next decade of loon conservation in Yellowstone will require adjustments for a mosaic of possible and probable effects of climate change. What might the effects be? Although difficult to fully predict at this writing, we anticipate changes in precipitation patterns, ice-out dates, lake levels, wildfire frequency and intensity, and disease outbreaks that may act in concert to affect loons, their prey, and their habitat in ways yet unforeseen. Only through continued monitoring of this isolated population will we understand how this iconic species will persist in a changing world. Furthermore, because this is a small, isolated, and well-monitored population at the extreme southern extent of the species range, lessons learned here may serve to inform conservation efforts in other population areas and National Park Service units.

Common loons historically were not a monitoring priority for YNP; with a small staff and budget, the YNP Bird Program did not have the resources to thoroughly assess the loon population on its own. Limited and inconsistent data collection has hampered analyses to understand population status, and while the status of these loons is likely stable, there have been indications of declines and territory vacancy in the recent past for unknown reasons. Until recently, it was also unknown how isolated the population was—a critical piece of knowledge now that loon monitoring is intensive, and vitally important should the population ever threaten to "blink out." Human disturbance during nesting and lake trout gill netting have emerged as management issues that were formerly underappreciated. Efforts to protect nests and coordinate netting locations are now routine. Finally, much of the success of the current program is due to the development of public-private partnerships as well as the regional working group. Although recent efforts and accomplishments have allowed us to determine the status of Yellowstone's breeding loon population, there is much to learn about the population dynamics and utilization of breeding and non-breeding habitat, as well as considerable gaps in our knowledge of these birds outside the breeding season. While the population shows a clear trend of growth over the past decade, the future of this population remains uncertain and at risk, due to its isolation, small size, and the threats to both reproduction and adult survivorship. It could be argued that the best option for these loons is to leave them alone, under the protection of Yellowstone and other agencies. However, humans have already interfered with their persistence in a variety of ways, and further human intervention is warranted to identify, understand, and correct these impacts. While it is likely that these loons will forever remain uncommon, continued collaborative efforts will help ensure they persist on the Yellowstone landscape.

Visit the Yellowstone's Birds website (https://press.princeton.edu/resources/yellowstones-birds-video-collection/v18) to watch an interview with Arcata Leavitt.

HARLEQUIN DUCKS

LUCAS SAVOY and SUSAN PATLA

One of the most beautiful birds in Yellowstone National Park (Yellowstone; YNP) is also one of the least well-known or studied wildlife species in the park (Fig. 19.1). The harlequin duck, by waterfowl standards, is a relatively small, stout, charismatic bird, most recognizable by the male's striking colorful and seemingly odd-patterned appearance. His plumage consists of an erratic array of slate gray, chestnut, blue, and bold white patterns. Fittingly, the origin of the term "harlequin" is in reference to the clown-like performers in Italian theater. In comparison, the females are drabber in appearance: gray-brown in color with white cheek patches. However, ask anyone who has had the privilege of observing harlequin ducks in their natural settings, and they will agree that the plumage of both the male and female seem perfectly designed to camouflage them in and out of the water. Harlequin ducks are among the family of waterfowl classified as sea ducks. As with most sea-duck species, the harlequin duck occupies interior freshwater habitats during the spring and summer breeding season, and then migrates back to the ocean, utilizing the near-shore marine ecosystem during the longer, non-breeding portion of the year.

N NORTH AMERICA, two separate harlequin duck populations occur, one on the Atlantic Coast and another on the Pacific Coast. The harlequin's breeding range is confined to the forested areas of the interior Pacific northwest United States and Canada, and the northern Atlantic Canadian provinces. For the Pacific Coast population, ducks that nest in Yellowstone represent some of the farthest-traveling breeding pairs.

During the breeding season, harlequins are secretive by nature, and paired individuals prefer the solitude of fast-flowing forested streams and rivers in remote mountainous terrain, in sites mostly secluded from human activity (Fig. 19.2). Prior to nesting, many of a harlequin's daily activities focus on foraging, often among the roughest sections of their breeding streams, diving for aquatic invertebrates that thrive among the rocks at the bottom of the streambed. Pristine waters are essential for supporting the abundant aquatic, protein-rich food sources required for breeding. Females especially need to be in optimal body condition for producing eggs. Adequate food resources are needed later in the summer for development of young as well.

While on the ocean during the non-breeding season, harlequin pairs become more communal and congregate in loose groups, numbering as few as tens and up to thousands of individuals. These wintering flocks utilize

HARLEQUIN DUCKS

▶ FIGURE 19.1 Male harlequin duck. Recent efforts in Wyoming, both in and out of Yellowstone National Park, have focused on radio-marking males and banding females to learn about migration patterns and other information valuable to protecting the population.
PHOTO BY RONAN DONOVAN

▼ FIGURE 19.2 Female harlequin duck. Compared to the male, females are marked drably to better conceal them throughout the summer while they raise young on their own.
PHOTO BY RONAN DONOVAN

near-shore, rocky coastlines, foraging in the tossing surf for submerged or intertidal invertebrates. They seek temporary refuge from the constant wave action by resting on exposed rocks. In some locations, harlequins are most visible during these non-breeding periods. Accessible ocean-view areas, which support harlequins, are popular locations for birders and photographers to observe this unique bird, which is otherwise cryptic and hard to locate during its breeding period in remote mountain streams. Although small, the harlequin duck is resilient and well-adapted for year-round residency in both salt and fresh turbulent waters.

The non-breeding wintering and molting locations for the western population of harlequins include the Pacific coastline and islands, extending from Oregon northward to Washington and Alaska, as well as British Columbia, Canada. The largest congregations of non-breeding harlequins are found in Alaska and British Columbia.

Their counterparts in the Atlantic tend to winter in small groups as far south as New Jersey, Rhode Island, and Massachusetts and in larger densities northward, in Maine and the Canadian Maritimes. The Pacific wintering population of harlequins is robust enough to allow a regulated hunting season managed through the Pacific Flyway Council, while fewer harlequins winter in the Atlantic and are not harvested.

Although the migratory harlequin duck is not constrained by state and provincial boundaries, the various respective wildlife agency managers have constructed their own differing management plans with the purpose of outlining conservation strategies to best protect the species within their jurisdiction. Ultimately, the overall population estimates of breeding and non-breeding harlequins by states and provinces dictate how they are managed. Alaska and British Columbia support the core population of breeding harlequins in western North America, where the combined estimated number of pairs may exceed 50,000, and these populations are considered secure. Much lower numbers of estimated pairs exist in Alberta (800–2,000 pairs), Washington (400 pairs), Oregon (72 pairs), Idaho (70 pairs), Montana (210 pairs), and Wyoming (70 pairs). In all these locations, harlequin ducks are designated "sensitive species" or "species of greatest conservation concern," and agencies have developed conservation plans to guide management efforts.

In Wyoming, the harlequin duck is one of the state's rarest breeding birds (the title of rarest belongs to the common loon; *see chapter 18*). As harlequins are specialists at remaining unnoticed and nest on remote breeding streams, inventorying this species is incredibly difficult and time-intensive. The current state-wide population estimate of 70 pairs is derived from a combination of anecdotal reports and formalized helicopter surveys initiated in 2002 and conducted every five years by the Wyoming Game and Fish Department in partnership with federal agencies. To detect and count breeding pairs in the pre-nesting season, when they are most visible, biologists survey remote streams and rivers from a low-flying helicopter, tracing segments of suitable habitat and recording harlequin observations. Field observations over the decades, plus aerial surveys, indicate that Yellowstone National Park supports a significant portion of Wyoming's breeding harlequin population, hosting upward of 16–24 pairs each year. Other important locations of breeding harlequins in Wyoming include Grand Teton National Park and the Bridger-Teton and Shoshone national forests. Harlequins have been reported occasionally from the Big Horn Mountains in Wyoming as well. Within Yellowstone National Park, the LeHardy Rapids viewing area is a consistent and accessible location to view harlequin pairs in early spring. Each year, upward of 10 or more pairs can be seen from mid-May to early June from the boardwalk as they slip in and out of the Yellowstone River rapids to feed and rest on the outcroppings of rocks. By the end of May, most harlequins will depart LeHardy Rapids, presumably to seek out smaller streams throughout Yellowstone, and perhaps beyond, to nest and raise young.

The female and male harlequins take on very different roles during the breeding period (Fig. 19.3). Male harlequins remain with the female in early spring on the streams and rivers to help select a nesting site, and to guard and mate with the female. However, once incubation is initiated, the males depart the streams and abandon their mates, migrating early back to the Pacific coast to undergo a feather molt (the process of shedding worn feathers and growing new). The female harlequin incubates and raises the young alone, escorting the ducklings in and out of the stream currents, feeding on insects and small aquatic invertebrates, and hiding from predators under low-hanging streambank vegetation. She and her young, which sometimes linger in their natal stream after the female adult departs, will undergo a migration of their own to the Pacific coast in late summer. No information existed until recently on the exact timing and routes taken by harlequins on their return to the West Coast.

In recent years, field studies of Yellowstone National Park and Wyoming's harlequin ducks have started to unlock some of the mysteries of the movements of these birds from their breeding areas to the Pacific coast. To

FIGURE 19.3 Harlequin ducks are difficult to survey but are associated with fast water and rapids, where males and females convene in spring to mate. After mating, males return to the wintering ground. PHOTO BY RONAN DONOVAN

best conserve Wyoming's harlequin ducks, biologists from the Wyoming Game and Fish Department, Biodiversity Research Institute, Grand Teton National Park, Yellowstone National Park, and Environment and Climate Change Canada teamed up to implement a harlequin duck-tracking study. This study focuses on identifying geographic linkages during the harlequin's migration, from the Greater Yellowstone Ecosystem and other interior nesting areas to their non-breeding areas on the Pacific coast. Funding for these studies was obtained through the Raynes Wildlife Fund, Wyoming Game and Fish Department, and Wyoming Community Foundation. Defining these geographic connections is important in identifying the harlequin's year-round habit requirements, which will provide wildlife managers crucial information for conserving this sensitive species. For example,

disturbances from increasing shipping activities or catastrophic events such as oil spills that may occur in the non-breeding coastal areas could have a profound effect on the small and sensitive breeding population within Yellowstone National Park and Wyoming.

During spring of the years 2016–2018, a team of biologists surveyed accessible backcountry streams and rivers to identify breeding pairs of harlequins in Yellowstone National Park and Grand Teton National Park. When pairs were located, the team, using well-tested techniques, safely captured the birds by flushing them downstream into a fine-mesh net (mist net) strung across the stream. Once in hand, the harlequins were weighed and measured, and blood and feather samples were collected for genetic and contaminant analyses. Each harlequin was uniquely tagged with a plastic, colored leg band containing a two-digit alphanumeric code, which allows the banded harlequins to be identified in future seasons from a distance, with the aid of binoculars or digital photography. Harlequins in Yellowstone and Wyoming are banded with an orange band with black lettering. To map the annual movements of harlequins, satellite transmitters were implanted in male harlequins by an experienced wildlife veterinarian who had performed over 1,000 previous implants on a variety of sea ducks. These transmitters were pre-programmed to provide varying levels of detail on the locations of the tagged harlequins for up to one year. The harlequin's locations, downloaded from satellites, are then digitally plotted and are available to the biologists in real time.

State-wide, over three seasons, a total of nine male harlequin ducks were tagged in Wyoming with a satellite transmitter, which included five harlequins from Yellowstone National Park. Biologists have cataloged interesting movements, from local movements among breeding streams and during migrations to non-breeding and wintering sites. One important discovery in Yellowstone was about movements prior to the nesting period; some harlequins frequently departed their streams at dusk to roost at night on large lakes, such as Yellowstone Lake. This overnight roost behavior is likely an attempt to seek out more expansive open water areas, which offer better protection from predators compared with the harlequin's narrow and confined, wooded breeding streams. Prior to nesting, harlequins in Yellowstone moved up and down their streams, traveling an average of 5-mile (8-km) sections of their breeding streams. Movement data also revealed that males typically depart breeding streams for the coast from June through early July. The initiation of male migration likely signals the approximate time the females begin incubating eggs, a process that will require the female to keep her eggs, generally a clutch size of five or six, warm and protected for approximately 28 days before they hatch. The duration of the male migration ranged greatly by individual. Some males executed a more direct migration route and arrived on the coastline in less than four days, while some birds stopped

more frequently and for longer time periods, resulting in a migration lasting upward of 38 days. During this migration, harlequins often utilized large river systems to stop and rest and tended to follow a route through the forested boreal habitats of northern Idaho, Washington, and British Columbia. Most Wyoming's harlequins traveled to the remote and rugged coastline of western Vancouver Island, British Columbia, while a few individuals utilized areas of Oregon and Puget Sound or the Salish Sea in Washington and British Columbia. These migration distances ranged from 466 to 1,118 miles (750 to 1,800 km). Three of the males tagged in Yellowstone National Park provided a complete year of tracking information, including their trip back to their breeding areas the following spring. Two of these males returned to the same breeding streams in the park they had used the previous year, while one male traveled to a new breeding area approximately 457 miles (735 km) away in British Columbia. Previous harlequin research has found that female harlequins generally return to their natal stream or area to breed and that the pairing process of male and females occurs on wintering areas. Therefore, it is likely that this male harlequin followed a new female to her breeding area, far from his previous breeding stream.

Although the migration tagging study has been completed, biologists will continue to catalog and interpret additional harlequin movement data and monitor breeding streams in Yellowstone to identify returning banded birds and to survey additional streams, looking for new pairs. This will allow biologists to refine future population estimates of the number of harlequin pairs that breed within the park and to document environmental changes that might affect future nest success. These combined efforts will help biologists and wildlife managers better understand their seasonal movements and year-round habitat requirements, and evaluate important aspects of Yellowstone and Wyoming's breeding population of harlequin ducks. The data are also being shared with a collaborative regional harlequin duck group, with other members conducting similar studies in Montana, Washington, Alberta, and British Columbia. When combined, these study findings will enhance our overall understanding on important regional linkages between western North America's harlequin breeding and non-breeding areas and will aid in future harlequin duck conservation. What we have learned to date provides us with a picture of the complexity and individual differences among harlequin ducks. We now know that the small population that lives in Yellowstone and Wyoming uses a variety of migration routes and wintering areas, giving us some confidence that harlequins will have the ability to adapt to future changes and continue to return to Yellowstone National Park for many future generations to enjoy.

HARLEQUIN DUCKS IN WYOMING—SUSAN PATLA

Harlequin ducks are one of the most difficult breeding bird species to monitor in Wyoming. Adults nest on remote mountain streams that are often inaccessible by foot or horse until July. By that time, males have returned to the Pacific Ocean, and drab-colored females with young are notoriously difficult to find. When my supervisor, Bob Oakleaf, head of Nongame for Wyoming Game and Fish, suggested that we use helicopter surveys, scheduled every five years, to count pairs in the pre-nesting period from mid-May to early June, I was skeptical.

I had two concerns. First, weather conditions during late spring can be highly erratic. Second, could we see pairs well enough to obtain accurate counts? Harlequins are masters at hiding. They can dive or scoot under vegetation quickly if disturbed. Would a low-flying helicopter spook the ducks?

To answer this second concern, we set up a control along two accessible streams that had been surveyed for a graduate study in the 1980s. If we could observe ducks there in expected numbers, we would have confidence in the method. Given our lack of options, it was worth a try. We had no data for many streams on national forest wilderness areas adjacent to and east of Yellowstone and Grand Teton national parks where nesting pairs were known to occur. Helicopter surveys could provide a way to map nesting distribution as well as obtain a state-wide population estimate.

Our first survey was in May 2002, with funding obtained through a new federal grant program for work on species of conservation concern. That first survey week did not start off well. Low clouds and snow obscured the high mountains. We had rented the helicopter for a week, and time was ticking. But the storm front finally moved on by day 3, opening blue sky.

That first year we surveyed over 340 miles (547 km) of stream segments. Bob, being the boss, sat in front next to the pilot with the big bubble view. I sat behind Bob looking for ducks that might resurface after the helicopter passed by. After five days, I could hardly move my head from constant twisting. But our success dampened my physical pains. We observed 63 harlequins, including 28 nesting pairs, 21 in locations not previously documented, greatly increasing our knowledge of the distribution and number of ducks in the state.

Since that first year, we have repeated these surveys four times. The harlequin breeding population in Wyoming appears stable. Low counts in drought years, we discovered, resulted from pairs delaying moves into breeding streams, as they waited in deeper-water areas for water levels to rise, likely to avoid predation by eagles and peregrines. Of note, we failed to find nesting pairs on many streams where habitat looked excellent. There is much yet to learn about harlequins, but the baseline data we collected will provide a yardstick to assess future changes in population number and distribution in the face of climate and habitat changes. And these surveys proved to be the most amazing aerial surveys of my career.

THE WATERFOWL OF YELLOWSTONE NATIONAL PARK

20

CARL D. MITCHELL

September 24th [1869] we arrived at Yellowstone Lake . . . the shallow water in its coves affords feeding ground for thousands of wild ducks, geese, . . . and swans.

N. P. LANGFORD 1905

ORTY SPECIES OF ducks, geese, and swans have been reported in Yellowstone National Park (Yellowstone; YNP) between 1869 and 2020. This chapter is about those waterfowl. Waterfowl are a fascinating group of birds widely distributed in the park and readily observed by visitors. Waterfowl are prized by birders, photographers, hunters, conservationists, and biologists. Ducks, especially breeding males in "nuptial plumage," often have colorful feathers, and all waterfowl species exhibit interesting behaviors. Each species is unique ecologically; they vary from one another in physical, physiological, and behavioral adaptations necessary for life in and around aquatic environments (Owen and Black 1990).

This chapter provides a brief introductory, non-technical overview of Yellowstone's waterfowl community, species ecology, habitat use, and interactions with other species. Detailed treatments are impossible because relatively little is known about most waterfowl in Yellowstone. In fact, comparatively little is known about waterfowl ecology in high-elevation habitats anywhere in North America. Most of what is known about waterfowl in these environments has been learned by studying the birds' large, intermountain basin wetland complexes that are very unlike Yellowstone's habitats. Nomenclature in this chapter follows Chesser et al. (2020).

WETLANDS AND OTHER ENVIRONMENTAL FEATURES

To understand the park's waterfowl, one has to understand wetland ecology. Wetlands are more than just surface water on the landscape. Wetlands are dynamic places that express geologic history, climate, weather, topography, hydrology, and soil. Yellowstone has about 228,766 acres (92,578 ha) of lakes, rivers, and marshes ranging in size from ponds of less than 0.1 acre (0.04 ha) to Yellowstone Lake at 215,080 acres (87,040 ha; Elliott and Hektner 2000). These wetlands are distributed across a diverse geological, edaphic (influenced by the soil), and topographic landscape. Geology ultimately determines the structure and function of soils, including physical, chemical, and mechanical properties. These properties in turn affect soil water-holding capacity, erosive capability, soil microbiota, and nutrient transfer to plants. Site-specific soils (e.g., loam, clay, sand, silt, gravel) and other characteristics (e.g., chemistry, nutrient cycling, productivity) and hydrology, along with elevation, precipitation, aspect, and slope, also determine which plant communities can occur on any given site (Weller 1981, Windell et al. 1986, Ringelman 1992). Topography and soil characteristics also determine the size, shape, depth, and duration of wetlands. These abiotic factors combine with

FIGURE 20.1 Male and female Barrow's goldeneyes in early spring on Yellowstone Lake, with the Red Mountains in the background. PHOTO BY LISA CULPEPPER

biotic features, like soil microflora and fauna, aquatic plants and invertebrates, and other animals that use wetlands, to create a suite of wetland habitats in the park that supports a diverse waterfowl community and wonderful opportunity for watching and studying wild waterfowl throughout their annual cycle.

The mean park elevation is about 8,000 feet (2,438 m) above sea level, and the park is therefore subject to heavy winter precipitation, low temperatures, and short growing seasons (Fig. 20.1). All these physical characteristics impose conditions on terrestrial and aquatic vegetation and the animals that depend on that vegetation. Long, cold winters mean thick ice forms on many waters, except where water is kept open by hydrothermal features, and result in prolonged cold and dark conditions underwater. Winter conditions influence water chemistry, ecological processes, the structure of wetland plant communities, and plant and animal phenology and productivity. All these factors affect how waterfowl use individual wetlands.

Palustrine wetlands (e.g., wet meadows, marshes, beaver ponds, and other small ponds), lacustrine wetlands (e.g., large or deep ponds and lakes), and riverine systems (e.g., streams and rivers) all can be found in the park (Elliott and Hektner 2000). These include a wide variety of ephemeral, temporary, seasonal, semi-permanent and permanent wetlands, thermal pools and springs, streams, rivers, ponds, and lakes of different sizes. Even roadside borrow ditches often are temporarily productive wetlands and may attract waterfowl. This array of wetland types also provides habitat for a multitude of different grasses, forbs (non-woody flowering plants other than grasses, sedges, and rushes; e.g., wildflowers), shrubs and trees, submerged and emergent aquatic plants, invertebrates, amphibians, fish, reptiles, birds, and mammals, many of which are dependent upon wetlands for all or most of their life histories (Weller 1981, Windell et al. 1986, Elliott and Hektner 2000). The result is that Yellowstone has a broad community of ducks, geese, and swans whose composition changes with the changing seasons.

Weller (1981) characterized most intermountain wetlands as "low production" areas for waterfowl, which may be true in relation to some of North America's most productive wetland areas, such as the prairie pothole region of the mid-continent. Nevertheless, high-elevation areas do produce significant numbers of waterfowl. Breeding puddle ducks and Canada geese are common in the park. Historically, Yellowstone has been (and hopefully will be again) a vital place for breeding and wintering trumpeter swans (*see chapter 17*) and for harlequin duck reproduction (*see chapter 19*).

Yellowstone also provides substantial habitat for migratory waterfowl. Skinner (1925) noted of Yellowstone Lake: "in the fall vast numbers of ducks, geese, and swans stop on their way south." Migratory species also benefit substantially from Yellowstone's size, protected status, and wetland diversity. Exactly which waterfowl species, in what numbers, and in which habitats are unknown.

One important trait especially germane to Yellowstone is that many waterfowl are "facultative" migrants. If adequate food, water, and shelter are available, they do not have to migrate. Because warm hydrothermal water flows throughout much of the park, Yellowstone provides a lot of productive open-water winter habitat for waterfowl (Skinner 1925, 1928a), despite the park's high elevation, bitter cold winters, and heavy snowfall. Rivers with open water in winter helped non-migratory trumpeter swans survive near extinction in the past (Banko 1960) and continue to support wintering swans, and other waterfowl, today.

THE WATERFOWL COMMUNITY—PRESENCE AND STATUS

Forty-eight species of ducks, geese, and swans occur in North America (Baldasarre 2014). Of these, about 27 species regularly can be found in the park during at least part of the year (Table 20.1). Most records of waterfowl abundance in Yellowstone are subjective rather than quantitative. There may be little difference between reports that say a species was "common" versus "abundant," for example. Despite the subjective nature of those reports, it is remarkable that there has been so little obvious change in species composition or relative abundance over about 150 years (Table 20.1). The facts that many early accounts were not primarily concerned with documenting wildlife, and that waterfowl tended to be less noteworthy to early explorers than forage for horses, Indigenous peoples, or large mammals (e.g., grizzly bears, elk, bison) probably have much to do with that. Even later commentators probably were more interested in recording sightings of large, charismatic fauna than in waterfowl.

There is documentation of 40 waterfowl species occurring in what is now Yellowstone National Park between 1869 and 2020, including two species of swan, four of geese, and 34 of ducks (Table 20.1). Of these, one swan,

TABLE 20.1 WATERFOWL (DUCKS, GEESE, AND SWANS) SPECIES PRESENCE AND RELATIVE STATUS IN YELLOWSTONE NATIONAL PARK, 1869–2020

SPECIES	1870–1880[A]	1902[B]	1915–1916[C]	1920–1930[D]	1930–1940[E]	1952[F]	1985[G]	1998[H]	2000–2010[I]	2020–2021[J]
Trumpeter swan	P	P	*	R*, Y	R*	R*	O*	C*	C*	U*
Tundra swan				RW		CM	OM	U	U	U
Canada goose	P	C*	A*	C*, Y	C	C*, Y	C*, Y	A*	A*	A*
Snow goose		X		P	RM	RM	R	R	R	RM
Ross's goose										RM
Greater white-fronted goose								X		
Wood duck	P	R, Y	R*	R	R	R	R			
Gadwall		C	C*	P	C*	C*, M	CM	C*	C*	C*
American wigeon		C		P	C*	C*, M	CM	C*	C*	C*
American black duck										X-
Mallard	C*	C*, W	PY	C*	A*, Y	A*, Y	A*	A*, PW	A*	
Blue-winged teal			P	C*	C	OM	U*	U*	U*	
Cinnamon teal	P	C*	C*	OM	C*	C*	C*			
Northern shoveler			P	O*	O*	O	U	U*	C*	

NOTE: Species status is indicated as A = abundant, C = common, O = occasional, P = present, Q = questionable record, R = rare, U = uncommon, and X = accidental. Season is indicated using * = breeding, W = winter, M = migrant, and Y = year-round resident.

one goose, and 18 ducks breed in the park, and most are "common"; three are considered particularly important breeders (trumpeter swans, harlequin ducks, and ruddy ducks) because the park is one of the few areas in the region where they breed. Common species are Canada goose, mallard, gadwall, American wigeon, cinnamon teal, northern shoveler, northern pintail, green-winged teal, ring-necked duck, lesser scaup, bufflehead, Barrow's goldeneye, common merganser, and ruddy duck (Table 20.1). These species are all common breeders in montane wetlands (Ringelman 1992). Most are also adaptable "generalist" species, capable of using various habitats across wide geographic areas.

Canada geese and mallards have always been common in the park but in recent years have become very abundant. Interestingly, only three other species have significantly changed their perceived status over the last 150 years. Northern shovelers were considered "occasional," "rare," or "uncommon" but have become "common breeders." Ring-necked ducks were considered "rare migrants" but now are a common breeding species (Table 20.1). The extent to which these are genuine biological

SPECIES	1870–1880[A]	1902[B]	1915–1916[C]	1920–1930[D]	1930–1940[E]	1952[F]	1985[G]	1998[H]	2000–2010[I]	2020–2021[J]
Northern pintail		C		P	O*	C*	CM	U*	U*	C*
Green-winged teal		C	C*	PY	*	C*	C*	C*, PW	C*	
Canvasback	C		P	OM	OM	RM	U*	U*	U*	
Redhead		C		P	OM	OM	RM	U*	U*	U*
Ring-necked duck			RM	R	O	C*	C*	C*		
Greater scaup		P	OM	C*			R			
Lesser scaup		P	O*, M	C*	C*	C*	C*, PW	C*		
Surf scoter		P					X			
White-winged scoter	X		P					X		
Black scoter			P	M	X					
Long-tailed duck				RW	R				X	
Harlequin duck			P	R*	R*	R*	R*	R*	U	
Bufflehead			P	CW	R*	O	U*	U*	C*	
Common goldeneye			W	PW	CW		W	U	U	CW

a Langford 1905, Comstock 1874; b Knight 1902, Mearns 1902; c Bailey 1915, 1916; d Skinner 1921, 1925, 1928a; Bailey 1926; e Bailey 1930, Fuller and Bole 1930, Kemsies 1930, Wright 1934, Kemsies 1935, Crook 1936; f Brodrick 1952; g Follett 1985; h Johnsgard 2009, 2013; i National Audubon Society 2002, 2005; Johnsgard 2009, 2013; j YNP Bird Program 2020, 2021

trends versus simply more careful observation is difficult to say, but it is suggestive that the park's waterfowl community has shifted slightly over time.

Several species, like wood ducks, are considered "rare." Species like trumpeter swans and harlequin ducks are uncommon, but regularly breed in Yellowstone and predictably can be observed by visitors. Other species are migrants that occasionally are seen during seasonal movements between breeding and wintering areas. These migrants include the tundra swan, snow goose, Ross's goose, canvasbacks, greater scaup, and hooded mergansers.

Waterfowl are highly mobile, and occurrence of several species is considered accidental, including the whooper swan, greater white-fronted goose, surf scoter, and white-winged scoter (Yellowstone National Park Bird Program 2020). McEneaney (2004) documented whooper swan presence one winter in Yellowstone. Although Skinner (1925) mentions seeing black ducks in 1922, he questioned whether the records were misidentified mallards. However, Crook (1936) positively identified black ducks in 1932. Curiously, although a small flock of captive, but free-flying, non-native mute swans bred on a private ranch only

30 miles (50 km) north of the park for many years (Reiswig 1986, Lenard et al. 2003), there is no record of this species being observed in Yellowstone, even as "accidental."

GENERAL LIFE HISTORY

It is important to emphasize that ducks are not all the same. There are many similarities between species, of course, but many are superficial. Different species have different plumage, courtship, reproductive, parental, migratory, and wintering behaviors, as well as life-history strategies, body size and physical structures, diets, gut morphologies, physiologies, habitat requirements at different annual periods (courtship, seasonal feeding, nesting, brood rearing, molting, etc.), and so on that differ (Johnsgard 1975, Owen and Black 1983, Baldasarre 2014). Ducks, geese, and swans have all evolved a range of morphological, ecological, and behavioral characteristics that are related to their life histories. Understanding a group of birds with so much biological and behavioral variation requires species-specific information to fully comprehend their behavior and biology. While many species seem to have very similar habits, slight differences in the timing of migration or nesting, nesting or foraging habitat, fine-scale food habits, and summer molting allow many species to simultaneously occupy the same wetlands.

Similarities include the obvious fact that all waterfowl require water (wetlands) for most of their life cycle. Many female waterfowl exhibit "natal philopatry"; that is, for nesting they return to the area near where they were hatched. Newly hatched waterfowl are cared for by one or both parents, and usually feed on aquatic invertebrates for one to several weeks before switching to adult diets. If they avoid what is often high mortality at young ages, most species of waterfowl have the potential to live long lives, 10–20 years or more. The fact that they have multiple feather molts renders their appearance very different between seasons, including during post-breeding alternate plumages (Johnsgard 1975, Baldasarre 2014). And cygnets, goslings, and ducklings change appearance over the summer as they grow, lose natal down, and begin to grow complex plumage.

Differences include their physical structures, including beak size and structure, always related to differing food

FIGURE 20.2 A common breeding duck in Yellowstone, a male American wigeon vocalizes in a light rain. PHOTO BY SCOTT HEPPEL

habits. For example, American wigeon feed on grasses and have relatively short, stubby beaks designed to pluck grass stems and seed heads (Fig. 20.2), while mergansers, who feed on small fish or large aquatic invertebrates, have long, thin bills with serrations to help grasp and hold slippery prey. Species with more general food habits, such as mallards, have bills suitable for collecting a variety of vegetable and insect foods. "Dabbling" ducks tend to use shallow waters for feeding and breeding, and have their legs placed farther forward on their bodies. In contrast, "diving" ducks tend to forage in deeper waters and have legs set farther back, more suitable for diving. Males may pair, or at least begin to form annual pair bonds, with females on wintering grounds, and will follow females back to the female's natal sites. In contrast, trumpeter swans may take years to develop what becomes a lasting pair bond. The adult bond may end before the young can fly (most ducks) or last for the life span of both partners (e.g., trumpeter swans). Some ducks nest in short upland cover (e.g., northern pintails) or dense vegetation (mallards), and some only over-water (e.g., canvasbacks). Brood care may be very short (e.g., ruddy ducks) or prolonged (e.g., Canada geese, trumpeter swans).

Body size has important effects on waterfowl life history (Baldasarre 2014). There are exceptions, but generally smaller species (e.g., teal) tend to arrive later and leave earlier on breeding grounds than larger species. Smaller species also have a different suite of potential predators, and smaller territories. They also usually have shorter breeding seasons (from arrival, nesting, incubation, fledging to fall departure). Smaller species also tend to have smaller eggs and clutch volumes, but these also represent a larger percentage of hen body size. This generally holds true for the entire range in size from small teal through medium-sized pintails, gadwalls, and mallards, to large Canada geese and even larger trumpeter swans.

As noted above, waterfowl may have different adult food habits, but generally most species (except mergansers) exhibit some trends. Females must eat enough protein and calcium-rich foods to improve their condition to lay eggs. Some individuals gain condition before arriving on breeding grounds ("capital breeders"), some improve condition after arrival at nesting sites ("income breeders"), and some do both, depending on the individual birds and their wintering and migratory habitat quality. The better condition females are in, the earlier they nest, the more eggs they lay, and the larger those eggs are (Owen and Black 1990). Larger eggs result in larger, healthier young, which in turn usually lead to higher survival rates. For species with short-enough breeding cycles, this also offers opportunities to renest if the first attempt is unsuccessful. Larger species like Canada geese and trumpeter swans have different breeding cycles (early nesting, smaller clutches of very large eggs, longer incubation, prolonged brood care, and time to fledge) and cannot renest.

For readers who want more detailed information on waterfowl, there are many excellent references on their biology, behavior, ecology, and management. These include Baldasarre (2014), Baldasarre and Bolen (2006), Kear (2005), Johnsgard (1968, 1975), Owen and Black (1983), Palmer (1976a, 1976b), and Birds of the World (https://birdsoftheworld.org) species accounts. There are literally thousands of books and technical papers on waterfowl in general, on specific waterfowl species, and on various aspects of their biology, ecology, and behavior. Trumpeter swans and harlequin ducks have their own chapters in this book (*see chapters 17 and 19*).

WATERFOWL ECOLOGY

The waterfowl in Yellowstone are diverse; their adaptations to different habitats and the enormous number and diversity of wetlands (Elliott and Hektner 2000) allow multiple species to coexist. However, studies on waterfowl in high-elevation habitats elsewhere in the Greater Yellowstone Ecosystem suggest that ducks, geese, and swans breeding in Yellowstone also have adaptations to local conditions. These conditions include cooler mean temperatures, highly variable local precipitation, short growing seasons, and deep winter snowpack. As alluded to above, all these characteristics affect local vegetation

THE WATERFOWL OF YELLOWSTONE NATIONAL PARK

types and structure, and micro-climates. Ponds in a state of "drawdown"—for example, when beaver dams breach during spring floods—do not recycle substrate nutrients in colder, wetter sites as fast as they would at warmer, lower-elevation sites (W. J. Kurtenbach, pers. comm.). Some specific waterfowl adaptations to high-elevation nesting sites that we have observed include slightly later nest-initiation dates, lower nesting densities, more nesting in vegetation that holds up to heavy snows, such as woody shrubs or in over-water sites that melt off sooner (e.g., emergent vegetation, muskrat houses), nesting in sites with more favorable micro-climates, and limited renesting (C. D. Mitchell, unpublished data). Ironically, climate change may improve some conditions for some breeding waterfowl in Yellowstone, although other possible changes may be harmful. These factors require additional attention.

Waterfowl also provide a variety of ecological linkages to areas outside of Yellowstone. They ensure a continuous genetic mix by recruiting breeding birds produced outside the park. Waterfowl produced outside of Yellowstone may also bolster local populations that may not produce enough young to maintain stable park populations. But migratory waterfowl may also import parasites,

FIGURE 20.3 (A) Bobcat sneaking up on mallards. (B) Coyote watching ducks in a winter marsh. (C) Coyote with dead mallard. PHOTOS BY LISA CULPEPPER

diseases, or pollutants (e.g., lead obtained by foraging in lead-contaminated sites outside the park). Many connections may be important to Yellowstone's waterfowl community but are largely speculative at present.

There are many important ecological interactions between Yellowstone waterfowl and many other plant and animal species. Notably, other chapters in this volume have referred to potential interactions between lake trout, cutthroat trout, Canada geese, bald eagles, and trumpeter swans (*see chapters 12 and 17*). Undoubtedly, waterfowl provide some prey for a variety of other birds, such as common ravens, California gulls, and raptors, especially golden eagles and bald eagles. Other park carnivores—including long-tailed weasels, mink, red fox, coyotes, wolves, bobcats, and possibly bears—undoubtedly benefit from having waterfowl eggs, young, and adults all available as potential prey (Fig. 20.3). This may be especially important in years with low vole or Uinta ground squirrel populations. Several studies have found waterfowl serve as alternate prey when vole populations are low (e.g., Brook et al. 2005, 2008). Conversely, piscivorous ducks like common mergansers might have local impacts on some fish populations in the park. Abundant Canada geese may compete for forage with grazing bison or elk in some areas.

On the other hand, grazers like bison may facilitate goose population growth by creating more favorable foraging conditions. Elk, bison, and other large ungulate grazing affects grassland structure, function, and plant productivity (Frank et al. 1998), and it is likely that large flocks of Canada geese do the same (Owen and Black 1990). It would be useful to know how bison and Canada goose grazing impacts plant communities, growth, and nutrition. Goose grazing could also affect populations of insects, other birds, or small mammals by altering grassland structure or productivity. Local grazing conditions might also affect Canada goose population growth.

Finally, beavers—and, to a lesser extent, muskrats—function as "ecosystem engineers." Beaver dams and resulting wetland and wet meadow habitats play an enormously important role in creating and maintaining waterfowl nesting, brood rearing, and foraging habitats (Ringelman 1991). Beaver ponds may even facilitate early breeding by Canada geese (Bromley and Hood 2013). Beaver ponds and dams affect local hydrology and wetland ecology and function (i.e., hydrology, surface flow, nutrient transport); they modify upland habitats adjacent to wetlands, and provide habitat for invertebrates, amphibians, reptiles, and fish, as well as waterfowl. Muskrats modify emergent plant structure, consume the same aquatic plants as waterfowl, and build houses that often serve as nest platforms.

WATERFOWL ISSUES

Few details are known about the ecology of most waterfowl while they are within Yellowstone National Park. The obvious exceptions are trumpeter swans (*see chapter 17*), Canada geese (Skinner 1925, 1928b), harlequin ducks (*see chapter 19*), and Barrow's goldeneye (Skinner 1925, 1937; Sawyer 1928). Even in these cases, information is limited, and much of it is dated. More complete and current data on all aspects of individual waterfowl species distribution, abundance, and ecology within Yellowstone is highly desirable.

Canada geese, in particular, pose several interesting and important questions. Population growth is apparent (*see above*; Fig. 20.4). Canada geese populations have increased almost everywhere within their range in recent years (Mowbray et al. 2020). More rigorous surveys that accurately and precisely count them, or at least provide reliable indices of population trends, are needed. It is extremely important to quantify any community impacts of geese increases, such as grazing consumption and nutrient enrichment, competition with large ungulates in or near wetland habitats, and to what extent Canada geese interact with other waterfowl species, as well as the predator-prey base. It would also be interesting to learn if there are any connections between park flocks and those outside Yellowstone in terms of Canada geese genetic structure or population growth.

 FIGURE 20.4 Canada geese have increased dramatically across Yellowstone, occupying nearly every wetland in the park and possibly impacting ecological relationships with other waterfowl. Two Canada goose families are depicted here. PHOTO BY LISA CULPEPPER

 FIGURE 20.5 A flock of Barrow's goldeneyes. PHOTO BY RONAN DONOVAN

THE FUTURE

Ideally, all native waterfowl species will persist in Yellowstone. Hopefully, some, like trumpeter swans, will increase their current abundance and distribution. In the absence of catastrophic change, waterfowl should continue to maintain their many ecological roles and provide enjoyment for visitors. With luck, natural historians and biologists will add to the knowledge of waterfowl and their ecology in the world's first national park. In turn, this newly acquired knowledge and understanding should be transmitted to visitors, park staff, and other waterfowl ecologists, so that everyone can gain a new appreciation for the ducks, geese, and swans of Yellowstone National Park (Fig. 20.5).

THE COLONIAL WATERBIRDS OF THE MOLLY ISLANDS

21

LAUREN E. WALKER and DOUGLAS W. SMITH

On the water the pelican is grace personified. With head bent back and close to his shoulders, and with his deep pouch tucked away between chin and throat, he moves majestically along like a ship under full sail.

M. P. SKINNER (1917)

INDIVIDUALLY KNOWN AS Sandy and Rocky, the Molly Islands are two nondescript outcrops of land in the southern end of Yellowstone Lake's Southeast Arm. Combined, the two islands measure less than 1.5 acres (0.6 ha; Fig. 21.1). Despite their tiny size, however, the Molly Islands meet many of the requirements for a successful waterbird breeding colony. They are remote and isolated, located far from the hubbub of Yellowstone National Park's (Yellowstone; YNP) highly visited front-country areas, and relatively difficult to reach for many of Yellowstone's terrestrial predators (bears, coyotes, etc.). They are surrounded by the fish-filled waters of Yellowstone Lake (more on this later), providing easy access to important food resources for nesting adults as well as hungry juveniles learning how to forage for themselves (Fig. 21.2). Finally, with little elevational gradient, they are

FIGURE 21.1 The Molly Islands (Rocky Island on the left and Sandy Island on the right) in the Southeast Arm of Yellowstone Lake during an aerial survey. NPS PHOTO

THE COLONIAL WATERBIRDS
OF THE MOLLY ISLANDS

low-lying and allow young flightless birds to easily reach the water when they are ready for their first swim. Accordingly, the Molly Islands comprise the sole breeding site for a suite of colonial waterbird species in YNP. For at least 150 years, American white pelicans, double-crested cormorants, California gulls, and Caspian terns arrived on the Molly Islands late each spring, shortly after the lake ice melted, to nest and raise their young (Fig. 21.3). Although the birds followed evolutionary and ecological precedent in choosing these islands to breed, their tenure on the islands has not been without perils, including actively antagonistic management practices, recreational disturbance, ecosystem change, and, today, volatile lake levels that threaten to wash away the colony altogether.

In the early 1900s, the National Park Service actively tried to exterminate the colony, blaming pelicans first for eating too many fish and then for harboring fish-killing parasites (Pritchard 1999). Between 1924 and 1931, hundreds of pelican eggs were destroyed and young pelicans killed in the name of protecting native fish populations and, ultimately, angling opportunities for park visitors (Fig. 21.4; Pritchard 1999). During this same period, however, the focus of the National Park Service began to shift and broaden, away from a management strategy that prioritized visitor opportunities to one that considered educational and research opportunities, as well as wildlife and its role in the ecosystem. In 1929, newly appointed National Park Service Director Horace Albright created

FIGURE 21.2 American white pelicans and a California gull squabble over a fish.
PHOTO BY SCOTT HEPPEL

FIGURE 21.3 American white pelicans and double-crested cormorants nesting on Rocky (A) and Sandy (B) islands. NPS PHOTOS

FIGURE 21.4 (A) Historically, the Molly Islands were not protected, as park staff felt American white pelicans ate too many cutthroat trout and spread disease. Human visits to the islands were common during nesting season, and eggs were sometimes destroyed. (B) California gulls respond to human visitors on the Molly Islands. Today, both islands are closed to boating and entry within a half-mile during the nesting season. NPS PHOTO/YELLOWSTONE ARCHIVES

the Branch of Research and Education, and in 1933, George Melendez Wright became chief of the newly created Wildlife Division. With these developments, and the public pressure from a growing conservation movement, the National Park Service stopped the culling of pelicans in Yellowstone in the early 1930s (Pritchard 1999).

Although they had ceased to be actively persecuted, the colonial birds on the Molly Islands continued to be plagued by human visitation to the islands and by nearby motorized boat traffic; both activities may disrupt birds during the sensitive nesting period in numerous ways. Disturbances such as these might flush incubating birds off eggs or away from young nestlings, leaving them exposed to the elements and to predation and, in the case of nestlings, without access to food. At a broader scale, human activities were impacting bird populations across the continent with pesticides such as DDT (dichloro-diphenyl-trichloroethane). Hoping to turn around descending

population trends, the park reduced recreational disturbance by limiting motorized boat activity on the lake and restricting visitation to the islands, and in 1972, the use of DDT was banned nationwide. By the early 1980s, however, park staff were noticing further declines in the Molly Islands' waterbird population, and today the colony continues to battle diverse and substantial challenges, from changing trophic (food-web) dynamics to climate change.

The primary food source for fish-eating birds on Yellowstone Lake has historically been the Yellowstone cutthroat trout. However, since the introduction of the non-native and invasive lake trout into the lake in the 1980s, populations of cutthroat trout have declined dramatically, both outcompeted and directly preyed upon by their larger cousins (Kaeding et al. 1996, Koel et al. 2005, Munro et al. 2005). Unfortunately, lake trout are not a simple substitute for cutthroat trout in the diet of most fish-eating birds on Yellowstone Lake. Unlike the native cutthroat, lake trout spend most of their time in relatively deep waters (Koel et al. 2005), out of reach of many avian piscivores. Furthermore, lake trout grow significantly larger than cutthroat, quickly becoming too large for consumption by even pelicans. While the food limitations impact the waterbird populations directly, the change in food availability has also forced a dietary shift in other more adaptable predators, namely bald eagles (Koel et al. 2019b; *see chapter 12*). Today, bald eagles on Yellowstone Lake are routinely spotted on the Molly Islands during the waterbird breeding season, waiting for vulnerable nestlings to wander away from their protective crèche.

In addition to changing predator regimes and prey availability, dramatic fluctuations in water levels on Yellowstone Lake, both within and between nesting seasons, may threaten the waterbird colonies on the Molly Islands (Fig. 21.5). Early in the breeding season, water-level rises may wash away nests or drown hatchlings. Later in the summer, flooding may corral older and more mobile chicks into relatively small areas, making them easy targets for aerial predators like eagles. Since the earliest records in the 1920s, the ice-off date on Yellowstone Lake has become more variable, the maximum elevation of the lake is higher (the peak of spring runoff), and the date of peak lake elevation is earlier (YNP, unpublished data). For birds that time their nesting according to instinct and generations of natural selection, these environmental changes make it difficult to effectively balance minimizing the risk of nest flooding with maximizing survival of their young.

To help understand these changes in water regimes and predators, Milton Skinner's book *The Birds of Yellowstone National Park* (1925) allows a useful historical comparison. Skinner wrote extensively about the Molly Islands and especially about the pelicans. Skinner first visited the islands in 1898 and noted stable numbers of pelicans through the 1920s, with an average of 250 breeding pairs and 150 young. Broad reproductive failures did occur, during "those extraordinary seasons when unusually high water caused the waves and cold spray to dash across their breeding ground" (Skinner 1925). Today, however, Skinner's "extraordinary seasons" are increasingly common.

FIGURE 21.5 Effects of lake-level fluctuations on the exposed area of Sandy Island, based on geo-referenced aerial photographs. An increase in lake level of 1.6 feet (0.5 m) can result in a loss of approximately 29,000 square feet (2,700 m²) of land area on Sandy Island.

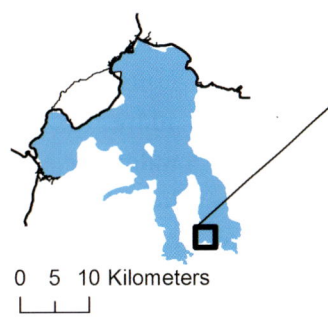

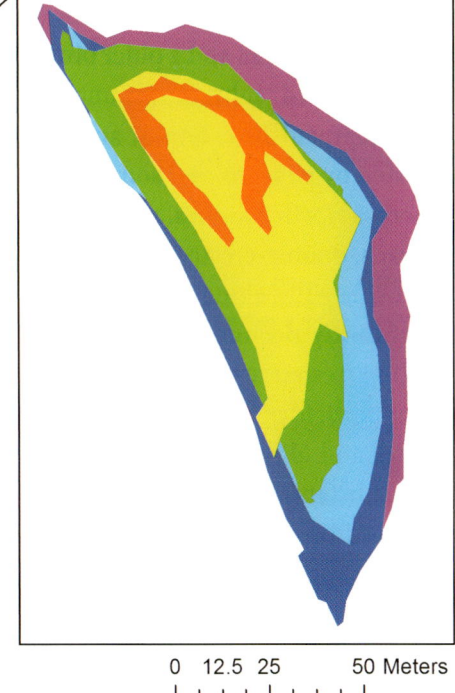

Color	Date	Lake Elevation (m)	Island Area (m²)
Pink	08/15/10	0.72	7006
Dark Blue	07/17/13	1.11	6011
Light Blue	05/28/17	1.24*	4739
Green	06/11/12	1.44	4076
Yellow	07/14/17	1.61*	2715
Red	06/28/11	1.93	567.4

* Lake elevation is estimated.

THE COLONIAL WATERBIRDS
OF THE MOLLY ISLANDS

Visits to the Molly Islands by Skinner in the 1920s and by the Bird Program in recent years provide additional insights into waterbird populations. Unlike the Bird Program's recent visits, Skinner often visited the islands while the birds were on nest. One such visit was on July 5, 1922, when he wrote about pelicans, California gulls, and Caspian terns, and mentioned not being able to find their nests. He later noted, however, finding young terns "too small to fly" on the islands, suggesting he made multiple visits during the breeding season. One photograph (Fig. 45 in Skinner 1925, but with no date) shows a large number of California gulls, which can no longer be found in such numbers. In 2022, coauthor Smith (observing from a campsite on shore, just north of the islands) observed a bald eagle perched on Rocky Island in the evening. The eagle was seen on Rocky again during a visit to the islands the next day. Both the response (nervousness, grouping together, and slowly moving away) by a flock of about 150 pelicans to this eagle during the evening observation and the numerous pelican carcasses found on the islands the following day were suggestive of previous predation. During a late-summer ground survey in 2019, biologists also observed signs of predation throughout the colony that were attributed to bald eagles, including 19 pelican carcasses (17 chicks and 2 older birds). During both visits, we also noted substantial amounts of fishing tackle (monofilament line, hooks, and spoons) that had been ingested by pelicans and somehow defecated out.

Across North America, populations of American white pelicans, double-crested cormorants, Caspian terns, and California gulls, along with numerous other bird species, significantly declined in the mid-20th century due to hunting, resource and habitat loss, and the widespread use of pesticides. The historic pattern in Yellowstone was similar and although populations continent-wide have since substantially recovered, today the status of colonial waterbird breeding on the Molly Islands remains tenuous. Historically, terns were always in relatively low numbers on the islands (Diem and Condon 1967). Although they are still seen on the lake each summer, they have not been observed nesting on the Molly Islands since 2005

(Fig. 21.6). California gulls, which numbered in the thousands in the 1940s, were reduced to between 30 and 40 individuals by 1966 (Diem and Condon 1967). Today, gulls are present in low numbers every summer, but researchers haven't detected any nesting attempts since 2016 (Fig. 21.7). While both cormorants and pelicans are still present each year and attempt to nest, pelican productivity (young fledged per nesting attempt) has declined substantially since the late 1970s, and the number of cormorant nesting attempts has declined steadily over the past 26 years, from 125 in 1994 to just 16 in 2019 (Fig. 21.7). Thus, the waterbird colony on the Molly Islands is now a priority for conservation and restoration.

Yellowstone continues to regulate the volume and speed of boat traffic in the Southeast Arm, and visits to the islands during the waterbird breeding season are forbidden, guidelines designed to minimize disturbance to the nesting birds. Furthermore, ongoing efforts by the Yellowstone Fisheries Program are gradually reducing the lake trout population in the Lake (Koel et al. 2019a). Biologists hope these efforts will subsequently allow native cutthroat trout populations to rebound and ultimately restore Yellowstone Lake's trophic web, including historically available foraging opportunities for eagles and pelicans alike. Although these ongoing efforts do not directly address threats to the waterbird colony from fluctuating water levels and climate change, Bird Program biologists hope they will help increase the success of the waterbird colony in the years with limited flooding. If the colony is more productive during the "good" years, this may perhaps buffer the population against the "bad," providing sufficient numbers of young to sustain the colony through years with high floods and low reproduction. Of course, even if that strategy works, it only works for the species that are still nesting on the Molly Islands, namely pelicans and cormorants. California gulls are still seen on the islands each summer and may be enticed to breed again. Further intervention, however, would be necessary to truly restore the colony to its former glory, with hundreds of pelicans, cormorants, gulls, and Caspian terns sharing two tiny islands on a vast lake.

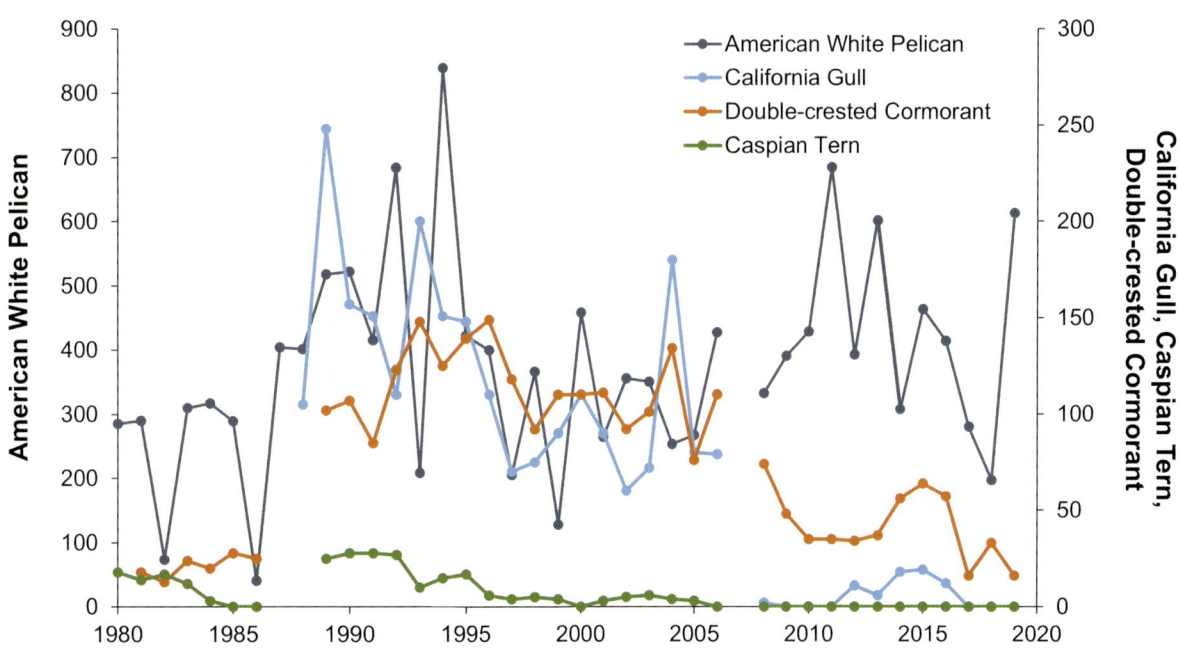

FIGURE 21.6 A Caspian tern skims the water surface. PHOTO BY SCOTT HEPPEL

FIGURE 21.7 Observed nest attempts by four colonially nesting waterbird species on the Molly Islands, from 1980 through 2019. Although intermittent records exist prior to 1980, particularly for pelicans, early observations were inconsistent, and records are too spotty to reliably reflect trends.

SHOREBIRDS OF YELLOWSTONE

KATHARINE E. DUFFY

A massive flock of semipalmated sandpipers departs its arctic breeding grounds on the coast of western Alaska in July, heading for wintering sites in South America. Dodging thunderstorms over northern Montana, a handful of these tiny sandpipers splits from the flock and flies straight south, eventually passing over Yellowstone. When the sandpipers reach Hayden Valley, mud flats along Alum Creek beckon. Other shorebirds forage here, so invertebrates must be plentiful, and the weary sandpipers descend.

ALTHOUGH SHOREBIRDS ARE not a major component of Yellowstone National Park's (Yellowstone; YNP) avian wildlife, shorebirds do occur in the park throughout the seasons, with some individuals remaining even in winter (Fig. 22.1). In the following discussion of shorebird species, I've indicated if the species is listed by the US Fish and Wildlife Service as a Bird of Conservation Concern (BCC) as well as the level of concern for each species as measured by the Assessment of the Conservation Status of Shorebirds (www.shorebirdplan.org; US Shorebird Conservation Partnership 2022), in which Level 1 indicates High Concern, Level 2 is Moderate Concern, and Level 3 indicates the species is of Least Concern.

inhabit some thermally heated marshes each winter. Because snipe tend to be secretive unless flushed from wetlands, their presence is most obvious in late spring and early summer as they perform their winnowing courtship flights overhead. As snow melts in spring, creating ephemeral wetlands, the winnowing of snipe is a common occurrence throughout Yellowstone. Wilson's phalaropes[3], the nesting shorebird species least-often encountered in YNP, nest in wetlands, apparently in low numbers, throughout the park. Females, with more colorful plumage than males, depart after egg laying, leaving males to incubate eggs and raise young until fledging. Sightings of Wilson's phalaropes most frequently occur during migration.

SHOREBIRDS NESTING IN YNP

Spotted sandpipers[3] commonly nest along rivers, ponds, and lakes throughout the park, just as these small shorebirds nest along waterways throughout much of the country. Killdeer[2] are conspicuous nesters in hydrothermal basins, with some inhabiting these thermal areas throughout winter, eating plentiful ephydrid flies that, in turn, eat microbial mats. Wilson's snipe[3] nest in marshes and

SHOREBIRDS ENCOUNTERED DURING MIGRATION

As certain shorebirds head to breeding sites as distant as arctic Canada and Alaska, they may stop in Yellowstone during their spring migration, mostly in April and May (Fig. 22.2). Fall migration of shorebirds can begin as early as June for phalaropes, with the peak for most species occurring during July and August and lasting into fall for

▲ FIGURE 22.1 While American avocets do not nest in Yellowstone, these large and distinctive shorebirds frequent the park during spring and fall migration. NPS PHOTO/JIM PEACO

▼ FIGURE 22.2 Observant birders searching for shorebirds on mudflats at the edge of park lakes and rivers sometimes find a solitary sandpiper, a species considered rare in the park in spring and fall. PHOTO BY HOWARD WEINBERG

certain species. Some shorebird species molt their colorful breeding plumage prior to their fall migration, while others molt at stopover sites during migration or where they spend the winter.

Anecdotal observations indicate these shorebird species are expected during migration (E. Hendrickson, pers. comm., J. Parker, pers. comm.):

American avocet[2] (Fig. 22.1)
Baird's sandpiper[3]
Least sandpiper[3]
Western sandpiper[2]
Long-billed dowitcher[2]
Solitary sandpiper[3] (Figure 22.2)
Willet[BCC,1]
Greater yellowlegs[3]

Based on frequency of observations, these shorebirds are less-common migrants, mostly seen during their fall migration (E. Hendrickson, pers. comm., J. Parker, pers. comm.):

Black-necked stilt[3]
Semipalmated plover[3]
Long-billed curlew (in 2019,
    a pair nested in Lamar Valley)[1]

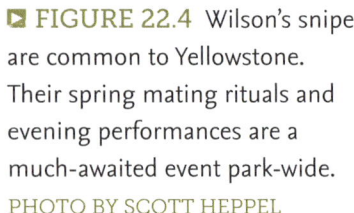

▲ FIGURE 22.3 Marbled godwits, willets, and other shorebirds seize the opportunity to forage in an ephemeral pond near Mammoth during spring migration. PHOTO BY HOWARD WEINBERG

◄ FIGURE 22.4 Wilson's snipe are common to Yellowstone. Their spring mating rituals and evening performances are a much-awaited event park-wide. PHOTO BY SCOTT HEPPEL

Marbled godwit[BCC,1]
Pectoral sandpiper[BCC,1]
Semipalmated sandpiper[1]
Lesser yellowlegs[BCC,1]
Red-necked phalarope[2]

Why pay attention to shorebirds? There's high conservation concern for six shorebird species listed above and moderate conservation concern for four species. Killdeer, seemingly ubiquitous throughout most of Canada and the lower 48 states, have been designated a species of moderate concern due to population declines, especially in Canada and the western United States.

Shorebirds that migrate exceptionally long distances have suffered devastating population declines worldwide. For these species, survival is a finely tuned balance of time, distance, weather, and especially abundant food, coupled with physiology. Mud flats crucial for feeding during migration are being actively lost to the effects of climate change (e.g., sea-level rise) and human manipulation that results in habitat loss (Weidensaul 2018, Rosenberg et al. 2019).

Although protected in the United States and Canada, many shorebird species continue to face illegal shooting during periods when they are outside protected areas. Legal protection doesn't shield shorebirds from other problems. For example, shorebirds that feed in agricultural evaporation ponds ingest a dangerous brew of toxins, from selenium to pesticides. Climate change presents additional threats, including mismatched timing of migration and food resources (https://whsrn.org/; Western Hemisphere Shorebird Reserve Network 2019).

A species with just a few concentrated gathering sites during any stage of its life cycle faces enormous risks: a natural or human-caused catastrophe where most of the population occurs could jeopardize the future of the species. Then sites attracting fewer individuals could attain more significance for perpetuation of the species. Every stopover site for long-distance migrating shorebird species has importance.

Does Yellowstone provide major stopover or staging sites for shorebirds? Probably not, based on current knowledge. If there's one special staging area or location that's critical to shorebirds, it hasn't yet been found. But during spring and especially fall migration, an array of shorebird species, sometimes in surprising numbers, forage in ephemeral wetlands and along mud flats adjacent to Yellowstone's many scattered lakes, ponds, streams, and rivers. Places like Alum Creek, where it enters the Yellowstone River in Hayden Valley, sometimes harbor individuals and flocks of several shorebird species—for example, Baird's sandpipers, greater yellowlegs, and least sandpipers. Pelican Bay, east of Fishing Bridge, often hosts shorebirds, along with numerous ducks. But shorebirds don't seem to linger in Yellowstone for long periods. They're here today and gone tomorrow.

Examples of fleeting occurrences of shorebirds in YNP include these sightings: Waves of red-necked phalaropes surprised observers in several parts of Yellowstone on May 19, 2019. An ephemeral pond west of the Gardner River hosted greater yellowlegs, willets, lesser yellowlegs, and marbled godwits on April 23, 2018 (H. Weinberg, pers. comm.). Black-necked stilts spent two days feeding in an ephemeral pond in the Upper Geyser Basin at Old Faithful (April 27, 2006; K. Duffy, pers. obs.), while a flock of about 20 marbled godwits dropped out of a snowstorm to land in another ephemeral pond near Old Faithful (April 26, 2009; K. Duffy, pers. obs.) but left within hours.

Observations of all shorebirds in the park, particularly of those other than the four species known to regularly nest in the park, may provide valuable data that can contribute to species conservation. If you spot shorebirds in Yellowstone, please submit your sightings to the park's Report a Wildlife Sighting website: www.nps.gov/yell/learn/nature/wildlife-sightings.htm. As with other citizen-science-led endeavors in Yellowstone, your input is imperative and can have a measurable impact on our understanding of how these birds use the park and how we manage for their continued presence across Yellowstone's landscape.

5 PASSERINES

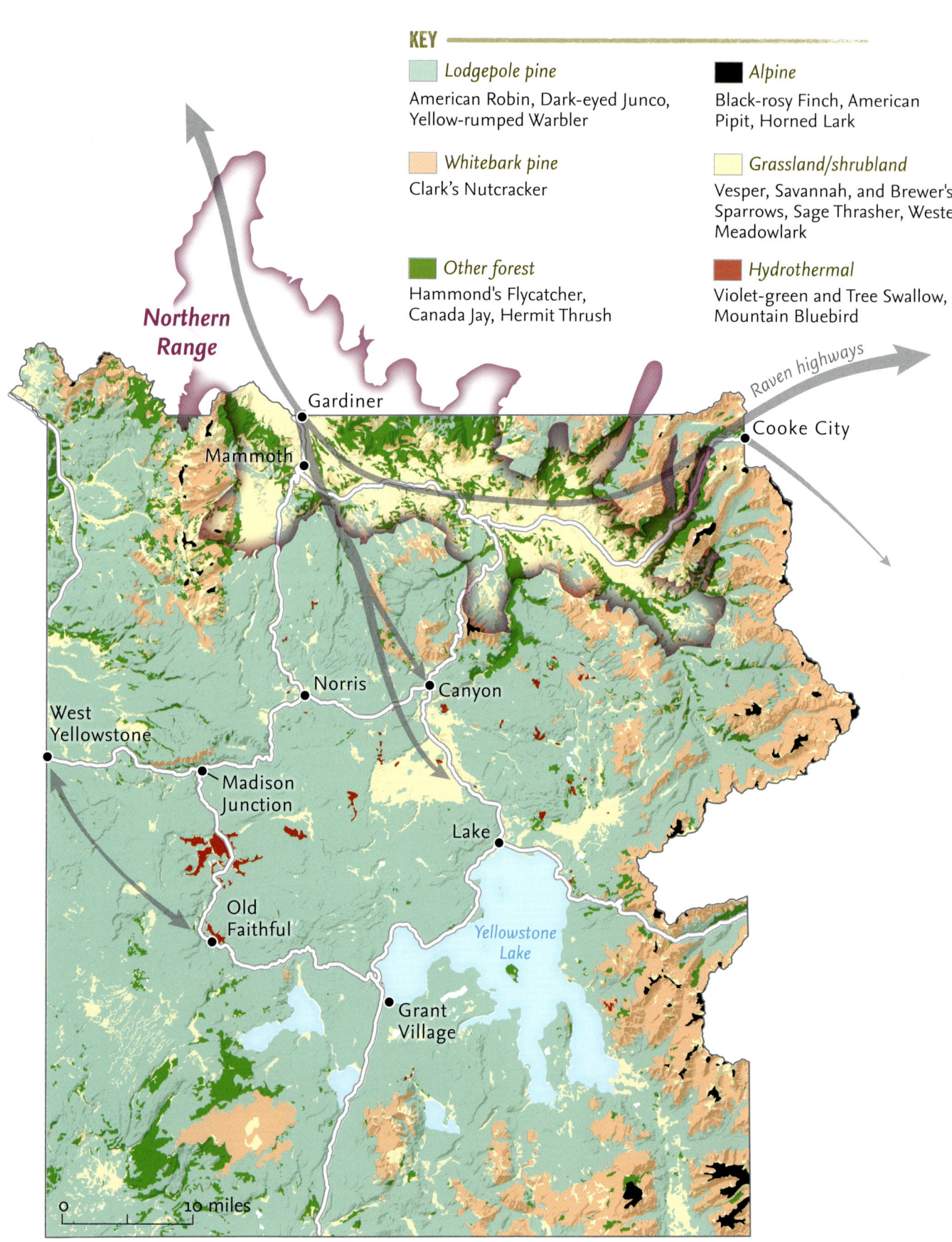

Lodgepole pine
American Robin, Dark-eyed Junco, Yellow-rumped Warbler

Whitebark pine
Clark's Nutcracker

Other forest
Hammond's Flycatcher, Canada Jay, Hermit Thrush

Alpine
Black-rosy Finch, American Pipit, Horned Lark

Grassland/shrubland
Vesper, Savannah, and Brewer's Sparrows, Sage Thrasher, Western Meadowlark

Hydrothermal
Violet-green and Tree Swallow, Mountain Bluebird

Northern Range

Raven highways

Gardiner

Cooke City

Mammoth

Norris

Canyon

West Yellowstone

Madison Junction

Lake

Old Faithful

Yellowstone Lake

Grant Village

0   10 miles

COMMON RAVENS IN YELLOWSTONE

MATTHIAS LORETTO, LAUREN E. WALKER, CAMERON HO, GEORGIA COLEMAN, DANIEL STAHLER, DOUGLAS W. SMITH, and JOHN M. MARZLUFF

We'd been waiting for hours. The trap was set before sunrise, a camouflaged net launcher baited with a very dead, road-killed mule deer carcass. What could be more enticing to a raven? Well, apparently, lots of things. Although we'd had many ravens fly over, some even circling back for a second look, not a single raven had landed to investigate our bait. Eventually we gave up, finally admitting that, once again, we'd been beaten. A frustratingly typical day trapping these cautious and wily birds.

LAUREN E. WALKER

COMMON RAVENS ARE smart, opportunistic omnivores that are ubiquitous in Yellowstone National Park (Yellowstone; YNP). This large, all-black bird is a member of the Corvidae family, a diverse group of intelligent birds comprised of magpies, jays, nutcrackers, and crows. Yellowstone is home to eight corvid species that, along with ravens, include the American crow, pinyon jay, Steller's jay, blue jay, Canada jay, black-billed magpie, and Clark's nutcracker (*see chapter 24*).

RAVEN BIOLOGY

Ravens, the world's largest songbird, are both predators and scavengers, utilizing these strategies to survive in a wide variety of climates and ecosystems (Dos Anjos et al. 2009). They are found in habitats as disparate as temperate rain forests and arctic tundra, and their diverse diet includes everything from seeds and insects to rodents, songbird eggs and nestlings, human garbage, and animal carcasses. Likely because of their tendency to scavenge carcasses, including human corpses, ravens are often portrayed in mythology as bad omens and harbingers of death. But, as the subject of diverse mythological and cultural significance across the globe, ravens are also viewed as clever tricksters, creators, and even providers (Marzluff and Angell 2005). Importantly, as documented in both the sociocultural and scientific literature, ravens have close ties to two mammalian species that have arguably shaped the Yellowstone Ecosystem more than any other: humans and wolves.

Ravens have a simple social system, especially in comparison with other corvids (Brown 1974). Non-breeding, non-territorial birds are vagrant, wandering hundreds of miles in search of food and roosting communally to share information concerning feeding locations (Marzluff et al. 1996). When they find a significant food resource, they often gather in large, but ephemeral, aggregations of unrelated birds (Parker et al. 1994) so as to overpower dominant local territory owners (Marzluff and Heinrich 1991, Heinrich and Marzluff 1995). In addition to learning about feeding locations at night roosts, vagrants (and territorial ravens as well) find food by homing in on the sounds of other ravens fighting over food (Heinrich and Marzluff 1991, Heinrich et al. 1993). The duration of this vagrant life varies among individuals and can last for over a decade (Loretto et al. 2016a). Upon ascending the dominance hierarchy (Heinrich 1994, Braun and Bugnyar 2012) and securing a lifelong mate and territory, ravens are thought to spend the rest of their lives mostly within an all-purpose territory (Webb et al. 2012). They nest annually, in late winter or early spring, and tend their fledged offspring for about one month before expelling them from the territory (Marzluff and Marzluff 2011).

Across the western United States, ravens have adapted to human development, foraging on roadkill and water treatment plant effluent, taking handouts from picnickers, and becoming regular visitors to trash cans and dumpsters in all settings from small towns to the largest cities (Restani et al. 2001, Marzluff and Neatherlin 2006, Webb et al. 2011). Historically, ravens have been viewed as agricultural pests, and have consequently faced human persecution. However, due to expanding human activity as well as federal protection by the Migratory Bird Treaty Act, ravens have increased in abundance and distribution across the western United States (Marzluff et al. 1994) and are rapidly recolonizing vacated portions of their range in the eastern United States (Hackworth et al. 2019). Similar patterns of range expansion and increasing abundance have also occurred in parts of Europe (e.g., United Kingdom; Sim et al. 2005, Risely et al. 2008). In areas away from people, however, ravens often rely on carrion provided by large predators. Yellowstone provides a unique scenario, where people abut wildlife, providing a reliable mix of food resources for ravens throughout the year. In the summer, many ravens can be found in campgrounds, picnic spots, and road pullouts, begging (or stealing) from the millions of visiting tourists. In the winter, when visitors are less common and human use of the park is focused on the northern range, ravens

are commonly seen scavenging ungulate carrion alongside wolves and other large carnivores. In fact, ravens have a well-documented relationship with wolves that allows them to maximize their knowledge of and exposure to wolf-killed carcasses; in many places, they are known to follow wolves directly, follow wolf tracks, and be attracted by wolf vocalizations (Mech 1966, Mech 1970, Harrington 1978, Stahler et al. 2002). Ravens are inherently neophobic (afraid of new experiences and objects), and the presence of wolves at a carcass alleviates that fear, enabling them to land more quickly and feed more readily (Heinrich 1988, Heinrich et al. 1995, Stahler et al. 2002). In Yellowstone, ravens are the most frequent scavenger at wolf kills and, other than coyotes, they remove more carcass biomass than any other scavenger species (Wilmers et al. 2003).

RAVENS AND WOLVES

Because of the close relationship between wolves and ravens across their range (Heinrich 1999), the reintroduction of wolves to the park in 1995 and 1996 (Fritts et al. 1995) likely had large impacts on the raven population in northern Yellowstone. Prior to 1995, ravens were largely dependent on winter-kill (ungulates that perish due to old age, injuries, harsh conditions, or lack of available food), as well as gut piles from hunter-killed ungulates outside the park boundaries. In fall and winter, food would have been abundant during hunting season and in late winter, but relatively scarce at other times. Raven numbers across Yellowstone's northern range probably varied dramatically from year to year, fluctuating with winter severity. Since wolves were reintroduced to the ecosystem and hunt ungulates year-round, ravens have been provided reliable and consistent winter foraging opportunities, allowing for more stable raven populations independent of winter harshness (Walker et al. 2018).

To learn more about how ravens have responded to the presence of wolves in northern Yellowstone, researchers from the Max Planck Institute of Animal Behavior in Germany and the University of Washington in Seattle are

FIGURE 23.1 Matthias Loretto (right) and John Marzluff (left) remove ravens from a net. We trapped the ravens using a net launcher at an elk outfitter site regularly visited by the birds. PHOTO BY ANDRIUS PASUKONIS

FIGURE 23.2 Matthias Loretto (right) attaches a solar-powered GPS transmitter to a raven as part of our ongoing study. The transmitters record and store the precise locations, temperature, and acceleration of the birds, uploading the information to a remote server via the cellular communication network.
PHOTO BY JOHN MARZLUFF

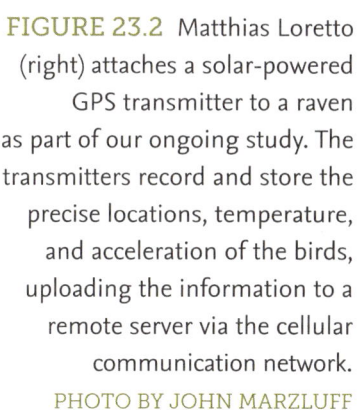

collaborating with park biologists to capture, band, and track ravens using GPS (global positioning system) transmitters (Fig. 23.1 and 23.2). After several months of trapping, biologists have captured and attached transmitters to 65 ravens and now regularly collect location data as these birds move throughout Yellowstone, the Greater Yellowstone Ecosystem (GYE), and the Rocky Mountains. Hundreds of thousands of raven locations have already been logged, providing the basis to address many long-standing and recent questions.

Between November 2019 and December 2020, around 43% of the 204 wolf kills monitored in the park attracted at least one of the GPS-tagged ravens; two attracted a dozen. About a quarter of these birds (22%) were regular wolfers, attending more than 10 kills. During the same period, only 25% of the 91 monitored cougar kills were visited by a tracked raven, with a maximum number of seven GPS-tagged ravens at a single kill. This apparent preference for wolf kills confirms previous observations from the long-term carcass monitoring of the Yellowstone wolf

project (Smith et al. 2020a) and likely has several causes: cougars are relatively secretive and solitary and spend more time in denser vegetation. Furthermore, after they make a kill, they often cache it, hiding it for later use. In contrast, wolf kills can be much easier for ravens to find because they generally hunt and make their kills in open landscapes. Ravens also cue into kill locations by following wolf howls (Harrington 1978).

Our observations of ravens utilizing wolf kills suggest that most birds locate such food bonanzas during the day, without reliance on information shared at communal roosts (Fig. 23.3). In comparison, research in Maine, where carnivores are rare and wolves entirely absent, revealed that ravens primarily found food bonanzas through information exchanged overnight at communal roost sites (Marzluff et al. 1996). In Yellowstone, ravens from nearby territories arrive quickly at the scene of a wolf kill, roost on their territories, and come and go from the kill independently until the food is consumed. Vagrant, non-territorial birds gather quickly at a kill that may occur near

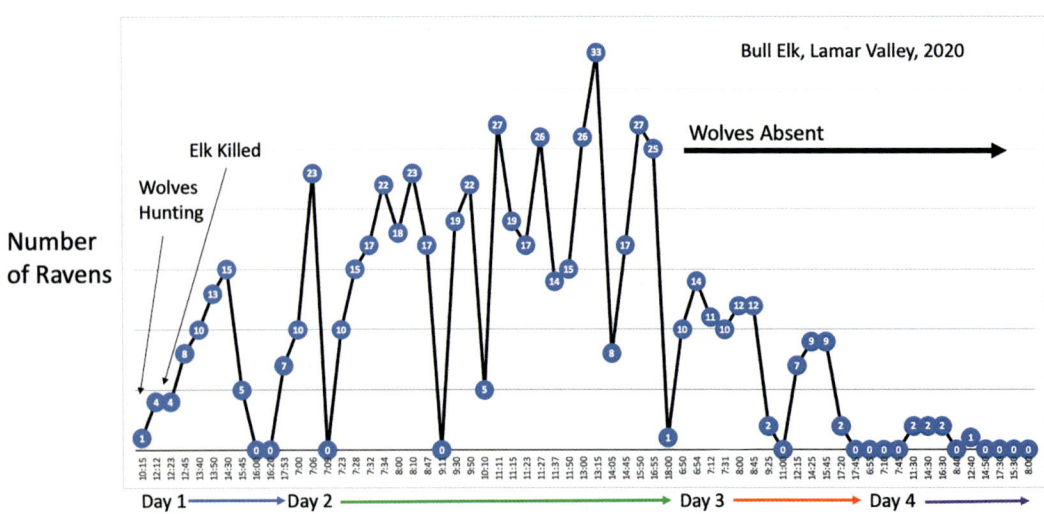

FIGURE 23.3 The ebb and flow of ravens at a wolf kill. We observed the Junction Butte wolf pack pursue and kill a bull elk at midday on February 24, 2020. Four ravens were associated with the wolves at this time, and their numbers gradually increased, mostly during the daylight hours but also over the night after the kill was made. This suggests most ravens were drawn to the food by the calls of ravens already at the carcass. Others may also have gained information about the kill site overnight, at a communal roost site.

where they are foraging, but some appear to find kills after long, seemingly direct flights to the area (Fig. 23.4). Vagrants remain for varying lengths of time near the kill, occasionally feeding and roosting nearby. However, others roost in the area with knowledgeable ravens but do not follow them to the kill. The low status of vagrant ravens, relative to territory holders, likely motivates vagrants to leave a kill where they are unable to obtain much food and instead exploit less-defensible anthropogenic resources. The presence of wolves is also important to how ravens utilize a kill; wolf defense of the carcass causes hourly ebbs and flows in the number of birds scavenging (Fig. 23.3) and dictates when ravens first feed. For example, at one kill site where ravens were present when an elk was killed at 12:23 p.m., wolf defense prevented birds from scavenging the carcass until 13:59.

Ravens also associate with wolves at their dens during the spring and summer. Wolf dens were visited by 13 GPS-tagged ravens during 2020, and most of these were nearby territory owners and young males. Research conducted on raven associations with wolves between 1997

and 1999 also found interactions at Yellowstone wolf dens (Stahler 2000). Although ravens were not marked in this early study, behavioral observations made during summer den monitoring found ravens (ranging from 1 to 10 individuals) were present during 55% of the daytime observation hours. These observations illustrated ravens' highly opportunistic foraging behavior, as well as the benefit of closely associating with wolves away from carcass sites themselves. Food rewards obtained by ravens at dens included carcass parts brought back by adults, regurgitated meat piles intended for wolf pups, and fresh scats deposited by pack members. In addition, ravens were observed directly interacting with young pups, including following pups at close distance around the den, pulling on their tails, engaging in short chase interactions, approaching close to nursing pups, and looking down into den holes while pups were inside. Ravens were also observed following wolves back to the den despite the availability of nearby carcasses, suggesting anticipation of food access from meat carries or regurgitations to pups. Stahler (2000) speculated that regular

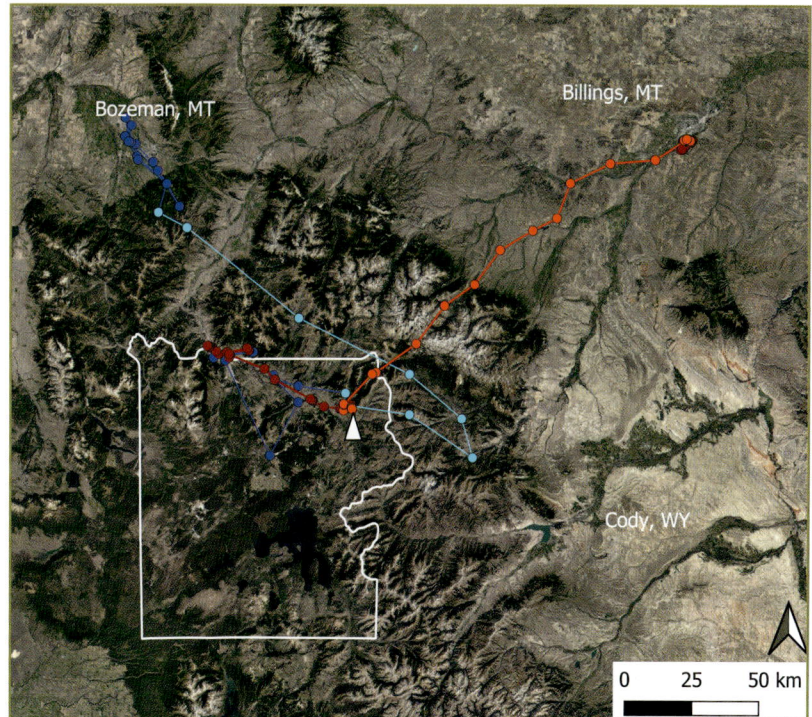

FIGURE 23.4 Frequently, ravens approach carcasses on directional flights over large distances. In this example, wolves killed an elk near Soda Butte (indicated by the tip of the white triangle) in Yellowstone National Park (white outline shows park boundaries) on March 2, 2020. One day later, a raven flew from Bozeman via Cooke City almost to Cody, turned back, and ended up at the kill site. GPS points and the interpolated trajectory from that day are shown in light blue; the movement from the two days before and after this trip are shown in dark blue. On March 5, 2020, another raven flew from Billings to the kill site, covering 96 miles (155 km), shown in light red. The movements two days before and after this trip are shown in dark red. MAPS CREATED IN QGIS 3.12.1, INCLUDING A GOOGLE SATELLITE BASEMAP

behavioral interactions between ravens and wolves at pup-rearing sites may serve as an influential mechanism for social learning for both species. From the first day of den emergence as a young pup, a wolf likely experiences close interactions with ravens daily for the duration of its life. These frequent encounters, particularly between socially impressionable pups and young ravens, likely facilitate learning opportunities about the risks and rewards of associating with one another. For ravens in particular, such learning could lead to familiarity with wolf response and intent. Ultimately, experience with differential outcomes of associating with wolves may enable ravens to learn how to avoid risks while interacting with a dangerous provider.

RAVENS' SPACE AND RESOURCE USE

Experienced wolf watchers may be surprised to learn that less than half of wolf kills over the past several years have been exploited by the GPS-tagged ravens, as one always observes ravens at a kill. However, our 65 tagged ravens represent only a small fraction of the large number of ravens foraging in the park. Additionally, it turned out that several of our ravens, mostly vagrant non-breeders but also territorial breeders, move far beyond the park boundaries. If one draws a polygon around all GPS-locations an individual had visited between November 2019 and February 2020, territorial breeders covered an average area of 1,470 square miles (3,806 km²), ranging from 75.3 square miles to 12,387 square miles (195 km² to 32,083 km²), while nomadic non-breeders utilized an average area of 10,506 square miles (27,210 km²), ranging from 88.4 square miles (229 km²) to 78,995 square miles (204,596 km²; Marzluff et al. 2022). However, vagrants and breeders do not use this space uniformly; instead, 95% of the time, ravens can be found in much smaller areas leading to an average size of 217 square miles (562 km²) for breeders and 1,076 square miles (2,787 km²) for non-

breeders. Regardless of the method to estimate space use, the area ravens in Yellowstone roam through is far greater than what has been observed in other locations (e.g., Loretto et al. 2016b, Loretto et al. 2017, Harju et al. 2018, Marchand et al. 2018). On average, ravens in Yellowstone move 24 miles (39 km) daily; the greatest distance traveled in a single day was more than 186 miles (300 km). Two non-breeders even moved over several months from Yellowstone National Park to Canada,

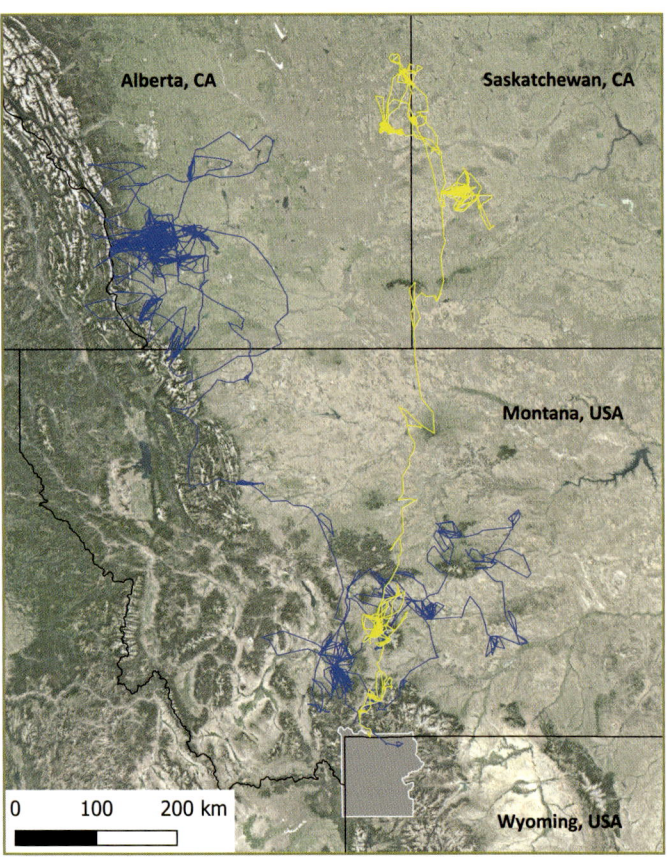

FIGURE 23.5 Movement trajectories of two non-breeding common ravens, a juvenile male (blue) and an adult male (yellow); both were trapped and banded in or close to Yellowstone National Park (gray area) and moved over several months to Alberta and Saskatchewan, Canada. These movements are the longest-recorded dispersal distances for common ravens (470 miles/757 km and 463 miles/745 km). MAP CREATED IN QGIS 3.12.1, INCLUDING A GOOGLE SATELLITE BASEMAP

which resulted in the longest-ever-recorded dispersal distances for ravens—470 miles (757 km) and 463 miles (745 km)—(Fig. 23.5).

Large differences in movement behavior occur not only between breeders and non-breeders, but also between seasons (Fig. 23.6), driven by changes in resource distribution. For instance, ravens exploited gut piles during the autumn and spring hunting seasons. They used dispersed natural resources, such as large insects, primarily during late spring through the summer. Wolf kills were most frequent in the Lamar Valley and in general throughout the northern range, and ravens concentrated their use of these resources in that area (Fig. 23.7). Throughout the Greater Yellowstone Ecosystem, ravens foraged on anthropogenic resources such as landfills, waste treatment centers, sewage ponds, road kills, farms, in cities, towns, and recreation sites (Fig. 23.7). Many of these food sources are highly predictable and almost always available. In winter, territorial Yellowstone ravens primarily foraged at anthropogenic food sources, often 30–60 miles (50–100 km) away from their territory, such as sewage-pond and waste-treatment centers in Gardiner, West Yellowstone, and Paradise Valley (Fig. 23.6), or landfills as far away as Cody, Wyoming, or Billings, Montana. Still, there is considerable individual variation in resource use. Some individuals used many resources equally, while others focused on specific resources, such as agriculture, dispersed natural foods, waste, and carcasses.

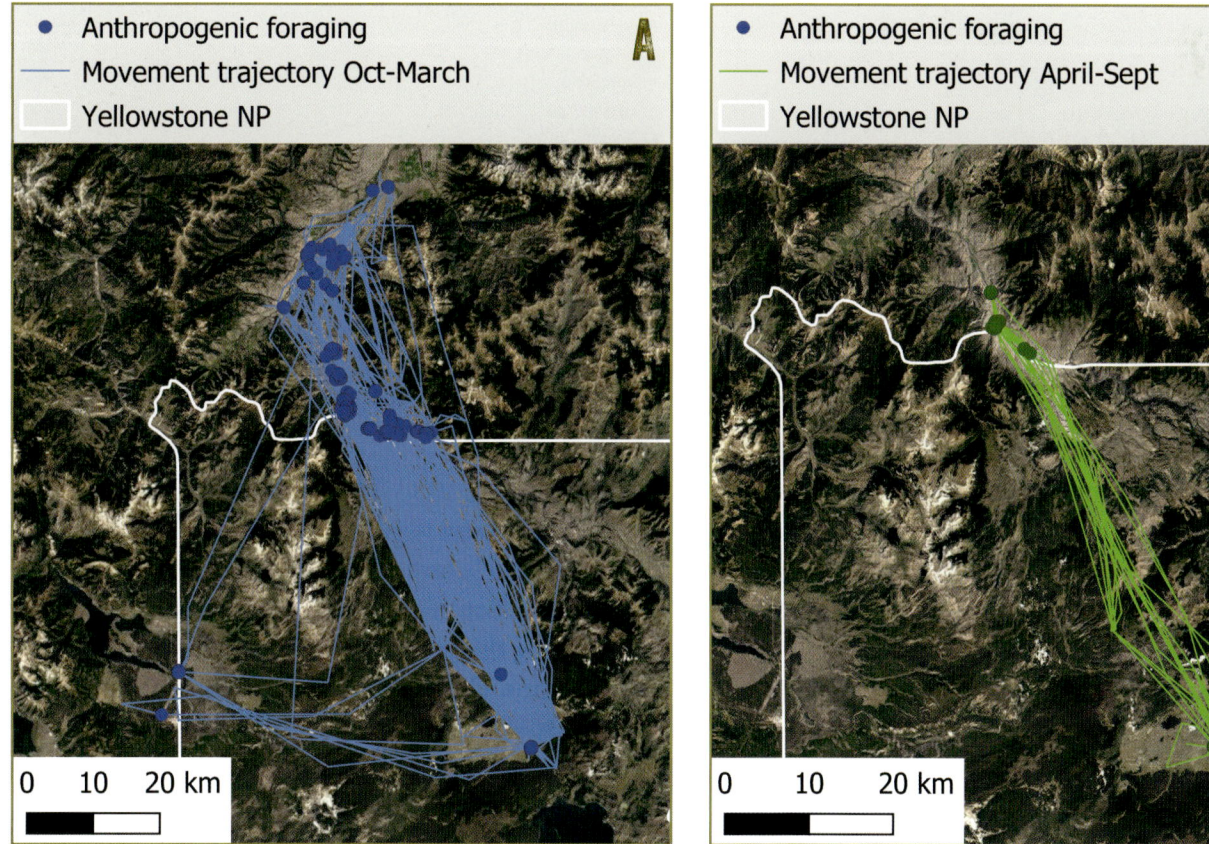

FIGURE 23.6 GPS trajectories of a breeding male common raven in Yellowstone National Park during (A) October 2019–March 2020 (blue lines) and (B) April–September 2020 (green lines). Dark blue and dark green points represent GPS locations of the common ravens that could be related to anthropogenic food sources.

227

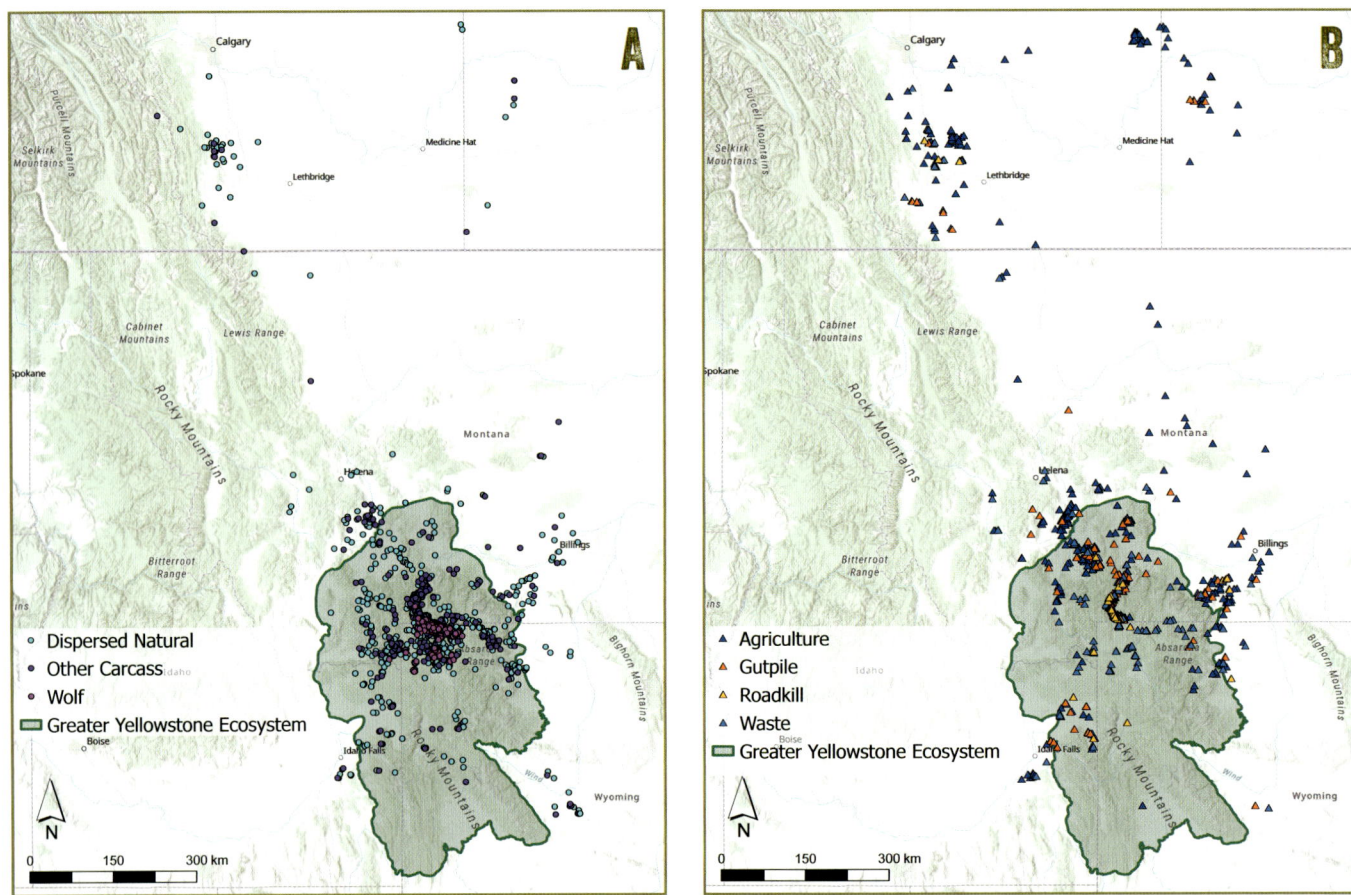

FIGURE 23.7 Spatial distribution of resource sites. (A) Natural resource sites in the Greater Yellowstone Ecosystem. (B) Anthropogenic resource sites in the Greater Yellowstone Ecosystem.

Ravens' great mobility in the Greater Yellowstone Ecosystem may be reflective of the relatively harsh Rocky Mountain winters, but also of the relatively low density of human occupation and therefore low density of anthropogenic food sources. Ravens moving between landfills must cover much larger areas than ravens in, for example, Central Europe (Loretto et al. 2017, Marchand et al. 2018).

While anthropogenic resources play an important role in foraging, ravens only rarely used anthropogenic structures for roosting at night (11% of roosts). In most cases, ravens roosted on trees (84%), sometimes on cliffs (3%), and only rarely in the open (about 1%; Fig. 23.8). From the movements of 42 GPS-tagged ravens between October 2019 and January 2021, we identified over 2,000 night roosts (Fig. 23.9). Most of the specific locations (2,021 of 2,095) were used for only one night, and during this period, each bird used, on average, 50 roosts. We mapped and identified frequently used roost locations by estimating the probability of the occurrence of roosting birds throughout the time. Inspection of the resulting roost-specific utilization distribution indicated that our ravens had a 30% chance of spending the night in one of seven distinct areas (an additional, more dispersed area with lower likelihood of use was frequented by many tagged birds; Fig. 23.10). Four of these areas included nightly gatherings of more than two ravens (High Use Areas 3, 6, 7, and 8), indicating they were likely communal roosts, with a mixture of both territorial breeders

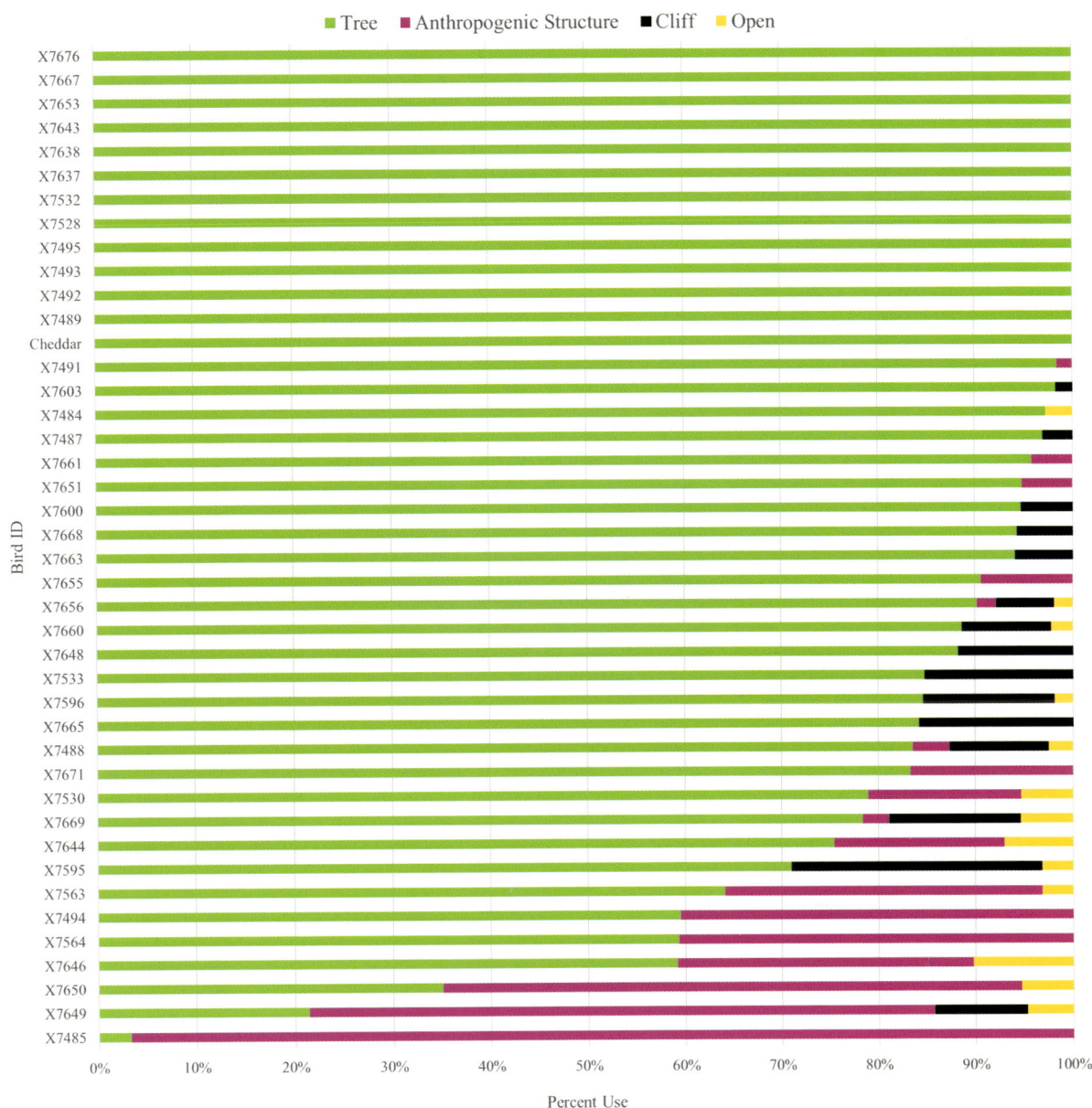

FIGURE 23.8 Variability of roost substrates used by individual ravens.

and vagrant non-breeders present throughout the duration of use. The other four areas were used only by territorial breeding ravens (High Use Areas 2 and 5) or by a territorial breeder (sometimes a pair) and one vagrant non-breeder raven (High Use Areas 1 and 4). A total of 27 individual ravens roosted at more than one high-use area (Fig. 23.9 a & b).

BEGGING AND STEALING

In Yellowstone, ravens commonly beg for food from humans visiting the park or living just outside the park. Territorial adult ravens in the Lamar and Hayden valleys routinely position themselves at picnic areas or pullouts and approach people. As the birds sidle up to their target,

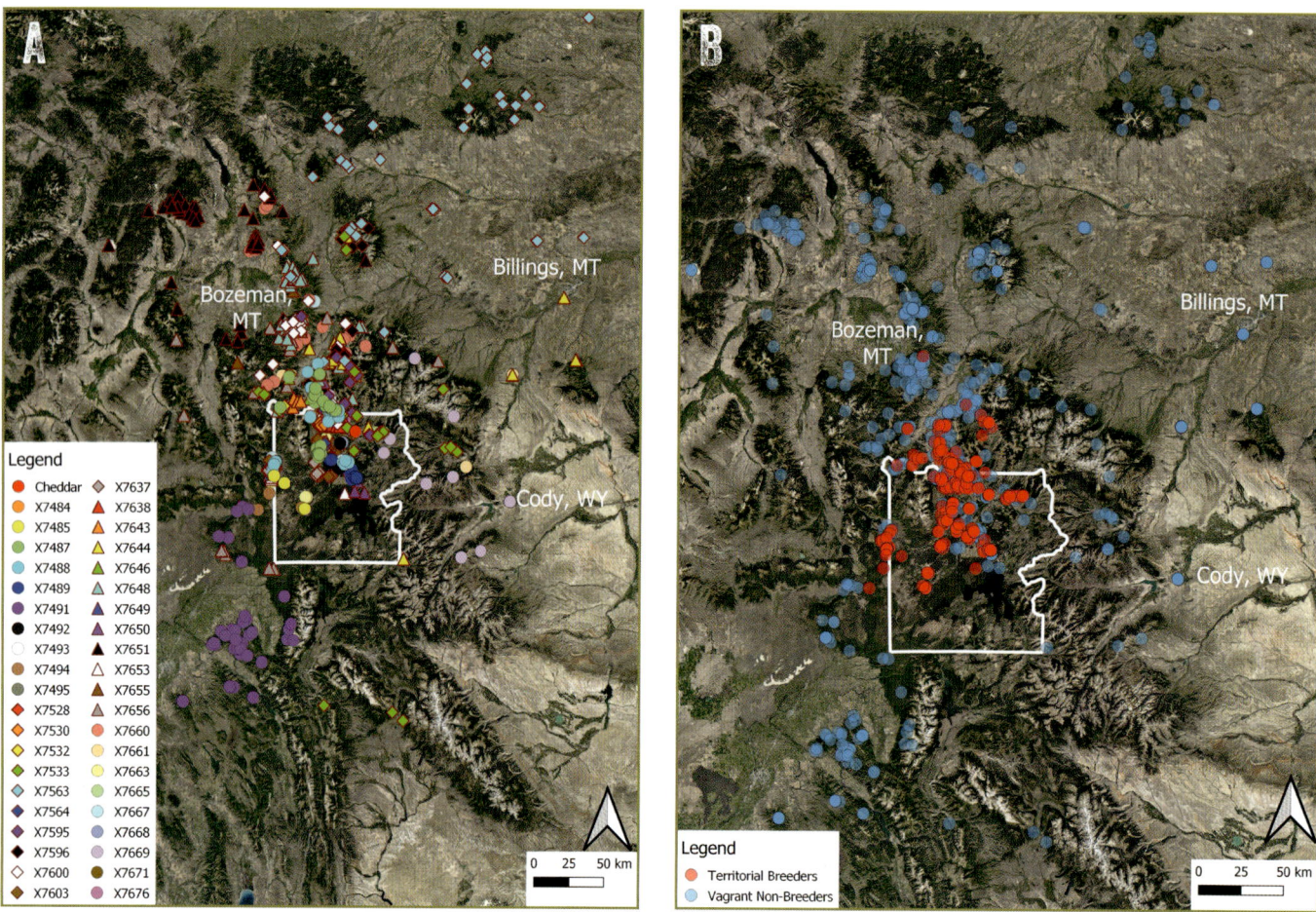

FIGURE 23.9 (A) All roost locations between October 2019 and January 2020, with different colors and symbols identifying individual ravens. (B) The same roost locations, colored by raven breeding status (red = vagrant non-breeder, blue = territorial breeder).

a car or picnic table, people toss them eggs, sandwiches, chips, and other scraps (Fig. 23.11). Although park regulations prohibit the feeding of wildlife in Yellowstone, ravens' proclivity to capitalize on human-sourced foods make this behavior in both ravens and humans nearly impossible to prevent. Outside of the park, one of our tagged birds routinely visited residents in Cooke City and the adjacent Beartooth Highway. This subordinate non-breeder (7523, aka Stevie) had a slight limp, which had attracted the attention and sympathy of residents, who provide her with eggs, pet food, chips, and scraps from the local tavern daily (Fig. 23.12).

Some ravens in Yellowstone raid the packs of tourists and the saddlebags of snowmobiles in search of edibles. In the early 2000s, ravens in the Old Faithful area learned to open the soft, zippered or Velcro-sealed bags popular on snowmobiles at the time, and stealing from snowmobiles was common (Fig. 23.13). Today, snowmobiles sport locking and hard-sided compartments that ravens have yet to learn how to open. This sort of arms race is typical among people and ravens and has driven cultural evolution in both species (Marzluff and Angell 2005).

▶ **FIGURE 23.10** High use roost areas within the study area. These high use areas are potential communal roosts or territorial roosts. We identified High Use Areas 1–7 as places where a tagged raven had a 30% or greater probability of spending the night. High Use Area 8 was visually identified as a potential communal roost site due to the high number of individuals spatially clumped in a small area at night.

▼ **FIGURE 23.11** A territorial breeding raven begs from a tourist at Tower Junction. The bird was rewarded with a slice of meat.
PHOTO BY JOHN MARZLUFF

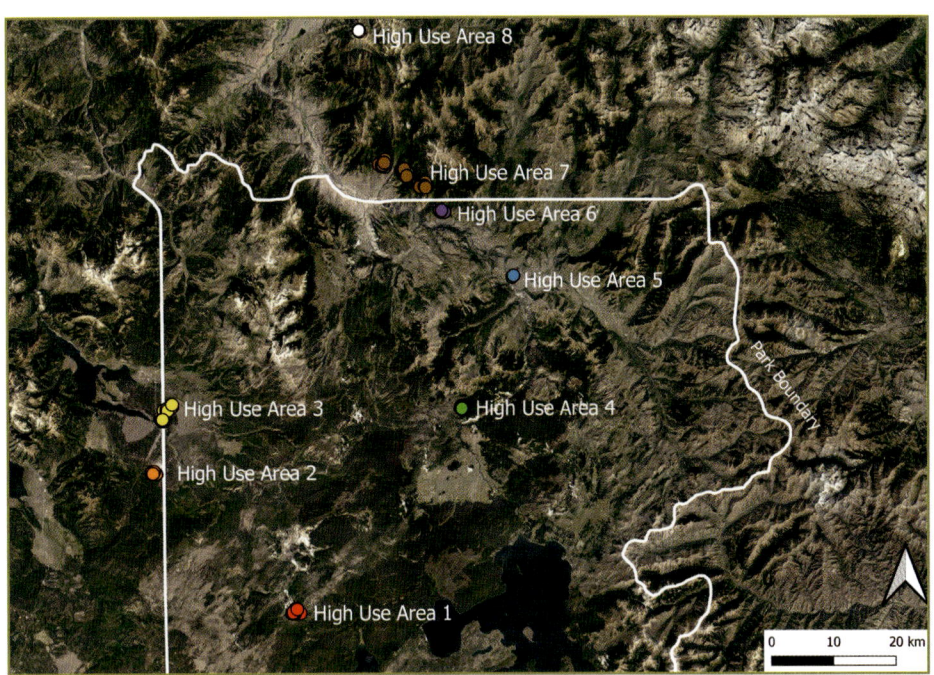

High Use Area 8
High Use Area 7
High Use Area 6
High Use Area 5
High Use Area 3
High Use Area 4
High Use Area 2
Park Boundary
High Use Area 1

0   10   20 km

▶ FIGURE 23.12 Stevie the raven eyes a Cheeto tossed to her by students at the Cooke City, Montana, school. Stevie begs from students at recess throughout the winter. PHOTO BY JOHN MARZLUFF

▼ FIGURE 23.13 (A–D) A raven systematically breaks into a snowmobile pouch, on the hunt for cookies. PHOTOS BY TOM MURPHY

NESTING BEHAVIOR AND PRODUCTIVITY

Ravens are among the earliest-nesting songbirds in the Northern Hemisphere. In Britain and Ireland, for example, pairs begin laying eggs as early as the first week in February (Ratcliffe 1997). There is, however, wide variation in the timing of nesting, with pairs nesting farthest north or at the highest elevations often breeding later than those in more temperate conditions. Ravens begin nesting in April in Greenland but as early as January in northern India (Goodwin 1986). In Yellowstone, ravens nest rather late relative to other locales. In 2021, we observed pairs building nests, copulating, and initiating incubation (which occurs with the laying of the last or penultimate egg in the clutch) from April 22 through at least April 28. Pairs in the lower elevations, around Mammoth Hot Springs and Gardiner, Montana, nest one to two weeks earlier than those in the colder or higher altitudes of the park's interior (Old Faithful, Lake, Hayden Valley), Blacktail

C

D

COMMON RAVENS IN YELLOWSTONE

Plateau, or Lamar Valley. In most locations, eggs hatch in mid- to late May, and conspicuous nestlings are begging hoarsely throughout June. Nestlings fledge in mid- to late June and remain on their natal territories with their parents at least until late July.

Ravens nest at a time when the climate in Yellowstone is not overly harsh or unpredictable and when foods will be plentiful for their nestlings and fledglings. Nesting in late April also reduces the energetic demands on incubating females that are frequently imposed by late-spring cold snaps and snows. Feeding nestlings in May and June, rather than February or March, allows parents to exploit abundant and seasonal foods, such as Uinta ground squirrels and terrestrial invertebrates. As fledglings gain independence in late summer, they can gorge on abundant grasshopper populations and, in short order, flock to gut

FIGURE 23.14 A female raven prepares to feed its nestlings in the Lamar Valley.
PHOTO BY CAMERON HO

piles left by recreational hunters each fall just outside the park boundaries.

Breeding ravens gather nest materials from within their territories. Early in the process, which may take several weeks, loud snaps and cracks can be heard from treed areas as dead branches, with finger-sized diameters, are broken loose by the birds' powerful beaks. Males and females gather materials, mostly as a coordinated unit, but lone birds may also work the nest. In Yellowstone, nests are 1.6–3.3 feet (0.5–1 m) from top to bottom. The outer nest layer is a woven basket of Douglas-fir, pine, sage, aspen, or cottonwood branches. Inside are layers of dried grass, shredded bark, rootlets, and finally a dense, insulating layer of bison wool.

As in other locales (Goodwin 1986, Ratcliffe 1997), Yellowstone ravens build their nests in trees, on cliffs, and on human artifacts. Of the 24 unique raven nest sites we found in 2020 and 2021, most were in coniferous or deciduous trees (6 in Douglas-fir, 6 in lodgepole pine, and 1 each in quaking aspen and narrowleaf cottonwood; Fig. 23.14). The typical arboreal nest is against the bole and in the upper half of the canopy of a tree on the edge of a small grove. Yellowstone nests were also built in cavities and on well-protected ledges of rocky cliffs (n = 8). One pair nested in the steel supporting system under the Highway 212 bridge across the Gardner River; another nested on a cell tower. Most pairs (9 of 11) re-used nests in 2021 that they occupied in 2020, regardless of success the prior year. In two cases, one member of the pair died between nesting seasons, and in both cases, the remaining partner (a male in one instance and female in the other) renested in the same location used previously.

Most ravens nesting in Yellowstone in 2020 and 2021 fledged their full brood of nestlings (2020: 9/14 nests; 2021: 11/20 nests; overall 59% of 34 nests). At the 20 successful nests where we were able to count the fledglings, each pair fledged an average of 3 offspring (range 1–4). As we monitored family groups in July, we observed fledglings chasing grasshoppers and being fed by their parents. We documented predation on one nest near Mud Volcano. We also observed the death of one fledgling at Yellowstone Lake after it was entangled in fishing line and hung from a high branch in a tree near its nest.

Nest failure is not uncommon for birds. We are just beginning to unravel how nest exposure, tenure with a mate, tagging, and reliance on human handouts affect nesting success of ravens in Yellowstone. During the breeding seasons of 2020 and 2021, half of the 22 nesting attempts by the ravens we tagged succeeded in producing fledglings (on average 2.9 per nest). During the same years, 67% of the 12 pairs of untagged birds we monitored fledged an average of 3.3 young, suggesting that tagging may slightly compromise nesting ravens. However, other aspects of the nest may confound this effect. For example, the failed nests were also more exposed to predators than those that succeeded. Two tagged females nested high in sparse trees near Old Faithful, and their chicks were preyed upon in both 2020 and 2021, most likely by an owl or eagle. Cliff nests exposed from at least one side are also more visible than nests in trees, and they failed at a high rate (60% of 10).

The raven project is planned to continue over the next several years and will focus on understanding the ways in which ravens exploit carnivores, the transition of individual ravens from non-breeder to breeder, the life span and lifetime reproductive success of individuals, and the importance of a diversity of invertebrate foods. In the meantime, modern technology offers the public unique insights into ongoing research—you can even follow the movements of our individual GPS-tagged ravens using the phone application Animal Tracker. Stay tuned for more insights into these fascinating birds!

Visit the Yellowstone's Birds website (https://press.princeton.edu/resources/yellowstones-birds-video-collection/v23) to watch an interview with John Marzluff.

24 CLARK'S NUTCRACKERS OF YELLOWSTONE NATIONAL PARK
CONSERVING AN ICONIC MUTUALISM

DIANA F. TOMBACK, THOMAS H. McLAREN, WALTER G. WEHTJE, LAUREN E. WALKER, and DOUGLAS W. SMITH

CLARK'S NUTCRACKER AND ITS NAMESAKE

The Clark's nutcracker was one of three new bird species collected in 1806, during the final year of the Lewis and Clark Expedition. Captain William Clark, the species' namesake, first sighted the bird near the Beaverhead River headwaters, Montana, on August 22, 1805, prior to the challenging westward crossing of the Bitterroot Mountains. He wrote in his journal, "I Saw to day Bird of the wood pecker kind which fed on Pine burs its Bill and tale white the wings black every other part of a light brown, and about the Size of a robin" (Lewis and Clark 2002, August 22, 1805, https://lewisandclarkjournals.unl.edu/item/lc.jrn.1805-08-22#lc.jrn.1805-08-22.02; Davis and Stevenson 1934). In the following year, on the return trip, Captain Meriwether Lewis collected specimens near the Clearwater River, in Idaho, and described the nutcracker in some detail in his May 28, 1806, journal entry. He recognized its relatedness to crows and wrote that he had viewed the bird previously "on the hights [sic] of the rocky Mountains," and "this bird feeds on the seed of the pine and also on insects. it resides in the rocky mountains at all seasons of the year, and in many parts is the only bird to be found" (Lewis and Clark 2002, May 28, 1806, https://lewisandclarkjournals.unl.edu/item/lc.jrn.1806-05-28#lc.jrn.1806-05-28.01; Davis and Stevenson 1934). No doubt the nutcracker vocalizations described by Lewis

as "loud squawling" and the bird's acrobatic flights from treetop to treetop, accentuated by flashy black and white wings and tails, caught his attention often as the expedition crossed the higher ranges of the Rockies. After the expedition returned, Clark's nutcracker was formally described and named by ornithologist Alexander Wilson (Davis and Stevenson 1934).

AN ICONIC MUTUALISM

Although Lewis and Clark's observations of nutcrackers may have been fleeting, more than two hundred years later, most of their observations hold true. Today we know that Clark's nutcracker not only feeds on the seeds of many western conifers but also serves as a keystone seed disperser for several, linking them ecologically (Fig. 24.1; Tomback and Kendall 2001, Tomback 2020). Seed dispersal is a consequence of nutcracker seed-caching behavior. In late summer and fall, nutcrackers harvest ripe conifer seeds and transport them in their expandable throat (sublingual) pouch anywhere from a few meters to as far as 20 miles (32 km) from source trees. Nutcrackers cache the seeds across mountain terrain varying from steep, rocky slopes and ridgelines to closed forest to post-fire landscapes, and from upper to lower treeline (Vander Wall and Balda 1977, Tomback 1978, Tomback 2001, Lorenz et al. 2011). Each nutcracker makes tens of thousands of seed caches each fall, typically depositing 1–15 seeds per

FIGURE 24.1

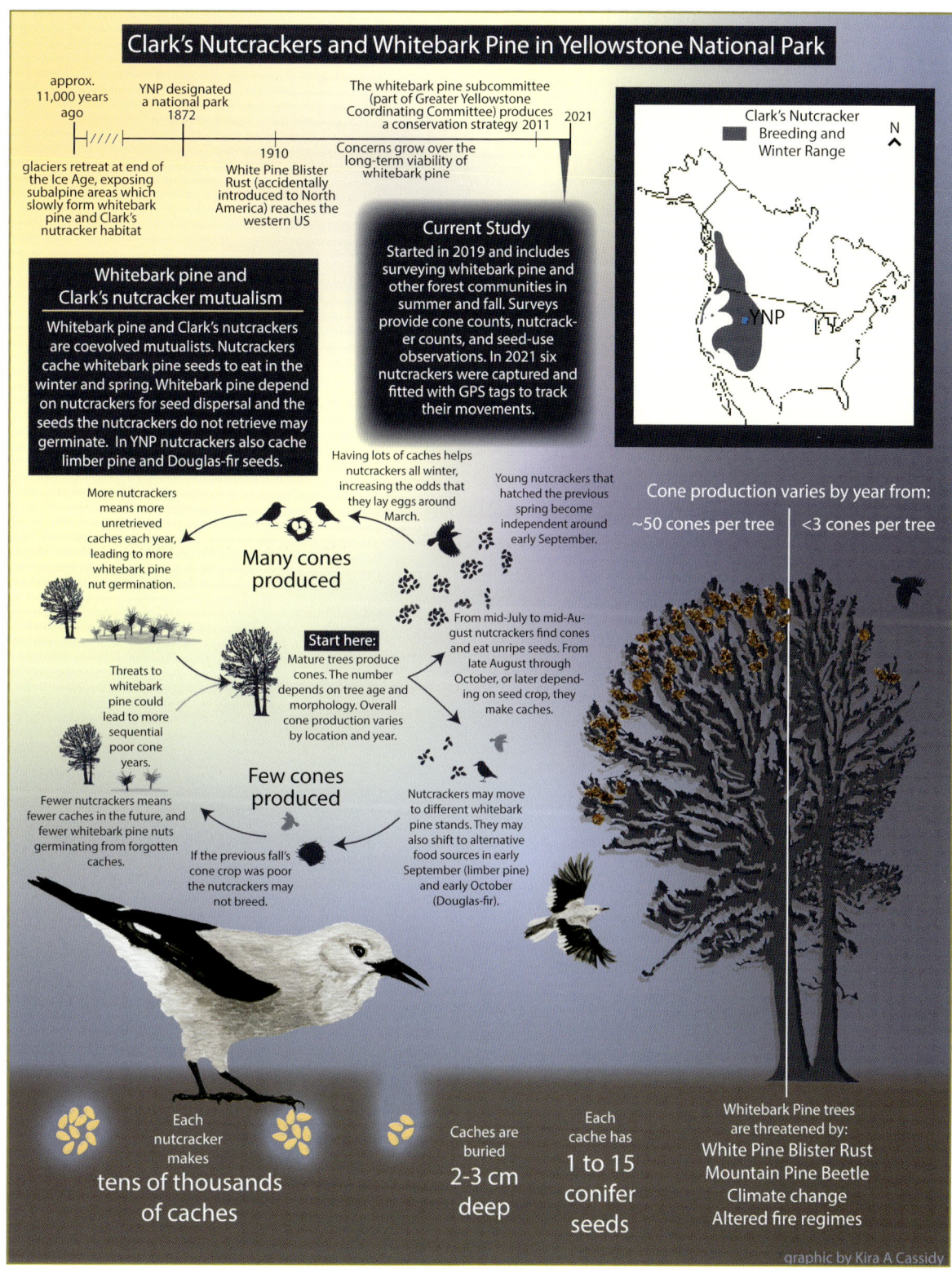

Clark's Nutcrackers and Whitebark Pine in Yellowstone National Park

approx. 11,000 years ago

YNP designated a national park 1872

The whitebark pine subcommittee (part of Greater Yellowstone Coordinating Committee) produces a conservation strategy 2011

2021

glaciers retreat at end of the Ice Age, exposing subalpine areas which slowly form whitebark pine and Clark's nutracker habitat

1910 White Pine Blister Rust (accidentally introduced to North America) reaches the western US

Concerns grow over the long-term viability of whitebark pine

Clark's Nutcracker Breeding and Winter Range

N

YNP

Current Study

Started in 2019 and includes surveying whitebark pine and other forest communities in summer and fall. Surveys provide cone counts, nutcracker counts, and seed-use observations. In 2021 six nutcrackers were captured and fitted with GPS tags to track their movements.

Whitebark pine and Clark's nutcracker mutualism

Whitebark pine and Clark's nutcrackers are coevolved mutualists. Nutcrackers cache whitebark pine seeds to eat in the winter and spring. Whitebark pine depend on nutcrackers for seed dispersal and the seeds the nutcrackers do not retrieve may germinate. In YNP nutcrackers also cache limber pine and Douglas-fir seeds.

More nutcrackers means more unretrieved caches each year, leading to more whitebark pine nut germination.

Having lots of caches helps nutcrackers all winter, increasing the odds that they lay eggs around March.

Young nutcrackers that hatched the previous spring become independent around early September.

Many cones produced

Threats to whitebark pine could lead to more sequential poor cone years.

Start here: Mature trees produce cones. The number depends on tree age and morphology. Overall cone production varies by location and year.

From mid-July to mid-August nutcrackers find cones and eat unripe seeds. From late August through October, or later depending on seed crop, they make caches.

Fewer nutcrackers means fewer caches in the future, and fewer whitebark pine nuts germinating from forgotten caches.

Few cones produced

If the previous fall's cone crop was poor the nutcrackers may not breed.

Nutcrackers may move to different whitebark pine stands. They may also shift to alternative food sources in early September (limber pine) and early October (Douglas-fir).

Cone production varies by year from:

~50 cones per tree | <3 cones per tree

Each nutcracker makes **tens of thousands of caches**

Caches are buried **2-3 cm deep**

Each cache has **1 to 15 conifer seeds**

Whitebark Pine trees are threatened by: **White Pine Blister Rust Mountain Pine Beetle Climate change Altered fire regimes**

graphic by Kira A Cassidy

ARTWORK BY KIRA CASSIDY

cache at depths of 0.8–1.2 inches (2–3 cm) in soil or rocky substrates and hiding some seeds in cracks and holes in trees or in downed logs. Cache retrieval, aided by the remarkable nutcracker spatial memory (Tomback 1980, Vander Wall 1982, Kamil and Balda 1985), usually begins some time in early to late winter near the start of the breeding cycle, when birds need to build up fat reserves and other food becomes difficult to find (Tomback 1978).

While nesting, nutcrackers feed themselves and their young with seeds from caches (Mewaldt 1956, Tomback 1978), enabling them to survive seasonal food scarcity. Unretrieved caches, on the other hand, may then germinate following snowmelt or summer rains, leading to forest regeneration (Tomback 1982, Tomback et al. 2001a). Given that nutcrackers place seeds above and below the treeline, elevational and distributional limits of whitebark pine and other conifers dispersed by nutcrackers may shift as climate changes.

Clark's nutcrackers are probably best known for their iconic coevolved mutualism with whitebark pine, a high-elevation, five-needle white pine. White pines are also known as softwood pines, a varied group including pinyon pines, bristlecone pines, and commercially valuable eastern and western white pines. Many of these white pines grow their foliage (needles) in clusters or fascicles of five. Whitebark pine has evolved nearly complete dependence on nutcrackers for seed dispersal (Fig. 24.2; Lanner 1980, Hutchins and Lanner 1982, Tomback 1982, Tomback and Linhart 1990). Whereas most conifers in the pine family Pinaceae, including spruce, fir, hemlock, and larch, have small, winged seeds, which are dispersed by wind from open cones, whitebark pine seeds are large and wingless, and the cones do not open. Instead, nutcrackers use their long, sturdy, and pointed beaks to break into cones and remove seeds (Tomback 1978). Furthermore, whitebark pine seed cones mature in early to mid-September,

FIGURE 24.2 (A) A Clark's nutcracker harvests seeds from a whitebark pine, using cones and the branch shoot for support during seed harvest. (B) A Clark's nutcracker harvesting seed on Dunraven Pass shows off a full throat pouch. PHOTOS BY DIANA TOMBACK

at least a month before the seed cones of most conifers (Krugman and Jenkinson 1974). This early maturation provides a timely food source after nutcrackers have depleted the previous year's seed caches and as they begin to prepare for another long winter and breeding season (Tomback 1978). Whitebark pine's tree and branch structure further facilitate the relationship with nutcrackers: the upper branches tend to be vertical, with horizontally oriented whorls of whitebark pine cones near the tips, providing secure landing platforms for the birds as they remove seeds from whitebark pine cones (Fig. 24.2).

CLARK'S NUTCRACKERS AND WHITEBARK PINE IN YELLOWSTONE NATIONAL PARK

In Yellowstone National Park (Yellowstone; YNP), visitors are treated to close views of Clark's nutcrackers industriously working the whitebark pine cones in trees bordering the Dunraven Pass and Chittenden Road hiking trails to the Mount Washburn fire lookout. In years when limber pine produces cones around park headquarters at Mammoth Hot Springs, picnicking visitors are entertained by small groups of nutcrackers flying back and forth, calling to their flockmates, harvesting seeds from the trees around the employee housing and administrative buildings, and then flying off with bulging throat pouches.

Park ranger and naturalist M. P. Skinner, who in 1916 published "The Nutcrackers of Yellowstone Park," in the ornithological journal *The Condor*, noted: "In the Yellowstone Park, the nutcrackers are everywhere," and "The vicinity of Mammoth Hot Springs and Fort Yellowstone seems to be a center of abundance both winter and summer." In fact, after three years of study, we would generally agree that nutcrackers can be seen anywhere in the park, often flying en route to seed resources or preferred forest communities to forage, make or retrieve seed caches, or to seek shelter. Curiously, Skinner—clearly more a wildlife expert than a botanist—in his paper mistakenly referred to the limber pine as "pinyon pine," although the ranges of pinyon pines end hundreds of miles to the south.

Clark's nutcrackers have been little studied in Yellowstone National Park, with only anecdotal observations to date (e.g., Skinner 1916). Although whitebark pine communities are one of several major forest types in the park (e.g., Renkin and Despain 1992), we believe they play a critical role in sustaining nutcracker populations and maintaining the seed dispersal mutualism. Whitebark pine, however, is declining in the Greater Yellowstone Ecosystem (GYE) from exotic disease, native pest outbreaks, and climate change impacts, including distributional shifts to higher elevations, reducing whitebark's total range (e.g., Tomback and Achuff 2010, Chang et al. 2014). In Yellowstone, it is possible that cumulative losses of whitebark pine will impact the population size and seasonal distribution of nutcrackers across the park's forest communities, but this requires study. Whitebark pine losses will also reduce food availability for many of Yellowstone's other seed-eating birds and small mammals, as well as for grizzly bears and black bears (e.g., Tomback and Kendall 2001). Bears raid red squirrel cone stores (middens) in early fall prior to hibernation and feast on the calorie-rich whitebark pine seeds.

Recognizing the downward trajectory range-wide for whitebark pine, the US Fish and Wildlife Service proposed to list the tree as threatened under the Endangered Species Act. This uncertain future highlights the need to better understand the potentially changing relationship between nutcrackers, whitebark pine, and alternative seed resources in Yellowstone National Park.

Because whitebark pine cone production varies across years and even among forests, Clark's nutcrackers are also well-adapted to use other seed resources (e.g., Tomback 1978, IGBST 2015). While nutcrackers prefer whitebark seeds, they harvest and cache seeds from other conifer species when necessary and sometimes migrate long distances in their search (Tomback 2020). In the GYE, which encompasses the park and surrounding federal lands, average whitebark pine cone production varies annually from fewer than 3 cones per tree to about 50 cones per tree. For whitebark pine stands sampled in Montana, Weaver and Forcella (1986) calculated an overall mean of about 75 mature seeds per cone. If whitebark pine cone production is high, estimates indicate that each nutcracker may store between 35,000 and 98,000 whitebark pine seeds per year in about 9,500 to 30,000 caches, respectively (Hutchins and Lanner 1982, Tomback 1982). Nutcrackers may then move on to other seed resources, especially in years of lower whitebark pine cone production, to cache additional seeds (e.g., Tomback 1978).

In the GYE, the most likely additional seed resources include Douglas-fir, which is widely distributed but has a small, winged seed and infrequent large cone crops, and limber pine, a relative of whitebark pine with large, wingless seeds but cones that open and slowly release seeds (e.g., Tomback et al. 2011, Williams et al. 2020). In the southern GYE, Schaming (2016) found that nutcrackers utilized Douglas-fir forests as breeding-season habitat and their seeds as food in early spring. Limber pine, which has a patchy and limited distribution in the park, also depends on nutcrackers for long-distance seed dispersal (Lanner and Vander Wall 1980, Tomback et al. 2011, Williams et al. 2020). Other seed-producing conifers in the GYE, notably Engelmann spruce and lodgepole pine, which comprises a major forest community type in Yellowstone National Park, are potential seed resources for nutcrackers, but their use has not been recorded previously (Tomback 2020).

THREATS TO WHITEBARK PINE AND THE ICONIC MUTUALISM

Additional seed resources may be key to retaining a Clark's nutcracker population in Yellowstone National Park in the not-too-distant future. Whitebark pine is declining across much of its range due to several factors: recent mountain pine beetle outbreaks; reduced cone production from crown damage and tree mortality caused by the exotic disease white pine blister rust (a fungal pathogen); altered fire regimes (and more severe fires); and climate change (Tomback et al. 2001b, Tomback and Achuff 2010). Limber pine is also declining from these threats, especially in the Rocky Mountains (Tomback and Achuff 2010, Goeking and Windmuller-Campione 2021). Although mountain pine beetle outbreaks have diminished over the last decade, white pine blister rust is slowly increasing in prevalence and is spreading across whitebark pine's range. It now represents a true threat to the long-term survival of the whitebark pine. For the GYE, data from monitoring transects indicate a mean whitebark pine infection rate of about 25% and increasing mortality from blister rust (Shanahan et al. 2016, 2017).

Previous research from the GYE north to Waterton Lakes National Park, Canada, found that nutcrackers were less likely to visit whitebark pine communities when tree damage and mortality resulted in low whitebark pine cone production. When cone production fell below about 400 cones per acre (1,000 cones per ha), the likelihood of nutcrackers visiting whitebark pine decreased steeply (McKinney et al. 2009, Barringer et al. 2012).

As whitebark pine further declines in Yellowstone National Park from blister rust mortality and additional wildfire, the total food available for Clark's nutcrackers also declines, leading to fewer birds and a potential feedback loop of less seed dispersal and fewer newly germinating whitebark pine seedlings. The likely imminent decline in nutcrackers (which may already have started) argues for the importance of initiating a park monitoring program.

STUDY OBJECTIVES

In July 2019, with funding and logistic support from Ricketts Conservation Foundation and Yellowstone National Park, Yellowstone Bird Program biologists and ecologists from the University of Colorado Denver initiated a long-term study of Clark's nutcrackers in the park. This intensive study seeks to answer the following questions:

Which forest community types in Yellowstone National Park are used by Clark's nutcracker, and how are they used (e.g., for seed resources, caching, shelter, etc.)? Does nutcracker use of forest communities differ between late summer and fall, and does this vary from year to year? And how do individual birds move within and outside the park? In short, we hope to learn how dependent the Yellowstone population of nutcrackers might be on the spatially limited and declining whitebark pine communities.

WHAT WE HAVE LEARNED AND HOW

Information and observations from the first three years of the study are building a picture of forest community use and spatial movements by nutcrackers. These data are being analyzed in more detail using statistical models (McLaren 2022), but we present some of the important initial findings here.

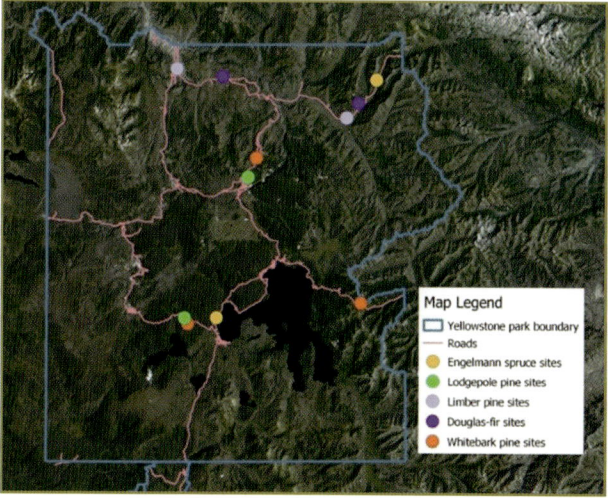

FIGURE 24.3 General locations of the 11 forest-community-study transects in Yellowstone National Park. The number of transects per community are as follows: whitebark pine (3), limber pine (2), lodgepole pine (2), Douglas-fir (2), and Engelmann spruce (2).

CONE PRODUCTION AND NUTCRACKER COUNTS

We identified the following major forest community types in Yellowstone National Park, in order from highest to lowest elevations: whitebark pine, spruce-fir-lodgepole pine, lodgepole pine, Douglas-fir, and limber pine. To investigate which forest communities are used by Clark's nutcrackers, we installed a series of transects (each 0.62 mile/1 km long) across the park with at least two per forest community type (Fig. 24.3, 24.4, and 24.5). Spaced along each transect we have five stations for nutcracker point counts and 11 tagged trees for cone counts during July of each year. The highest-elevation transects are in whitebark pine communities, up to 9,380 feet (2,860 m), and the lowest-elevation transects are in small stands of limber pine as low as 6,160 feet (1,880 m). We performed one complete set of nutcracker point counts four times between mid-July and October to determine bird presence during different cone-ripening stages.

We found that whitebark and limber pine seeds and cones mature at the end of August or in early September, whereas the seeds and cones of spruce, lodgepole pine, and Douglas-fir do not mature (and open) until the end of September or in early October. Although whitebark pine cone production varied among transects and from year to year, some cone production occurred in all three years of the study, with relatively large cone crops at Dunraven Pass in 2020 and 2021 (Fig. 24.6). Given the large size and thus the high energy value of whitebark pine seeds (e.g., Tomback 1982), even the comparatively low cone numbers represented an important energy source for nutcrackers. Nutcracker numbers were relatively high in whitebark pine throughout the study. Limber pine, which has similarly large, energy-rich seeds, had high cone production on one transect and relatively high numbers of nutcrackers in all three years. Douglas-fir produced large cone crops in 2020 and 2021, and nutcrackers were reliably present in all three years. Lodgepole pine produced few new cones in each year, and Engelmann spruce had

CLARK'S NUTCRACKERS OF YELLOWSTONE NATIONAL PARK

A

B

CLARK'S NUTCRACKERS OF YELLOWSTONE NATIONAL PARK

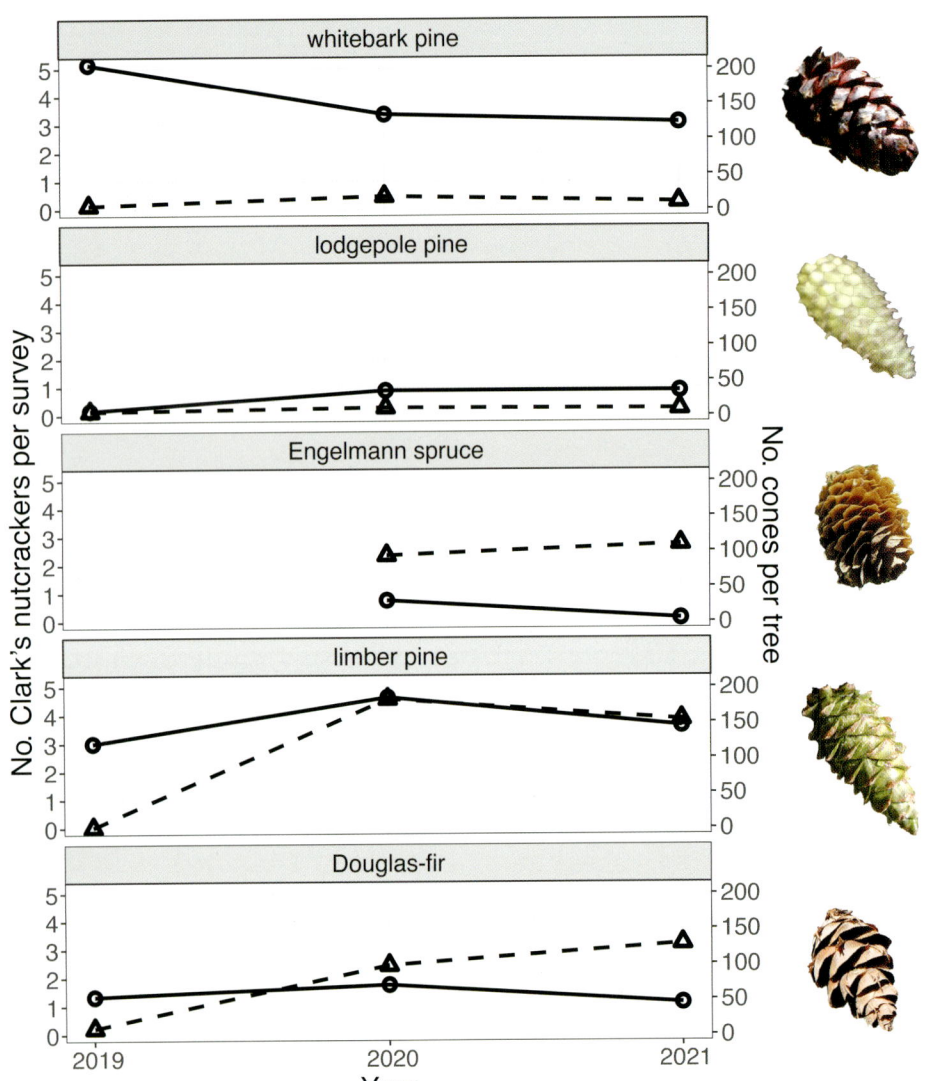

FIGURE 24.6 Relationship between nutcracker survey numbers and cone production in each year of the study. The nutcracker survey numbers were calculated as follows: for each survey, the number of nutcracker visual observations for the five point counts were added to provide a total count. Then the total for each transect survey was averaged across all the transect surveys for a given forest-community type for that year. Visual observations included all birds using each forest-community type during the point counts (i.e., no overflights). Cone counts from the 11 tagged trees along each transect were averaged, providing an index of the size of the annual cone crop for that forest-community type. The number of transects per forest-community type is provided in Fig. 24.3.

high cone counts, but both community types had few to no nutcracker observations each summer and fall.

SEED HARVESTING AND CACHING

During point counts and based on walking between points over the three years, we noted nutcrackers harvesting seeds, although the numbers were not large. We saw whitebark pine seeds harvested on 11 occasions, with six of these observations during September transect visits. We observed limber pine seeds harvested on three occasions (twice in September), Douglas-fir seeds harvested twice (September and October), and lodgepole pine seeds harvested three times (July and October)—a first-time observation for this conifer. We observed seeds cached on or near whitebark pine transects twice, limber pine transects four times, and a Douglas-fir transect and spruce transect once each.

ROAD TRANSECT OBSERVATIONS

In November 2019, we established a long roadside survey transect approximately 30 miles (50 km) long with 10 point-count stops at car pullouts. The transect began at Mammoth Terrace in the west and ended just west of the park's northeast entrance. The pullouts included the following forest community types: limber pine/Douglas-fir, Douglas-fir, lodgepole pine, subalpine fir, Engelmann spruce, and Engelmann spruce/subalpine fir. In the first two years of the study, one complete road transect survey was completed each month from November through March. At each pullout, we followed the same point-count protocol used on the forest-community transects but also included nutcracker vocalizations as evidence of bird presence.

During the first road transect period (2019–2020), we performed a total of 50 point counts, and for the second period (2020–2021), 40 point counts. During the first period, we saw or heard nutcrackers on 22% of point counts; for the second period, we saw or heard nutcrackers on 52% of point counts. We observed nutcrackers during 30% of the limber pine/Douglas-fir point counts, 38% of Douglas-fir counts, 33% of lodgepole pine counts, 33% of subalpine fir counts, 56% of Engelmann spruce counts, and 33% of Engelmann spruce/subalpine fir counts. The road-transect counts indicated nutcrackers were present in all road-transect forest community types during the late fall and winter seasons.

GPS-TAGGED NUTCRACKERS

In March 2021, we established five feeding/trapping locations across the northern road in the park. Our goal was to trap, leg-band, and place a GPS (global positioning system) tag on six nutcrackers in total. Three nutcrackers were trapped in Douglas-fir habitat, and another three in a mixed spruce-fir-lodgepole pine community near the east side of the park (Fig. 24.7). The GPS transmitters (Lotek Pinpoint Argos) were 3-gram battery-powered satellite transmitters that provided GPS locations on different schedules we programmed for each bird, ranging from once every two days to once every two weeks. Given the different rates of GPS fixes, battery life ranged from less than one month to more than 10 months after tagging. We undertook this pilot study to see if satellite transmitters could provide useful data on nutcracker movements throughout the year.

Despite the small number of tagged birds, the nutcrackers provided important information on spatial movements. All captured birds had brood patches, indicating they were breeding in the Douglas-fir and mixed spruce-fir-lodgepole pine communities where they were caught. Where in most songbird species only the nesting female develops a brood patch—an unfeathered area of heavily vascularized skin below the breast, which is applied to the eggs during incubation—in Clark's nutcrackers, both sexes develop a brood patch (Mewaldt 1952). Given that nutcrackers nest in late winter, which is extremely early for songbirds, brood patches in both sexes allow the mated pair to take turns incubating eggs or brooding nestlings to allow the other bird to retrieve seed caches to feed itself and the young.

GPS locational data indicated the three birds trapped in the Douglas-fir forest community generally stayed close to the capture site early on, making short flights, most likely to retrieve caches or forage. The one bird whose transmitter continued working into the fall months traveled as far as 7 miles (11 km) north of the park on multiple occasions into whitebark pine forest; the bird was likely harvesting and caching whitebark pine seeds on these trips. The three birds captured in the spruce-fir-lodgepole pine forest community west of the Northeast Entrance all moved to higher elevations near Barronette Peak in late April, possibly to retrieve whitebark pine caches that became accessible with the retreat of the winter snowpack. The two remaining birds with functioning transmitters then traveled outside the park in late June to early July. One nutcracker flew 30 miles (48 km) east into the Absaroka Mountains, stopping about 20 miles (32 km) outside the park, where it remained. The other nutcracker traveled about 75 miles (120 km) to the southwest, spending the summer and fall in the Centennial Mountains, 20 miles (32 km) west of the park and northwest of Island Park, Idaho, and began visiting a backyard feeder in the area in November 2021.

In summary, the three birds with transmitters that continued working into the fall traveled outside Yellowstone National Park to harvest whitebark pine seeds. This confirms that Clark's nutcrackers in Yellowstone National Park range beyond the park, frequenting whitebark pine communities throughout the Greater Yellowstone Ecosystem to supplement the limited local availability. Notably, although our sample size was small, we did not observe any of our tagged birds foraging in stands of limber pine or Douglas-fir, which would have allowed them to remain closer to the breeding areas where we caught them.

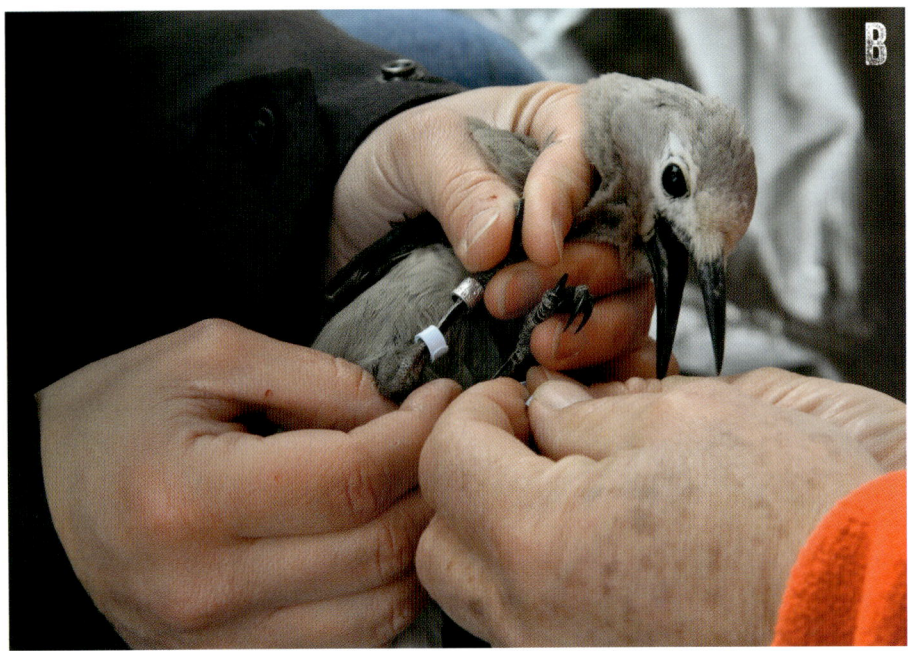

FIGURE 24.7 (A) A radio transmitter safely affixed to a Clark's nutcracker. (B) Banding a Clark's nutcracker, the traditional and most reliable way to mark birds. NPS PHOTOS/DOUGLAS SMITH

CLARK'S NUTCRACKERS IN YELLOWSTONE NATIONAL PARK: PRESENT AND FUTURE

The results from the first three years of this study indicate that whitebark pine, especially, but also limber pine, are the most important food resources for nutcrackers in Yellowstone National Park (Fig. 24.8). Nutcrackers also harvested and cached Douglas-fir seeds later in fall. We have, for the first time, evidence that nutcrackers may eat and possibly cache the small seeds of lodgepole pine on occasion. The GPS tracking information suggests some nutcrackers travel widely, and even outside the park, in search of whitebark pine seeds. We do not yet know whether individuals that leave the park return to the park or travel farther beyond. We hope to answer these and other

FIGURE 24.8 Critical to high-elevation communities due to their seed caching of whitebark pine, which leads to the tree's regeneration, the Clark's nutcracker is a year-round resident with an extraordinary memory for finding stashed seeds. It is also strikingly beautiful. PHOTO BY SCOTT HEPPEL

questions over time. Our current plans are to continue this study, and GPS-tag additional birds, for several more years.

ANNUAL USE OF FOREST COMMUNITIES

The point-count data and associated observations indicated that during July, nutcracker family groups (two parents and recently fledged young) were most frequently encountered in whitebark pine communities and, to some extent, in Douglas-fir communities. In whitebark pine communities, nutcrackers retrieved caches but also harvested unripe seeds from cones to feed begging juveniles, starting in mid-July. In early September, nutcrackers began harvesting whitebark and limber pine seeds and flying off with full throat pouches for seed caching. In fall months, some nutcrackers may travel widely to find whitebark pine communities with cone production.

In September and October, we observed extensive overflights of nutcrackers with full throat pouches from the whitebark pine forests on Mount Washburn, near our Dunraven Pass transect, south toward lodgepole pine communities, where they likely cached seeds. During October, we observed nutcrackers in Douglas-fir communities harvest seeds and then cache them in nearby rocky areas. We know that by mid-March, nutcrackers were breeding in Douglas-fir and mixed spruce-fir-lodgepole pine communities, where they may have access to seed caches made the previous fall. In March 2020, we observed nutcrackers retrieve seed caches from rocky outcrops near a Douglas-fir forest community transect.

THE FUTURE

The decline of Clark's nutcrackers in response to the decline of whitebark pine potentially initiates a downward spiral in both species in Yellowstone National Park. Ideally, we hope our study, and any future population monitoring, may help prevent the loss of the iconic mutualism between Clark's nutcracker and whitebark pine. Information from the GPS-tagged nutcrackers indicates that the spatial scale at which nutcrackers use whitebark pine communities transcends park boundaries. Any plans to manage and restore whitebark pine would be most effective if implemented at the scale of the Greater Yellowstone Ecosystem. Restoration of whitebark pine will not only ensure a future for Clark's nutcrackers, but also that their keystone seed-dispersal services will continue to benefit forest communities throughout the park and beyond.

Visit the Yellowstone's Birds website (https://press.princeton.edu/resources/yellowstones-birds-video-collection/v24) to watch an interview with Diana Tomback.

SONGBIRDS

LAUREN E. WALKER, KATHARINE E. DUFFY, MARY BETH ALBRECHTSEN, and DOUGLAS W. SMITH

With a cheerful song and a bright blue suit, mountain bluebirds are a welcome sight to bird biologists, resident ranchers, vacation rental owners, and visitors alike when they arrive back in northern Wyoming in late March. A colorful contrast to the gray winter landscape, bluebirds herald the beginning of the end of the long winter season. Despite their optimism, however, bluebirds often find themselves battling cold and snowy weather for months before the fleeting summer finally arrives (Fig. 25.1). Throughout the year, over 100 songbird species can be found in Yellowstone, and many, like bluebirds, migrate hundreds or even thousands of miles to breed here each spring and summer. On wintering grounds, along migration corridors, and in the park, the songbirds of Yellowstone National Park (Yellowstone; YNP) face a myriad of challenges.

ACROSS NORTH AMERICA, many songbird species are declining or imperiled due to habitat loss or fragmentation, changing climate and fire regimes, and invasive species. Recent estimates suggest we have lost as many as 3.2 billion songbirds since the 1970s, a period over which other avian guilds, including raptors and waterfowl, have increased (Rosenberg et al. 2019). In many respects, Yellowstone is a haven for songbirds—a refuge from significant human infrastructure and other land-use changes. However, Yellowstone provides its own unique set of challenges. The park is located on a high-elevation plateau with a short dry summer and a potential for snow any month of the year. The landscape is dominated by coniferous forests and sagebrush steppe grasslands, with only small veins of deciduous riparian habitat that provide forage and nesting sites for the greatest diversity of songbirds (Despain 1990). Furthermore, the evolving management of Yellowstone's wildlife, fires, and human visitors since the park's inception has impacted Yellowstone's songbird populations significantly. For example, early in the park's history, invasive grasses were intentionally introduced to portions of the northern range to provide forage for bison and elk (Goodacre 1933, Whittlesey 1995). Those invasive grasses persist today, impacting the composition and structure of the grasslands, and thus the habitat quality for grassland songbirds. The extirpation and more recent reintroduction of large predators to the park had cascading effects to ungulate populations and the habitats on which they graze and browse, further impacting the riparian zones depended upon by many of Yellowstone's songbirds (Beschta and Ripple 2016). Since the park's inception, fire-management policy has fluctuated significantly between "put them all out" and "let them all burn," landing today on a more flexible in-between

strategy that allows for fire control near human infrastructure but acknowledges the important role of wildland fire in shaping functional and resilient ecosystems (Zimmerman 2009). However, as wildfires become more frequent and more severe due to climate change, this policy may still allow vulnerable habitats to disappear (Westerling et al. 2011). Finally, visitation to the park has increased dramatically in the past decades, from 2 million visits a year in 1980 to nearly 4.9 million in 2021, Yellowstone's busiest season on record. More visitors ultimately means more cars on the roads and more hikers on the trails, resulting in increased ambient noise along roadways and in commonly visited areas of the park, as well as widened hiking and social trails across the park. Taken as a whole, these management decisions and policies that have shaped Yellowstone and created the park as it is today have also had deep running ramifications for the composition and structure of the park's vegetation. These impacts to the vegetation additionally impact wildlife, including songbird populations. Moving forward, climate change is predicted to bring additional, unique challenges to the park's managers and songbirds alike.

Like most passerine species across North America, most of Yellowstone's summer songbirds migrate north each spring and south each fall, spending their winters in warmer, more forgiving climates. Similarly, many of the songbirds that nest in areas north of the park pass through the park each spring and fall, making these shoulder seasons particularly exciting for sighting rare species or large groupings of otherwise uncommon or solitary birds. Only a few hardy songbird species, including corvids (jays, crows, and ravens), nuthatches (both red- and white-breasted), mountain chickadee, Townsend's solitaire, and Bohemian waxwing, choose to winter in Yellowstone. Thus, in addition to variability across habitat types, songbird diversity and abundance shift considerably between seasons.

FIGURE 25.1 A mountain bluebird bravely faces a Wyoming spring. PHOTO BY TOM MURPHY

To assess songbird populations in the park and document their status in the face of unknown future management and climate-change impacts, the Yellowstone Bird Program monitors songbirds across seasons and habitat types, using a variety of survey methods. We utilize (1) point counts in the summer to assess diversity and abundance of breeding songbirds along roadways, and in willow, grasslands and sagebrush, and mature forests; (2) line transects in the fall to see which songbirds utilize the park during fall migration; and (3) a songbird-banding station, located in a willow-lined riparian corridor on the northern range, to gain information about songbird numbers, species diversity, productivity, and survival in the summer and fall. In turn, the status of different songbird populations gives us insight into the condition and health of the broader Yellowstone ecosystem.

A WORD ABOUT METHODS

POINT COUNTS

Much bird research, both in general and in Yellowstone, is conducted utilizing a survey method known as point counts. In general, point counts involve standing in a known location, for a set time, and recording the birds you see and/or hear within a specified distance (Fig. 25.2). Some survey protocols call for 3-minute counts (e.g., Breeding Bird Survey counts; BBS), while others are longer. Our breeding season, habitat-specific surveys are 10 minutes long. BBS counts allow for all birds to be included within a 1,312-foot (400-m) radius, while observers conducting

FIGURE 25.2 Howard Weinberg, a biological-science technician with the Yellowstone Bird Program, conducts a point count in the willows of the Lamar Valley.
NPS PHOTO/MARY BETH ALBRECHTSEN

our habitat-specific surveys record all birds detected out to 131 feet (40 m) or 328 feet (100 m) for grasslands. Because point counts are conducted from the same locations, year after year, they can detect changes in songbird abundance or diversity simply by comparing the numbers or types of birds that are seen. More complicated versions of point counts involve recording the distance to each bird detected, which ultimately allows for the calculation of bird density.

During the breeding season, the dawn chorus is the best time of day to conduct point counts—songbirds are active and vocal, and males proclaim their territories with loud song. In Yellowstone, this means being at your survey site by 5:30 a.m. For the biologists, this often means getting out of bed by 3:00 or 3:30 a.m. For some, it's a dreaded part of the field season. Others relish the early mornings, with a quiet chill in the air, just you and the birds (and in Yellowstone, your field partner, and maybe some elk, and hopefully not a bison, moose, or bear). Either way, point counts are a rite of passage for most songbird biologists.

LINE TRANSECTS

Rather than remaining stationary at a single point, line transects, as their name suggests, are conducted by walking along a predetermined path. The observer then records birds seen and heard out to a specified distance. The Bird Program utilizes line transects for our fall migration surveys to help ensure that flocks of songbirds, grouping together in preparation for migration, won't be missed just because they fall outside the allotted time of a point count. Our protocol is relaxed, with observers allowed to take whatever time they want or need, and birds are recorded up to distances of several hundred meters, depending on the habitat type. Fall surveys are also not limited to early mornings; during this time of year, birds aren't singing much but are active and foraging throughout the day to gain enough energy to fly long migratory distances and enough calories to fatten up in preparation for colder weather.

BIRD BANDING

At our banding station, 10 mist nets 39 feet (12 m) long and 6.5 feet (2 m) high are set up through gaps in the willows. When it's done right, and the weather cooperates, the nets are largely invisible to the birds, and they are caught as they fly back and forth between small clumps of vegetation. When the sun gets high and the nets are no longer in the shade, or if it gets windy, as commonly happens in the late morning or early afternoon in Yellowstone, the nets become more visible. Additionally, birds aren't generally as active later in the day; because of that and combined with the effects of sun and wind on the nets, we tend to catch fewer birds as the day goes on.

Once every 10 days during the summer breeding season, we set up our nets at sunrise and leave them open for six hours, checking every 20 minutes to see what we've caught. In what we refer to as our fall migration period (although it's almost entirely in what most of the country calls "summer"), we are more relaxed, setting up nets once a week at a leisurely hour (usually around 8 a.m.) and closing when our capture rate becomes unbearably slow. When we catch a bird, we carry it in a small cotton bag to the banding table, where it is measured, examined, and given a small aluminum bracelet before being released (Fig. 25.3). The metal bands, issued by the United States Geological Survey's (USGS) Bird Banding Lab, each have a unique nine-digit number that, if the bird is eventually recaptured, provides information about where and when the bird was initially banded, and what condition it was in.

BIRDS IN SPECIFIC HABITATS

GRASSLANDS AND SAGEBRUSH STEPPE

"It just kept singing!" This key observation is enough to give away the identity of the mystery bird being described

FIGURE 25.3 Panel of the banding-station process. (A) Biological-science technician Dylan Sanborn retrieves a bird out of a mist net. (B) Banding and measuring a yellow warbler. (C) Biological-science technicians Brenna Cassidy (left) and Mary Beth Albrechtsen (right) discuss a yellow warbler, evaluating molt and body condition. (D) Senior author Lauren Walker releases a spotted towhee, an uncommon Yellowstone capture. NPS PHOTOS: (A–C) BY DOUGLAS SMITH, (D) BY JAKE FRANK

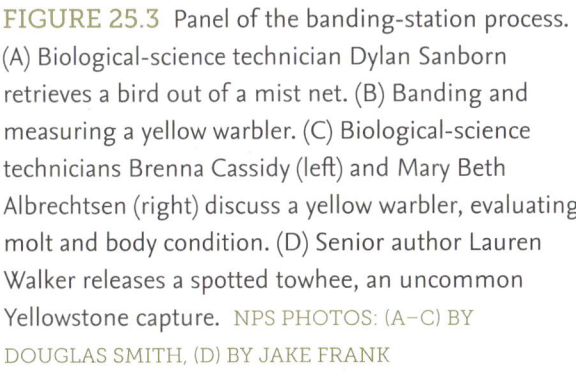

AVIAN COMMUNITY RESPONSES TO BISON GRAZING IN THE NORTHERN RANGE OF YELLOWSTONE NATIONAL PARK—DANIELLE A. FAGRE, WILLIAM M. JANOUSEK, and VICTORIA J. DREITZ

First light in the northern range of Yellowstone National Park is a special time of day in a special place. The sagebrush is wet with dew, making it even more fragrant, steam is rising in the distance from an unseen thermal feature, bison grunt and grumble on the valley floor, and the birds sing with unbridled enthusiasm. You wouldn't know it just from watching, but there is an interesting relationship between bison grazing and the grassland and shrub-steppe songbirds, and Yellowstone National Park is one of few places where we can learn about it. We set out to understand this interaction, and so had the privilege of counting songbirds on many early mornings in the northern range.

This might seem like a surprising or far-fetched interaction. After all, what could bison grazing possibly have to do with songbirds? The key is through the vegetation in songbird habitat. Grassland and shrub-steppe songbirds respond to grazing by large herbivores because grazing alters vegetation structure. Grazing reduces vegetation height (Tastad 2013), increases bare ground (Augustine et al. 2012, Lwiwski et al. 2015), and affects the spatial variability of vegetation (Adler et al. 2001), making it patchier, like a mosaic. When animals like bison graze some areas more frequently than others, it creates different habitat conditions to which songbirds respond, with each species seeking out its preferences in vegetation. Horned larks, for example, seem to prefer areas with short, sparse vegetation

(Rotenberry and Wiens 1980), while clay-colored sparrows prefer areas with a shrub component (Madden et al. 2000). Historically in North America, bison were much more widespread and would have helped create those conditions. We wanted to know whether modern-day bison grazing influences how songbirds use their habitat. Our study investigated the importance of bison grazing to grassland and shrub-steppe songbird species richness (the number of different species in the study area) and abundance (the number of individuals of a species), bringing us one step closer to understanding their habitat needs and guiding conservation for their preservation.

Our study took place in the northern range of Yellowstone National Park, Wyoming, and the National Bison Range (NBR), Montana. There were about 3,500 bison that used Yellowstone's northern range, where our study area was about 344 square miles (890 km²). In contrast, there are about 350 bison in the NBR, which is about 29.3 square miles (76 km²). We counted songbirds by walking transects in which two observers walked in a single file. The first observer stated all birds they saw, while the second observer recorded the observations and any additional birds they saw (Nichols et al. 2000). We also measured bison grazing intensity by counting bison patties (Tastad 2013) along transects.

We found that bison grazing intensity in both study sites has a substantial influence on the number of grassland and shrub-steppe songbird species present

by a curious visitor. Sometimes described (perhaps a bit unfairly), as a "washed-out robin," the sage thrasher more than makes up for a somewhat plain appearance with its incredible vocal prowess (Kaufman 1996). The gray, streaky bird sings a loud and melodious song that can continue for minutes on end without breaks, providing a beautiful acoustic backdrop to Yellowstone's sagebrush steppe.

Sage thrashers are considered sagebrush obligates (they require sagebrush habitat to successfully reproduce and for their populations to flourish), and as their habitat shrinks across the west, so do their populations. In Yellowstone, these birds persist in small numbers, but like many of their grassland and sagebrush neighbors, they remain vulnerable.

FIGURE 25.4 Avian species occupancy responses to bison-grazing intensity, measured by bison patty counts in the National Bison Range and the northern range of Yellowstone National Park. Individual species are represented by different colors and noted by their four-letter code: BRSP = Brewer's sparrow, CCSP = clay-colored sparrow, GRSP = grasshopper sparrow, GTTO = green-tailed towhee, HOLA = horned lark, LASP = lark sparrow, SATH = sage thrasher, SAVS = savannah sparrow, VESP = vesper sparrow, and WEME = western meadowlark.

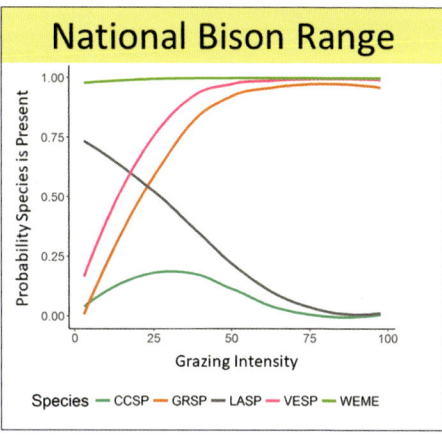

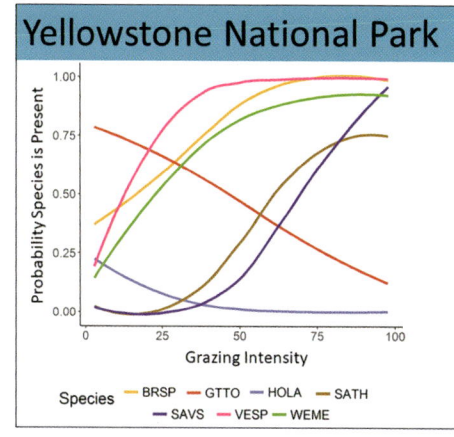

(Fig. 25.4). As bison grazing intensity increases, there are more species present. This pattern peaks at intermediate bison grazing intensities, which supported the highest number of species. In Yellowstone National Park, Brewer's sparrow, sage thrasher, savannah sparrow, vesper sparrow, and western meadowlark were all more likely to be present in plots with higher bison grazing intensity, but green-tailed towhees and horned larks were less likely to be present. In NBR, grasshopper sparrows, vesper sparrows, and western meadowlarks were more likely to be present in plots with higher bison grazing intensity, but clay-colored sparrows and lark sparrows were less likely to be present. We also found that the abundance of vesper sparrows and western meadowlarks increased with higher bison grazing intensity at both study sites, but western

meadowlark abundance responded more positively to the same level of bison grazing intensity in Yellowstone compared to NBR.

Our overall findings show that modern-day bison grazing is associated with habitat conditions for multiple songbird species and supports songbird community diversity. This suggests that having a broad spectrum of grazing intensities across the landscape would help ensure ideal habitat conditions are present for numerous grassland and shrub-steppe songbird species. This relationship is important to understand because this group of songbirds has suffered long-term, nationwide declines. The more we understand about their habitat needs, the better we can safeguard their continued existence, both within and outside of Yellowstone National Park.

Grassland songbirds are among the most imperiled across North America, as many grassland habitats have been invaded by non-native plants, fragmented by infrastructure, or converted to agriculture or rangeland for cattle (Vickery et al. 1999). While Yellowstone grasslands and sagebrush steppe host a variety of invasive weeds, they remain relatively intact and provide forage for dynamic populations of ungulates, including bison (McGarvey and Geremia 2022). In 2016 and 2017, research conducted by University of Montana graduate student Danielle Fagre revealed preliminary insights into the relationship between bison grazing and grassland-bird communities (Fagre 2018). Notably, Danielle's work highlighted a sizable gap in the Bird Program's data collection, and

FIGURE 25.5 Grassland scene. (A) In winter, resident ravens, magpies, and eagles (both species) visit virtually every wolf kill due to winter food limitation. (B) In spring, migrants fill Yellowstone's grasslands, including western meadowlarks, Brewer's and Vesper's sparrows, Brewer's blackbirds, brown-headed cowbirds, red-tailed hawks, and, uncommon but unmistakable due to their exquisitely melodious and variable repertoire of songs, sage thrashers. Wolf kills are much less valuable due to abundant warm-weather food sources. ARTWORK BY JACK DELAP

in 2018, we added grassland and sagebrush steppe surveys across the northern range to our summer and fall repertoire.

Within the greater grassland habitat type, many subtypes exist in Yellowstone. From Gardiner's open-shortgrass habitat dominated by the invasive cheatgrass and hoary alyssum, to the sagebrush steppe of the Blacktail Plateau and the tallgrass of the Lamar Valley, the grasslands of the northern range are variable. Likewise, each grassland type provides foraging and nesting habitat for a unique group of grassland birds. In shortgrass areas with a high intensity of bison grazing and abundant non-natives, the most abundant songbird species during the breeding season are the horned lark, Brewer's blackbird, and western meadowlark. In taller but non-native grasslands, meadowlarks and savannah sparrows are most common. In contrast, sagebrush steppe with fewer non-natives and less grazing pressure host abundant vesper and Brewer's sparrows, along with occasional sagebrush specialists like the sage thrasher (Fig. 25.5). In general, these trends also hold true during fall migration, although some species become more broadly abundant across grassland types, such as Brewer's sparrow, vesper sparrow, and savannah sparrow. Additional species like the mountain bluebird and yellow-rumped warbler also utilize Yellowstone's grasslands in the fall, although they are not generally detected during our breeding-season surveys. In the winter, open grasslands become host to a different type of food resource—carrion. Bison and elk die of winter-kill (they perish in late winter due to old age, injuries, particularly harsh conditions, or lack of available food) and by wolf predation, and their carcasses provide a much-needed source of calories for ravens, magpies, and other hungry scavengers (Fig. 25.5). The year-round patterns in grassland songbird abundance and diversity, better revealed through more years of songbird survey data, may reveal further insights into habitat differences across the grasslands of the northern range and ultimately may help inform management guidelines for both bison populations and invasive plants.

WILLOWS AND RIPARIAN AREAS

As we walk up to the net at a riparian bird-banding station, I get a small rush of adrenaline as I spot a flash of yellow. Head down, caught in the upper trammel of the net, it's a yellow warbler. The most common capture at our songbird banding station, these birds are beautiful—bright yellow with vibrant red streaks on their breast—and every individual that we can band, measure, and evaluate for breeding condition helps us better understand how this songbird, and its willow habitat, are faring in Yellowstone.

After being bagged and banded, and having tolerated numerous measurements, as well as a thorough inspection of feather and body condition, the warbler is subjected to a few final photographs before I open my hand to let him go. He perches for a moment on a finger and then is gone. Back to his life, his nest, doing what warblers do. We hope to see him again next year.

LAUREN E. WALKER

Covering less than 1% of the park, willow-lined riparian corridors are relatively rare within Yellowstone's vast acreage. Throughout the West, however, riparian zones often provide habitat for most of the local avian diversity (Ohmart 1994, Dobkin et al. 1998, Berger et al. 2001). In the park, they are vital breeding and foraging areas for a multitude of songbirds. Willow stands are exceptional in the role they play for songbirds, yet they are incredibly dynamic, at the whim of a variety of abiotic (Tercek et al. 2010) and biotic factors, including vegetation removal by ungulates (Beschta and Ripple 2016).

The extirpation of many of Yellowstone's large predators in the early half of the 20th century allowed prey populations of browsing ungulates to flourish, foraging and moving across the landscape without fear (Laundré et al. 2001). Without large predators, riparian vegetation was heavily browsed, and songbird diversity and abundance decreased in over-browsed riparian corridors

(Jackson 1992). Large-predator recovery began in Yellowstone in the mid-1990s, concurrent with changes to state management outside the park, and populations of elk and other ungulates declined and browsing pressure on riparian vegetation was reduced substantially (Ripple et al. 2001, Fortin et al. 2005, Beschta and Ripple 2016). Furthermore, recent restoration efforts and rebounds in Yellowstone's beaver populations may also aid riparian vegetation recovery (McColley et al. 2012). Willow stands, and the songbird populations that rely on them, have slowly begun to recover (Olechnowski and Debinski 2008, Baril 2009, Baril et al. 2011), and today songbird diversity and abundance in willow stands are useful bellwethers of the health of these habitats.

Willow-songbird communities were first studied in the park in 2005 by Montana State University graduate student Lisa Baril (Baril 2009, Baril et al. 2011). Continuing that early work, the Bird Program began conducting annual breeding-season surveys in riparian areas in 2008, utilizing point-count survey techniques to survey three types of willow stands. Tall willow stands, such as those found in Willow Park, are tall and dense and were never over-browsed. Suppressed willows are heavily browsed and are patchy and short, like the willows found near the confluence of Soda Butte Creek and the Lamar River. Finally, released willow stands, like those along Blacktail Deer Creek, were formerly over-browsed but are recovering, in part due to management and research actions. Over the past 12 years, biologists have observed at least 49 different songbird species in willow habitat during the summer, with yellow warblers the most abundant overall. Although biologists saw greater overall songbird abundance in some survey years, species diversity did not vary greatly through time. In contrast, both songbird abundance and species diversity differ among willow-stand types. Tall and released willow stands host greater species diversity than suppressed stands, and willow specialists like willow flycatchers, gray catbirds, and yellow and Wilson's warblers (Fig. 25.6a,b) are all more abundant in taller willow-stand types. In contrast, suppressed willows appear to provide habitat for generalist and grassland species. We found Brewer's blackbirds and savannah sparrows (Fig. 25.6c) most commonly in suppressed willow stands, while house wrens, European starlings, and green-tailed towhees were only observed in this stand type.

The patterns of songbird distribution in willow stands across the northern range highlight the degree of willow recovery since these surveys were initiated, as predator populations have recovered, and ungulate browse patterns have shifted. Today, released willows, which were once over-browsed, are now largely characterized by tall but dispersed willow shrubs, exhibiting some structural characteristics, like each of the tall and suppressed willow stands. These structural similarities contribute to songbird species overlap, and some species, like song sparrows (Fig. 25.6d) and dark-eyed juncos, are similarly abundant across all three willow-stand types during the summer months. Additionally, several riparian songbirds that were absent from over-browsed, suppressed willow stands 12 years ago, when these surveys began, including yellow warblers (Fig. 25.6a), willow flycatchers, and song sparrows (Fig. 25.6d), are now broadly abundant as willow stands have recovered. In Lisa Baril's graduate work, between 2005 and 2007, Wilson's warblers during the breeding season were found solely in tall willow stands (Baril 2009, Baril et al. 2011). Today, these small but flashy birds can now be observed in all our surveyed willow stands (Fig. 25.6b).

Willow stands are equally important for songbirds in the fall, when these habitats are used by an even broader diversity of species than during the breeding season. During fall migration, we detect numerous birds that we would normally expect in forest (e.g., mountain chickadee, yellow-rumped warbler, and orange-crowned warbler) or grassland (e.g., mountain bluebird, Brewer's sparrow, and vesper sparrow) habitats. Furthermore, some expected willow species shift notably in relative abundance. For example, we detect fewer yellow warblers than Wilson's warblers in the fall, whereas the opposite pattern is seen in the breeding season.

While point-count surveys help to reveal patterns in songbird diversity and abundance between years and between willow-stand types, a songbird banding station

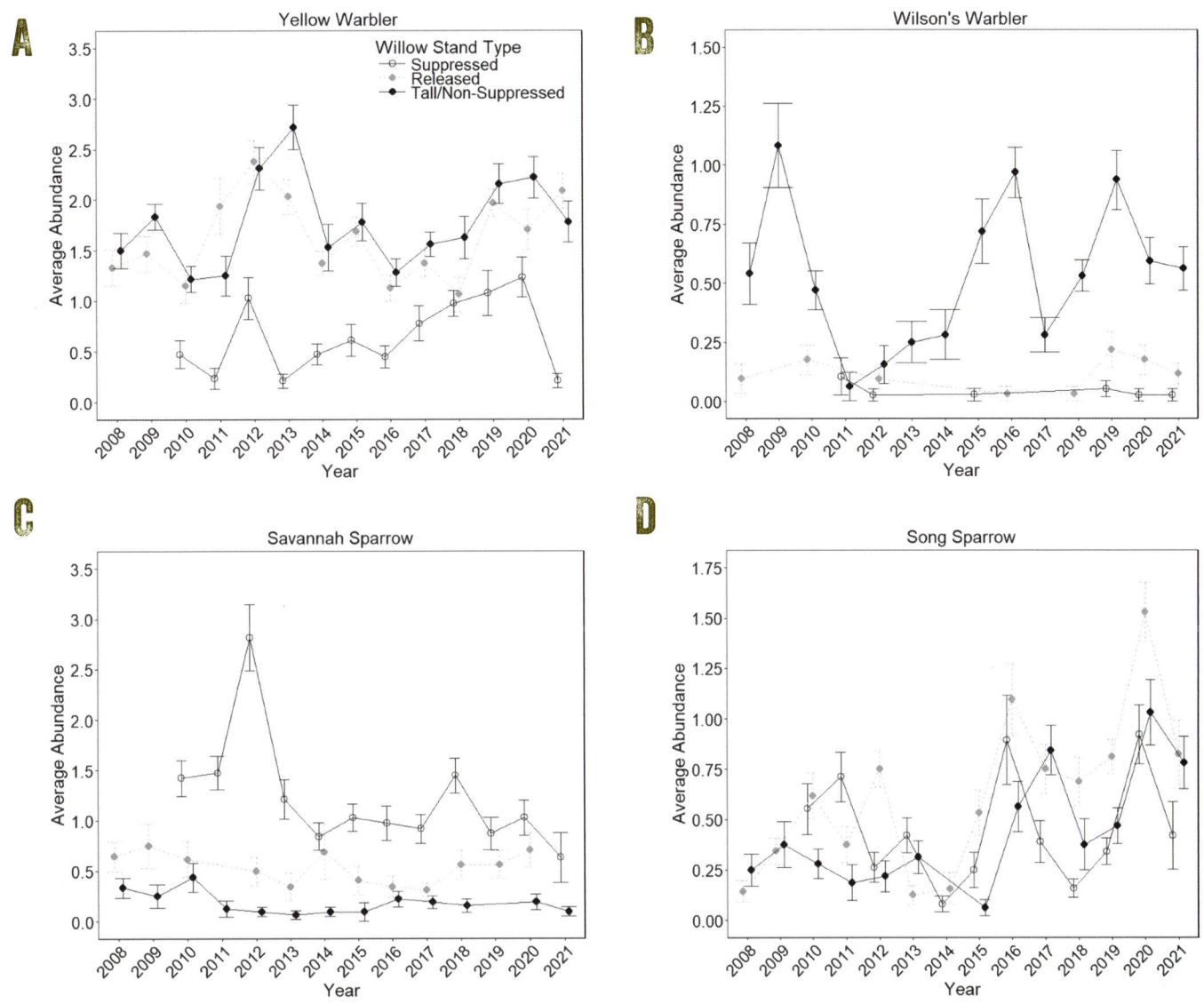

FIGURE 25.6 Average abundance of (A) yellow warbler, (B) Wilson's warbler, (C) savannah sparrow, and (D) song sparrow in three types of riparian/willow communities.

may additionally provide information about songbird reproduction and survival, as well as within-year seasonal differences in diversity and abundance. For the first time in Yellowstone's history, in 2018 we established a bird-banding station in a single "released" willow stand on the northern range. In 2018 and 2019, we followed the protocols of an international bird-banding program known as MAPS (Monitoring Avian Productivity and Survivorship;

www.birdpop.org/pages/maps.php), which compiles data from thousands of bird-banding stations across North America. In 2020 and 2021, we were constrained by Covid-19 restrictions and were unable to follow the regimented MAPS protocol. We did, however, continue banding-station operations through the breeding season as we were able. By banding during the breeding season, between June and early August, we are better able to determine how many birds of which species are nesting in our approximately 5-acre (2-ha) area of study. By catching young birds after they fledge their nests, we can get a sense of productivity, or how successful the breeding season was. And, if we're lucky and those birds return to breed again in the same area in future years, we will be able to estimate annual rates of survival of those young and adult songbirds.

By mid-August, young birds are becoming independent, and songbirds shift their focus from nesting to preparing for their fall migration. A few weeks later, local birds begin to leave, and birds that breed farther north may pass through, heading south. Banding during this period can reveal dramatically different patterns in songbird diversity and abundance than during the summer breeding season, and to help assess use of

riparian habitats by juveniles and migrating songbirds, we continue weekly banding operations through the end of September, weather permitting.

During the first four years of banding-station operation, the Yellowstone Bird Program captured and banded 1,093 passerine and near-passerines (e.g., woodpeckers). Across 22 banding days during the breeding season and 20 days during migration, we banded 453 and 640 songbirds, respectively, belonging to at least 40 different species (Fig. 25.7). Yellow warblers were the most common capture during the breeding season, comprising nearly 21% of all summer captures. While most (63%) yellow warblers recorded were captured in the summer, we also captured 55 during fall migration. Many other common summer residents, however, were almost completely absent from

FIGURE 25.7 Two common songbirds of Yellowstone's riparian zones: (A) yellow warbler and (B) warbling vireo. PHOTOS BY GREG ALBRECHTSEN

fall banding records. More than 92% of recorded American robins and warbling vireos and 88% of gray catbirds were captured in the breeding season. In contrast, Wilson's warblers were rarely caught during the summer but are the most common species recorded in the fall. Migrating Wilson's warblers represented nearly 25% of all fall captures, and over 95% of all Wilson's warblers captures. Ruby-crowned kinglets, white-crowned sparrows, chipping sparrows, and yellow-rumped warblers were also most common during migration, and we captured more than 75% of individuals of these species in the fall.

In total, we captured and banded 434 juvenile birds at the Yellowstone banding station during our first four seasons. Seventy-seven hatch-year birds of 15 species were caught during the breeding season, indicating they were nesting in or nearby our riparian study area. The most common hatch-year bird was the yellow warbler (34 birds, 44% of all juveniles in the breeding season); young gray catbirds (13%) and song sparrows (10%) were also relatively common during the breeding season.

From 2019 through 2021, we recaptured 35 birds banded in a previous season, including six birds initially banded as juveniles. Three individuals banded in 2018, all as adults, were recaptured in 2021 and thus were at least five (one MacGillivray's warbler) and six years old (two warbling vireos). The return and recapture of riparian birds in future years will reveal additional and important information about adult and juvenile survival in Yellowstone's willow habitats.

MATURE FORESTS AND RECENT BURNS

The rising, swirling notes effervesce from the dense and shady vegetation surrounding us. Hiking the slow and steady climb up the Pebble Creek trail, what we've heard is the musical but haunting song of a Swainson's thrush—for anyone who has ever paid attention, it is magical and unforgettable. Common in spruce-fir forests across northern North America, these robin relatives are sensitive to human disturbance and vulnerable to habitat degradation by logging practices, wildfire, and climate change (Mack and Yong 2020). In Yellowstone, a changing fire regime that disproportionately targets mature forest may threaten this species' persistence.

Historically, forest-management guidelines across the country aimed to put out wildfires as quickly as possible. Today Yellowstone's forests are largely allowed to burn, so long as people and important human infrastructure are not threatened. Climate warming, however, may result in more frequent and severe fires in Yellowstone (Westerling et al. 2011, Rocca et al. 2014), and by their very nature, mature forests take a long time to replace. When a mature forest burns, it may take several hundred years to return to its pre-burn state, and if fires become more frequent, it may never get there. Furthermore, more frequent fires may adversely affect forest resilience, preventing the buildup of a sufficient seed bank, and creating an alternative stable state with more open woodlands or even non-forest habitat (Westerling et al. 2011, Turner et al. 2019). To document the songbird diversity that relies on mature forests in the park, in 2017 the Bird Program began conducting point-count surveys in three forest types that vary in structure and conifer diversity. Using the results of these, along with surveys of burned forests conducted between 2010 and 2017, we hope to highlight the importance of maintaining a diversity of forest types in Yellowstone at various stages of growth and post-fire recovery.

More diverse forests, with a greater variety of tree species as well as understory cover, generally support more birds; they provide important nesting habitat and foraging opportunities for more individual birds from a more diverse suite of species (Fig. 25.8). Although the exact composition depends on a variety of landscape qualities, including elevation, precipitation, soil composition, and fire history, Yellowstone's forests are largely a mix of Douglas-fir, subalpine fir, Engelmann spruce, and lodgepole pine (Despain 1990). Since the initiation of annual forest surveys in 2017, Bird Program biologists have observed 37 different species of passerines and woodpeckers utilizing forest stands in the breeding season. In surveys of lodgepole pine forests, biologists find approximately

FIGURE 25.8 American three-toed woodpecker has a larva (immature form) of a wood-boring beetle in its beak. PHOTO BY HOWARD WEINBERG

half the number of total birds, and fewer than half the number of bird species, compared with more diverse forests of pine, spruce, and fir (Fig. 25.9).

Avian species composition also varies between forest types. As forests age and mature, they generally provide better habitat for cavity-nesting species, like woodpeckers and swallows, that often require larger trees and standing snags (Carey et al. 1991). While some species like the mountain chickadee (Fig. 25.10) and ruby-crowned kinglet are equally abundant across the surveyed forest types, others become more abundant with increasing forest complexity

(e.g., dark-eyed junco and yellow-rumped warbler; Fig. 25.10). In fact, many of Yellowstone's forest songbirds breed only in more mature forest stands with mixed tree composition. Flycatchers (dusky, Hammond's, and olive-sided), woodpeckers (hairy woodpecker and northern flicker), and forest specialists, like the golden-crowned kinglet and western tanager, are all absent from lodgepole-dominant forests during the breeding season. In the fall, however, forest songbirds seem less picky; diversity somewhat evens out, and species like brown creeper and golden-crowned kinglet become more equally abundant across forest types.

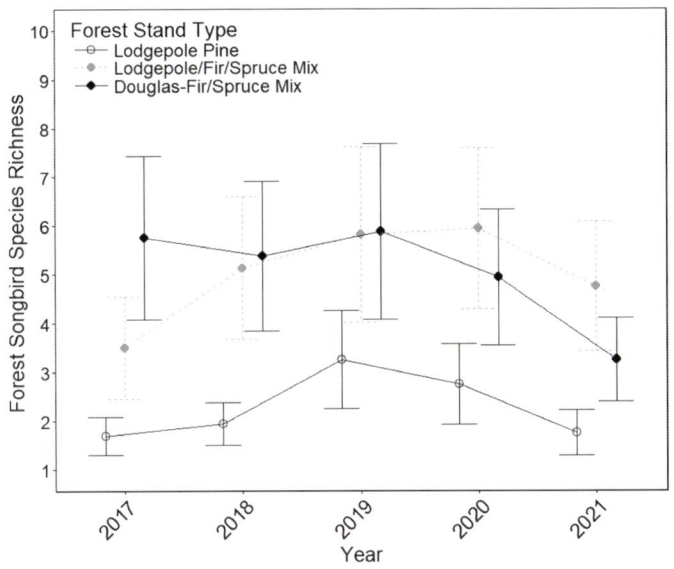

FIGURE 25.9 Average songbird species richness in three forest types, 2017–2021.

▼ FIGURE 25.10 Average abundance of (A) mountain chickadee and (B) yellow-rumped warbler in three forest types, 2017–2021.

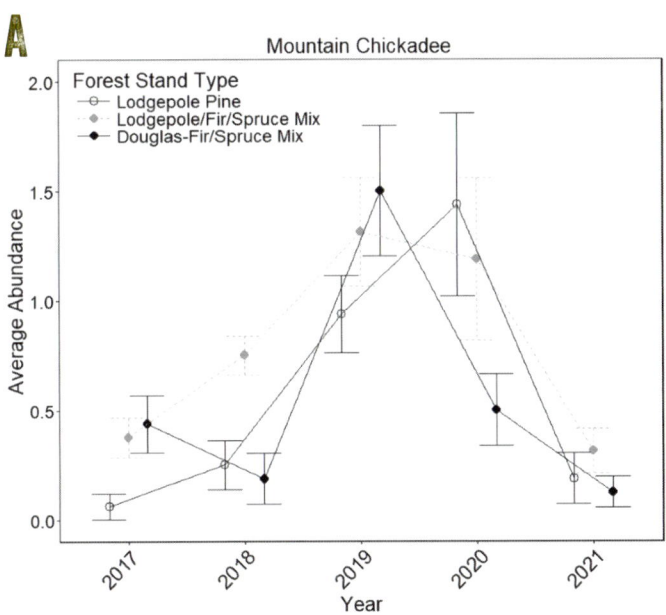

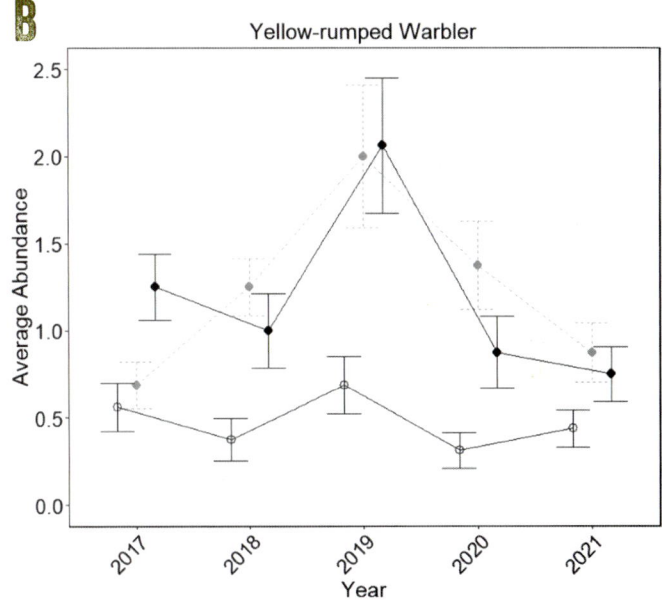

EFFECTS OF WILDFIRE

Dynamic landscape processes such as wildfire are important engineers of forest composition and structure. Immediately after fire, burned areas have ample snags and insects for foraging and may provide ideal habitat for many cavity nesters and post-fire specialists, like the black-backed woodpecker. However, the conditions are short-lived, and Bird Program surveys indicate that woodpecker activity peaks approximately two years after the burn and then begins to decline again. In much of Yellowstone, initial forest regrowth after fires is dominated by the serotinous-coned lodgepole pine (Turner et al. 2019). Therefore, although fire may initially benefit some bird species, regrowth conditions that lean toward lodgepole monocultures often represent a decline in habitat quality for breeding songbirds.

SONGBIRDS, CLIMATE CHANGE, AND LOOKING FORWARD

Within Yellowstone, climate-change effects on ecosystem processes are largely unknown, especially with respect to birds. Detecting changes in phenology (the timing of seasonal events such as migration or the onset of breeding) or shifts in species' breeding or wintering ranges will help inform management decisions and add to our understanding of the impacts of climate change on Yellowstone's songbird community.

To assess phenological shifts in response to climate change, the Bird Program has recorded the spring arrival dates of common migrant songbirds to Yellowstone's northern range since 2005. Although arrival dates have varied for many species, the arrival of American robins has become notably earlier over the past 15 years. Birds may initiate their spring migration earlier due to an early spring (or a warmer winter; Swanson and Palmer 2009) or a shorter migration, as more birds are able to winter in areas local to their breeding grounds (Brown and Miller 2016).

As the climate changes, a songbird species' breeding or wintering range may expand, retract, or shift across its former expanse (e.g., VanDerWal et al. 2013, Gillings et al. 2015). In the Greater Yellowstone Ecosystem, these shifts could result in changes to both the abundance of individual species and overall songbird diversity. Observations from Bird Program surveys, banding-station records, rare bird reports from park visitors (www.nps.gov/yell/learn/nature/wildlife-sightings.htm), and citizen-science efforts such as eBird (www.ebird.org) are vital for documenting these changes through time. For example, while many of the songbird species captured at the banding station are relatively common in Yellowstone, we have also captured several species considered rare or unusual to the area, including the northern waterthrush and spotted towhee. With the progression of climate change, both species are predicted to expand their breeding ranges into Yellowstone (Wu et al. 2018), and an increase in observations or capture records in future years would help confirm those expectations.

FOR THE BIRDS

The next time you are out in the park, standing with a pack of wolf watchers in the Lamar Valley or stuck in traffic as a herd of bison mosey lazily across the road, take a moment to appreciate the songbirds that no doubt surround you. Look for their bright colors or flitting movements, and listen for their cheery songs and ever-present calls. Much can be learned about the birds, and the state of the park, from a few moments of observation.

As a program of the National Park Service (NPS), the Yellowstone Bird Program abides by the NPS mission and aims to monitor and document the status of Yellowstone's bird populations. As biologists, however, we study birds because they intrigue and inspire us. The professional and personal pursuits intersect with the common goals of ensuring that the birds and the ecosystem persist, for their own right and for the enjoyment of visitors to the park. Many of the current songbird research and monitoring projects in Yellowstone have begun relatively recently, in only the past few years. With continued focus, however, these efforts will increasingly contribute meaningful insights into the park's songbird populations, as the years fly by.

Visit the Yellowstone's Birds website (https://press.princeton.edu/resources/yellowstones-birds-video-collection/v25) to watch an interview with Lauren Walker.

SPECIES NAMES USED IN TEXT

American avocet (*Recurvirostra americana*)

American bison (*Bison bison*)

American black bear (*Ursus americanus*)

American black duck (*Anas rubripes*)

American coot (*Fulica americana*)

American crow (*Corvus brachyrhynchos*)

American dipper (*Cinclus mexicanus*)

American kestrel (*Falco sparverius*)

American mink (*Mustela vison*)

American pine marten (*Martes americana*)

American pipit (*Anthus rubescens*)

American robin (*Turdus migratorius*)

American salmonfly (*Pteronarcys dorsata*)

American three-toed woodpecker (*Picoides dorsalis*)

American white pelican (*Pelecanus erythrorhynchos*)

American wigeon (*Mareca americana*)

American woodcock (*Scolopax minor*)

Baird's sandpiper (*Calidris bairdii*)

Bald eagle (*Haliaeetus leucocephalus*)

Bank swallow (*Riparia riparia*)

Barn swallow (*Hirundo rustica*)

Barrow's goldeneye (*Bucephala islandica*)

Bighorn sheep (*Ovis canadensis*)

Biting flies (*Stomoxys* spp.)

Bitterroot (*Lewisia rediviva*)

Black cottonwood (*Populus trichocarpa*)

Black rosy-finch (*Leucosticte atrata*)

Black-backed woodpecker (*Picoides arcticus*)

Black-billed magpie (*Pica hudsonia*)

Black-capped chickadee (*Poecile atricapillus*)

Black-headed grosbeak (*Pheucticus melanocephalus*)

Black-necked stilt (*Himantopus mexicanus*)

Black-tailed jackrabbit (*Lepus californicus*)

Blue jay (*Cyanocitta cristata*)

Bobcat (*Lynx rufus*)

Bohemian waxwing (*Bombycilla garrulus*)

Boreal owl (*Aegolius funereus*)

Brewer's blackbird (*Euphagus cyanocephalus*)

Brewer's sparrow (*Spizella breweri*)

Brine fly (*Ephydra cinerea*)

Brown creeper (*Certhia americana*)

Bufflehead (*Bucephala albeola*)

Burrowing owl (*Athene cunicularia*)

California condor (*Gymnogyps californianus*)

California gull (*Larus californicus*)

Calliope hummingbird (*Selasphorus calliope*)

Canada goose (*Branta canadensis*)

Canada jay (*Perisoreus canadensis*)

Canvasback (*Aythya valisineria*)

Caspian tern (*Hydroprogne caspia*)

Cassin's finch (*Haemorhous cassinii*)

Chestnut-collared longspur (*Calcarius ornatus*)

Chipping sparrow (*Spizella passerina*)

Chukar (*Alectoris chukar*)

Cinnamon teal (*Spatula cyanoptera*)

Clark's grebe (*Aechmophorus clarkii*)

Clark's nutcracker (*Nucifraga columbiana*)

Clay-colored sparrow (*Spizella pallida*)

Cliff swallow (*Petrochelidon pyrrhonota*)

Common loon (*Gavia immer*)

Common merganser (*Mergus merganser*)

Common raven (*Corvus corax*)

Common yellowthroat (*Geothlypis trichas*)

Cooper's hawk (*Accipiter cooperii*)

SPECIES NAMES USED IN TEXT

Cougar (*Puma concolor*)

Coyote (*Canis latrans*)

Dark-eyed junco (*Junco hyemalis*)

Deer mouse (*Peromyscus maniculatus*)

Desert cottontail (*Sylvilagus audubonii*)

Domestic cat (*Felis catus*)

Domestic chicken (*Gallus domesticus*)

Domestic dog (*Canis lupus familiaris*)

Double-crested cormorant (*Nannopterum auritum*)

Douglas-fir (*Pseudotsuga menziesii*)

Dusky flycatcher (*Empidonax oberholseri*)

Dusky grouse (*Dendragapus obscurus*)

Eared grebe (*Podiceps nigricollis*)

Eastern kingbird (*Tyrannus tyrannus*)

Eastern white pine (*Pinus strobus*)

Elk (*Cervus canadensis*)

Engelmann spruce (*Picea engelmannii*)

European starling (*Sturnus vulgaris*)

Ferruginous hawk (*Buteo regalis*)

Flammulated owl (*Psiloscops flammeolus*)

Franklin's gull (*Leucophaeus pipixcan*)

Gadwall (*Mareca strepera*)

Golden eagle (*Aquila chrysaetos*)

Golden-crowned kinglet (*Regulus satrapa*)

Grasshopper sparrow (*Ammodramus savannarum*)

Gray catbird (*Dumetella carolinensis*)

Gray partridge (*Perdix perdix*)

Gray wolf (*Canis lupus*)

Gray-crowned rosy-finch (*Leucosticte tephrocotis*)

Great blue heron (*Ardea herodias*)

Great gray owl (*Strix nebulosa*)

Great horned owl (*Bubo virginianus*)

Greater prairie-chicken (*Tympanuchus cupido*)

Greater sage-grouse (*Centrocercus urophasianus*)

Greater scaup (*Aythya marila*)

Greater white-fronted goose (*Anser albifrons*)

Greater yellowlegs (*Tringa melanoleuca*)

Green-tailed towhee (*Pipilo chlorurus*)

Green-winged teal (*Anas crecca*)

Grizzly bear (*Ursus arctos horribilis*)

Hairy woodpecker (*Dryobates villosus*)

Hammond's flycatcher (*Empidonax hammondii*)

Harlequin duck (*Histrionicus histrionicus*)

Hermit thrush (*Catharus guttatus*)

Hooded merganser (*Lophodytes cucullatus*)

Horned grebe (*Podiceps auritus*)

Horned lark (*Eremophila alpestris*)

Horsehair lichen (*Bryoria fremontii*)

House finch (*Haemorhous mexicanus*)

House wren (*Troglodytes aedon*)

Indian paintbrush (*Castilleja* spp.)

Island fox (*Urocyon littoralis*)

Killdeer (*Charadrius vociferus*)

Lake trout (*Salvelinus namaycush*)

Lark sparrow (*Chondestes grammacus*)

Lazuli bunting (*Passerina amoena*)

Least sandpiper (*Calidris minutilla*)

Lesser scaup (*Aythya affinis*)

Lesser yellowlegs (*Tringa flavipes*)

Limber pine (*Pinus flexilis*)

Lincoln's sparrow (*Melospiza lincolnii*)

Lodgepole pine (*Pinus contorta*)

Long-billed curlew (*Numenius americanus*)

Long-billed dowitcher (*Limnodromus scolopaceus*)

Long-billed murrelet (*Brachyramphus perdix*)

Long-eared owl (*Asio otus*)

Long-tailed weasel (*Mustela frenata*)

MacGillivray's warbler (*Geothlypis tolmiei*)

Mallard (*Anas platyrhynchos*)

Marbled godwit (*Limosa fedoa*)

Marsh wren (*Cistothorus palustris*)

Meadow vole (*Microtus pennsylvanicus*)

Merlin (*Falco columbarius*)

Montane vole (*Microtus montanus*)

Moose (*Alces alces*)

Mountain bluebird (*Sialia currucoides*)

Mountain chickadee (*Poecile gambeli*)

Mountain cottontail (*Sylvilagus nuttallii*)

Mountain goat (*Oreamnos americanus*)

Mountain pine beetle (*Dendroctonus ponderosae*)

Mule deer (*Odocoileus hemionus*)

Muskrat (*Ondatra zibethicus*)

Mute swan (*Cygnus olor*)

Narrowleaf cottonwood (*Populus angustifolia*)

North American beaver (*Castor canadensis*)

North American river otter (*Lontra canadensis*)

Northern flicker (*Colaptes auratus*)

Northern goshawk (*Accipiter gentilis*)

Northern harrier (*Circus hudsonius*)

Northern pintail (*Anas acuta*)

Northern pocket gopher (*Thomomys talpoides*)

Northern pygmy-owl (*Glaucidium gnoma*)

Northern rough-winged swallow (*Stelgidopteryx serripennis*)

Northern saw-whet owl (*Aegolius acadicus*)

Northern shoveler (*Spatula clypeata*)

Northern waterthrush (*Parkesia noveboracensis*)

Olive-sided flycatcher (*Contopus cooperi*)

Orange-crowned warbler (*Leiothlypis celata*)

Osprey (*Pandion haliaetus*)

Passenger pigeon (*Ectopistes migratorius*)

Pectoral sandpiper (*Calidris melanotos*)

Peregrine falcon (*Falco peregrinus*)

Pileated woodpecker (*Dryocopus pileatus*)

Pine siskin (*Spinus pinus*)

Pinyon jay (*Gymnorhinus cyanocephalus*)

Pinyon pine (*Pinus* spp.)

Piping plover (*Charadrius melodus*)

Prairie dog (*Cynomys* spp.)

Prairie falcon (*Falco mexicanus*)

Pronghorn (*Antilocapra americana*)

Quaking aspen (*Populus tremuloides*)

Racoon (*Procyon lotor*)

Rats (*Muridae* spp.)

Red crossbill (*Loxia curvirostra*)

Red fox (*Vulpes vulpes*)

Red phalarope (*Phalaropus fulicarius*)

Red-backed vole (*Clethrionomys* spp.)

Red-breasted nuthatch (*Sitta canadensis*)

Red-necked grebe (*Podiceps grisegena*)

Red-necked phalarope (*Phalaropus lobatus*)

Red-tailed hawk (*Buteo jamaicensis*)

Red-winged blackbird (*Agelaius phoeniceus*)

Ring-necked duck (*Aythya collaris*)

Ring-necked pheasant (*Phasianus colchicus*)

Rock wren (*Salpinctes obsoletus*)

Rocky Mountain juniper (*Juniperus scopulorum*)

Ross's goose (*Anser rossii*)

Rough-legged hawk (*Buteo lagopus*)

Rubber rabbitbrush (*Ericameria nauseosa*)

Ruby-crowned kinglet (*Corthylio calendula*)

Ruddy duck (*Oxyura jamaicensis*)

Ruffed grouse (*Bonasa umbellus*)

Rufous hummingbird (*Selasphorus rufus*)

Rusty blackbird (*Euphagus carolinus*)

Sage thrasher (*Orescoptes montanus*)

Sandhill crane (*Antigone canadensis*)

Savannah sparrow (*Passerculus sandwichensis*)

Semipalmated plover (*Charadrius semipalmatus*)

Semipalmated sandpiper (*Calidris pusilla*)

Sharp-shinned hawk (*Accipiter striatus*)

Sharp-tailed grouse (*Tympanuchus phasianellus*)

Snow goose (*Anser caerulescens*)

Snowshoe hare (*Lepus americanus*)

Solitary sandpiper (*Tringa solitaria*)

Song sparrow (*Melospiza melodia*)

Sooty grouse (*Dendragapus fuliginosus*)

Sora (*Porzana carolina*)

Spotted sandpiper (*Actitis macularia*)

Spotted towhee (*Pipilo maculatus*)

Spruce (*Picea* spp.)

Spruce budworm (*Choristoneura fumiferana*)

Spruce grouse (*Canachites canadensis*)

Steller's jay (*Cyanocitta stelleri*)

Surf scoter (*Melanitta perspicillata*)

Swainson's hawk (*Buteo swainsoni*)

Swainson's thrush (*Catharus ustulatus*)

Townsend's solitaire (*Myadestes townsendi*)

Tree swallow (*Tachycineta bicolor*)

Trumpeter swan (*Cygnus buccinator*)

Tundra swan (*Cygnus columbianus*)

Turkey vulture (*Cathartes aura*)

Uinta ground squirrel (*Urocitellus armatus*)

Varied thrush (*Ixoreus naevius*)

Vesper sparrow (*Pooecetes gramineus*)

SPECIES NAMES USED IN TEXT

Violet-green swallow (*Tachycineta thalassina*)

Virginia rail (*Rallus limicola*)

Warbling vireo (*Vireo gilvus*)

Western grebe (*Aechmophorus occidentalis*)

Western meadowlark (*Sturnella neglecta*)

Western sandpiper (*Calidris mauri*)

Western tanager (*Piranga ludoviciana*)

Western white pine (*Pinus monticola*)

Whitebark pine (*Pinus albicaulis*)

White-breasted nuthatch (*Sitta carolinensis*)

White-crowned sparrow (*Zonotrichia leucophrys*)

White-faced ibis (*Plegadis chihi*)

White-tailed jackrabbit (*Lepus townsendii*)

White-tailed ptarmigan (*Lagopus leucura*)

White-winged scoter (*Melanitta fusca*)

Whooper swan (*Cygnus cygnus*)

Wild turkey (*Meleagris gallopavo*)

Willet (*Tringa semipalmata*)

Williamson's sapsucker (*Sphyrapicus thyroideus*)

Willow (*Salix* spp.)

Willow flycatcher (*Empidonax traillii*)

Willow ptarmigan (*Lagopus lagopus*)

Wilson's phalarope (*Phalaropus tricolor*)

Wilson's snipe (*Gallinago delicata*)

Wilson's warbler (*Cardellina pusilla*)

Wood duck (*Aix sponsa*)

Yellow warbler (*Setophaga petechia*)

Yellow-bellied marmot (*Marmota flaviventris*)

Yellow-headed blackbird (*Xanthocephalus xanthocephalus*)

Yellow-rumped warbler (*Setophaga coronata*)

Yellowstone cutthroat trout (*Oncorhynchus clarkii bouvieri*)

BIBLIOGRAPHY

Ackerman, J. T., C. A. Eagles-Smith, M. P. Herzog, C. A. Hartman, S. H. Peterson, D. C. Evers, A. K. Jackson, J. E. Elliot, S. S. Vander Pol, and C. E. Bryan. 2016. Avian mercury exposure and toxicological risk across western North America: A synthesis. *Sci Total Environ* 568:749–769.

Adler, P. B., D. A. Raff, and W. K. Lauenroth. 2001. The effect of grazing on the spatial heterogeneity of vegetation. *Oecologia* 128:465–479.

Alerstam, T., and A. Hedenstrom. 1998. The development of bird migration theory. *J Avian Biol* 29:343–369.

Alt, K. L. 1980. Ecology of the breeding bald eagle and osprey in the Grand Teton–Yellowstone National Parks complex. MS thesis, Montana State University.

Anctil, A., A. Franke, and J. Bêty. 2014. Heavy rainfall increases nestling mortality of an arctic top predator: Experimental evidence and long-term trend in peregrine falcons. *Oecologia* 174:1033–1043.

Andersen, D. E. 1995. Productivity, food habits, and behavior of Swainson's hawks breeding in southeast Colorado. *J Raptor Res* 29:158–165.

Anderson, D. W. 1972. Eggshell changes in certain North American birds. In *Proceedings—XVth International Ornithological Congress*, edited by K. H. Voous. Leiden, Netherlands: Brill Publishers.

Anthony, A. W. 1903. Migration of Richardson's grouse. *Auk* 20:24–27.

Augustine, D. J., D. T. Booth, S. E. Cox, and J. D. Derner. 2012. Grazing intensity and spatial heterogeneity in bare soil in a grazing-resistant grassland. *Rangel Ecol Manag* 65:39–46.

Bailey, V. O. 1915. Field notes, Wyoming, Montana, and Yellowstone National Park, July 10–October 2, 1915. Accessed January 16, 2021. www.biodiversitylibrary.org/item/237392#page/1/mode/1up.

———. 1916. Field notes, Yellowstone National Park, February 29–April 14, 1916, and California, September 4–October 3, 1916. Accessed January 16, 2021. www.biodiversitylibrary.org/item/237393#page/1/mode/1up.

———. 1926. Field notes, Yellowstone National Park, Michigan, and Minnesota, July 3–September 17, 1926, and California, October 20–December 1, 1926. Accessed January 16, 2021. www.biodiversitylibrary.org/item/237526#page/1/mode/1up.

———. 1930. *Animal Life of Yellowstone National Park*, Springfield, IL, and Baltimore, MD: Charles C. Thomas.

Baldasarre, G. 2014. *Ducks, Geese, and Swans of North America*, vol. 2, Baltimore, MD: Johns Hopkins University Press.

Baldasarre, G., and E. G. Bolen. 2006. *Waterfowl Ecology and Management*, 2nd ed., Malabar, FL: Krieger Publishing Company.

Banko, W. E. 1960. *The Trumpeter Swan: Its History, Habits, and Population in the United States.* Washington, DC: US Government Printing Office.

Baril, L. M. 2009. Change in deciduous woody vegetation, implications of increased willow (*Salix* spp.) growth for bird species diversity, and willow species composition in and around Yellowstone National Park's Northern Range. MS thesis, Montana State University.

Baril, L. M., A. J. Hansen, R. Renkin, and R. Lawrence. 2011. Songbird response to increased willow (*Salix* spp.) growth in Yellowstone's northern range. *Ecol Appl* 21:2283–2296.

Baril, L. M., D. W. Smith, T. Drummer, and T. M. Koel. 2013. Implications of cutthroat trout decline for breeding ospreys and bald eagles at Yellowstone Lake. *J Raptor Res* 47:234–245.

Baril, L. M., D. B. Haines, D. W. Smith, and R. J. Oakleaf. 2015. Long-term reproduction (1984–2013), nestling diet, and eggshell thickness of peregrine falcons (*Falco peregrinus*) in Yellowstone National Park. *J Raptor Res* 49:347–358.

Baril, L. M., D. B. Haines, L. E. Walker, and D. W. Smith. 2017a. Autumn raptor migration in Yellowstone National Park, 2011–2015. *Can Field-Nat* 131:303–311.

Baril, L. M., D. W. Smith, D. B. Haines, L. E. Walker, and K. Duffy. 2017b. *Yellowstone Raptor Initiative: 2011–2015 Final Report.* YCR-2017–04. Yellowstone National Park, WY: National Park Service.

Barringer, L. E., D. F. Tomback, M. B. Wunder, and S. T. McKinney. 2012. Whitebark pine stand condition, tree abundance, and cone production as predictors of visitation by Clark's nutcracker (*Nucifraga columbiana*). *PLOS One* 7 (5):e37663. Doi:0.1371/journal.pone.0037663.

Barrows, M. 1936. *Trumpeter Swans of Yellowstone National Park, Summer 1936.* Internal report. Yellowstone National Park, WY: National Park Service.

BIBLIOGRAPHY

Bates, W. J., and M. O. Morretti. 1994. Golden eagle (*Aquila chrysaetos*) population ecology in eastern Utah. *Gt Basin Nat* 54:248–255.

Bechard, M. J. 1982. Effect of vegetative cover on foraging site selection by Swainson's hawk. *Condor* 84:153–159.

Bechard, M. J., C. S. Houston, J. H. Sarasola, and A. S. England. 2020a. Swainson's hawk (*Buteo swainsoni*), version 1.0. In *Birds of the World*, edited by A. F. Poole. Ithaca, NY: Cornell Lab of Ornithology. doi.org/10.2173/bow.swahaw.01.

Bechard, M. J., T. R. Swem, J. Orta, P.F.D. Boesman, E.F.J. Garcia, and J. S. Marks. 2020b. Rough-legged Hawk (*Buteo lagopus*), version 1.0. In *Birds of the World*, edited by S. M. Billerman. Ithaca, NY: Cornell Lab of Ornithology. doi.org/10.2173/bow.rolhaw.01.

Bent, A. C. 1961. *Falco peregrinus anatum*: Duck hawk. In *Life Histories of North American Birds of Prey*. New York: Dover Publications.

Berger, J., P. B. Stacey, L. Bellis, and M. P. Johnson. 2001. A mammalian predator-prey imbalance: Grizzly bear and wolf extinction affect avian neotropical migrants. *Ecol Appl* 11:947–960.

Beschta, R. L., and W. J. Ripple. 2016. Riparian vegetation recovery in Yellowstone: The first two decades after wolf reintroduction. *Biol Conserv* 198:93–103.

Bildstein, K. L., K. D. Meyer, C. M. White, J. S. Marks, and G. M. Kirwan. 2020. Sharp-shinned hawk (*Accipiter striatus*), version 1.0. In *Birds of the World*, edited by S. M. Billerman, B. K. Keeney, P. G. Rodewald, and T. S. Schulenberg. Ithaca, NY: Cornell Lab of Ornithology. doi.org/10.2173/bow.shshaw.01.

Billerman, S. M., B. K. Keeney, P. G. Rodewald, and T. S. Schulenberg. 2022. *Birds of the World*. Cornell Laboratory of Ornithology, Ithaca, NY. Last modified 2022. Accessed Sept 2019–June 2020. https://birdsoftheworld.org/bow/home.

Bird, D. M. 2009. The American kestrel: From common to scarce? *J Raptor Res* 43:261–262.

Birdlife International. 2018. *Gavia immer*. The IUCN Red List of Threatened Species 2018: e.T22697842A132607418.

Bortolotti, G. R. 1984. Trap and poison mortality of golden and bald eagles. *J Wildl Manage* 48:1173–1179.

Bradley, M., R. Johnstone, R. Court, and T. Duncan. 1997. Influence of weather on breeding success of peregrine falcons in the Arctic. *Auk* 114:786–791.

Braun, A., and T. Bugnyar. 2012. Social bonds and rank acquisition in raven nonbreeder aggregations. *Anim Behav* 84:1507–1515.

Brodrick, H. J. 1952. *Birds of Yellowstone National Park*, Yellowstone National Park, Yellowstone, WY.

Bromley, C. K., and G. A. Hood. 2013. Beavers (*Castor canadensis*) facilitate early access by Canada geese (*Branta canadensis*) to nesting habitat and areas of open water in Canada's boreal wetlands. *Mamm Biol* 78:73–77.

Brook, R. W., D. C. Duncan, J. E. Hines, S. Carrière, and R. G. Clark. 2005. Effects of small mammal cycles on productivity of boreal ducks. *Wildlife Biol* 11:3–11.

Brook, R. W., M. Pasitschniak-Arts, D. W. Howerter, and F. Messier. 2008. Influence of rodent abundance on nesting success of prairie waterfowl. *Can J Zool* 86 (6):497–506.

Brown, D., and G. Miller. 2016. Band recoveries reveal alternative migration strategies in American robins. *Anim Migr* 3:35–47.

Brown, J. L. 1974. Alternate routes to sociality in jays—with a theory for the evolution of altruism and communal breeding. *Am Zool* 14:63–80.

Bull, E. L., and J. R. Duncan. 2020. Great gray owl (*Strix nebulosa*), version 1.0. In *Birds of the World*, edited by S. M. Billerman. Ithaca, NY: Cornell Lab of Ornithology. doi.org/10.2173/bow.grgowl.01.

Cade, T. J. 1969. The northern peregrine populations. In *Peregrine Falcon Populations: Their Biology and Decline*, edited by J. J. Hickey. Madison: University of Wisconsin Press.

———. 1982. *Falcons of the World*. Ithaca, NY: Comstock/Cornell University Press.

Cade, T. J., J. H. Enderson, and J. Linthicum. 1996. *Guide to Management of Peregrine Falcons at the Eyrie*. Boise, ID: Peregrine Fund.

Carey, A. B., M. M. Hardt, S. P. Horton, and B. L. Biswell. 1991. *Spring Bird Communities in the Oregon Coast Range*. Pacific Northwest Research Station, Olympia, WA: US Department of Agriculture and US Forest Service.

Chang, T., A. J. Hansen, and N. Piekielek. 2014. Patterns and variability of projected bioclimatic habitat for *Pinus alvicaulis* in the Greater Yellowstone Area. *PLOS One* 9 (11):e111669.

Chapman, F. M. 1901. A Christmas Bird Census. In *Bird-Lore*, vol. 3. New York: MacMillan Co.

Chesser, R. T., S. M. Billerman, K. J. Burns, C. Cicero, J. L. Dunn, A. W. Kratter, I. J. Lovette, N. A. Mason, P. C. Rasmussen, J. V. Remsen Jr., D. F. Stotz, and K. Winkler. 2020. Sixty-first supplement to the American Ornithological Society's check-list of North American birds. *Auk* 137 (3):ukaa030.

Childs, F. W. 1934. *Trumpeter Swans of Yellowstone National Park and Adjacent Areas*. Internal report. Yellowstone National Park, WY: National Park Service.

Clark, S. G. 2021. *Yellowstone's Survival: A Call to Action for a New Conservation Story*. New York: Anthem Press.

Cockrell, L. E. 2014. Landsat Evaluation of Trumpeter Swan Historical Nesting Sites in Yellowstone National Park. MS thesis, Eastern Kentucky University.

Comstock, T. B. 1874. The Yellowstone National Park. *Am Nat* 8:65–79.

Condon, D.D.L. 1941. *Preliminary Report on the Trumpeter Swan of Yellowstone National Park*. Yellowstone National Park, WY: National Park Service.

Craighead, D., and R. Smith. 2002. *Breeding Raptor Census in Grand Teton National Park, 2002*. Annual Report 26:21–25. Laramie: University of Wyoming National Park Service Research Center.

Crandall, R. H., B. E. Bedrosian, and D. Craighead. 2015. Habitat selection and factors influencing nest survival of golden eagles in south-central Montana. *J Raptor Res* 49:413–428.

Crook, C. 1936. Further additions to the list of the birds of Yellowstone National Park. *Wilson Bull* 48 (2):136–137.

Davis, W. B., and J. Stevenson. 1934. The type localities of three birds collected by Lewis and Clark in 1806. *Condor* 36:161–163.

Despain, D. G. 1990. *Yellowstone Vegetation: Consequences of Environment and History in a Natural Setting.* Boulder, CO: Roberts Rinehart Publishers.

Diem, K. L., and D. D. Condon. 1967. *Banding Studies of Water Birds on the Molly Islands, Yellowstone Lake, Wyoming.* Yellowstone National Park, WY: Yellowstone Library and Museum Association.

Diffendorfer, J. E., J. C. Stanton, J. A. Beston, W. E. Thogmartin, S. R. Loss, T. E. Katzner, D. H. Johnson, R. A. Erickson, M. D. Merrill, and M. D. Corum. 2021. Demographic and potential biological removal models identify raptor species sensitive to current and future wind energy. *Ecosphere* 12 (6):e03531.

Dobkin, D. S., A. C. Rich, J. A. Pretare, and W. H. Pyle. 1995. Nest-site relationships among cavity-nesting birds of riparian and snowpocket aspen woodlands in the northwestern Great Basin. *Condor* 97:694–707.

Dobkin, D. S., A. C. Rich, and W. H. Pyle. 1998. Habitat and avifaunal recovery from livestock grazing in a riparian meadow system of the northwestern Great Basin. *Conserv Biol* 12:209–221.

Dos Anjos, L., S.J.S. Debus, S. C. Madge, and J. M. Marzluff. 2009. Family *Corvidae* (crows). In *Handbook of Birds of the World*, edited by J. del Joyo, A. Eliott, and D. A. Christie. Barcelona: Lynx Editions.

Dudley, N. 2008. *Guidelines for Applying Protected Area Management Categories.* Gland, Switzerland: IUCN Publication Services.

Duerr, A. E., T. A. Miller, M. Lanzone, D. Brandes, J. Cooper, K. O'Malley, C. Maisonneuve, J. A. Tremblay, and T. Katzner. 2015. Flight response of slope-soaring birds to seasonal variation in thermal generation. *Funct Ecol* 29:779–790.

Elbroch, M. L., C. O'Malley, M. Peziol, and H. B. Quigley. 2017. Vertebrate diversity benefiting from carrion provided by pumas and other subordinate, apex felids. *Biol Conserv* 215:123–131.

Elliott, C. R., and M. M. Hektner. 2000. *Wetland Resources of Yellowstone National Park.* Yellowstone National Park, WY: National Park Service.

Enderson, J. H. 1969. Population trends among peregrine falcons in the Rocky Mountain region. In *Peregrine Falcon Populations: Their Biology and Decline*, edited by J. J. Hickey. Madison: University of Wisconsin Press.

Estep, J. A. 1989. *Biology, Movements, and Habitat Relationships of the Swainson's Hawk in the Central Valley of California, 1986–87.* Department of Fish and Game, Wildlife Management Division, Nongame Bird and Mammal Program.

Evers, D. C. 1993. A replicable capture method for adult and juvenile common loons on their nesting lakes. In *Proceedings—1992 Conference by the North American Loon Fund*, edited by L. Morse, S. Stockwell, and M. Pokras. Washington, DC: US Fish and Wildlife Service.

———. 1994. Activity budgets of a marked common loon (*Gavia immer*) nesting population. *Hydrobiologia* 279–280:415–420.

———. 2007. *Status Assessment and Conservation Plan for the Common Loon (Gavia immer) in North America: 2007.* BRI Report 2007–20. Hadley, MA: US Department of the Interior and US Fish and Wildlife Service.

———. 2018. The effects of methylmercury on wildlife: A comprehensive review and approach for interpretation. In *Encyclopedia of the Anthropocene*, edited by D. A. Dellasala and M. I. Goldstein. Amsterdam: Elsevier. doi.org/10.1016/B978-0-12-809665-9.09985-7.

Evers, D. C., J. D. Kaplan, M. W. Meyer, P. S. Reaman, A. Major, N. Burgess, and W. E. Braselton. 1998. Bioavailability of environmental mercury measured in common loon feathers and blood across North America. *Environ Toxicol Chem* 17:173–183.

Evers, D. C., and K. M. Taylor. 2014. *Journey with the Loon.* Minocqua, WI: Willow Creek Press.

Fagre, D. A. 2018. Avian Community Responses to Bison Grazing in North American Intermountain Grasslands. MS thesis, University of Montana.

Farmer, C. J., L. J. Goodrich, E. R. Inzunza, and J. P. Smith. 2008. Conservation status of North America's birds of prey. In *State of North America's Birds of Prey. Series in Ornithology*, edited by J. Smith. Washington, DC: American Ornithologists' Union.

Farmer, C. J., and J. P. Smith. 2009. Migration monitoring indicates widespread declines of American kestrels (*Falco sparverius*) in North America. *J Raptor Res* 43:263–273.

Ferrer, M., and J. A. Donazar. 1996. Density-dependent fecundity by habitat heterogeneity in an increasing population of Spanish imperial eagles. *Ecology* 77:69–74.

Fieberg, J., and L. Börger. 2012. Could you please phrase "home range" as a question? *J Mammal* 93:890–902.

Follett, D. 1985. Birds of Yellowstone and Grand Teton National Parks. Yellowstone, WY: Yellowstone Library and Museum Association, and National Park Service.

Fortin, D., H. L. Beyer, M. S. Boyce, D. W. Smith, T. Duchesne, and J. S. Mao. 2005. Wolves influence elk movements: Behavior shapes a trophic cascade in Yellowstone National Park. *Ecology* 86:1320–1330.

Frank, D. A., D. J. McNaughton, and B. F. Tracy. 1998. The ecology of the earth's grazing ecosystems. *BioScience* 48 (7):513–521.

Franklin, A. B. 1987. Breeding Biology of the Great Gray Owl in Southeastern Idaho and Northwestern Wyoming. MS thesis, Humboldt State University.

Fritts, S. H., E. E. Bangs, J. A. Fontaine, W. G. Brewster, and J. F. Gore. 1995. Restoring wolves to the northern Rocky Mountains of the United States. In *Ecology and Conservation of Wolves in a Changing World*, edited by L. D. Carbyn, S. H. Fritts, and D. R. Seip. Edmonton, AB: Canadian Circumpolar Institute.

BIBLIOGRAPHY

Fuller, A. B., and B. P. Bole, Jr. 1930. Observations on some Wyoming birds. *Sci Pub Clevel* 1:37–80.

Fuller, M. R., W. S. Seegar, and L. S. Schueck. 1998. Routes and travel rates of migrating peregrine falcons *Falco peregrinus* and Swainson's hawks *Buteo swainsoni* in the western hemisphere. *J Avian Biol* 29:433–440.

Furniss, M. M., and R. Renkin. 2003. Forest entomology in Yellowstone National Park, 1923–1957: A time of discovery and learning to let live. *Am Entomol* 49:198–209.

Gale, R. S., E. O. Garton, and I. J. Ball. 1987. *The History, Ecology and Management of the Rocky Mountain Population of Trumpeter Swans.* Montana Cooperative Wildlife Research Unit, Missoula, MT: US Fish and Wildlife Service.

Gillings, S., D. E. Balmer, and R. J. Fuller. 2015. Directionality of recent bird distribution shifts and climate change in Great Britain. *Global Change Biol* 21:2155–2168.

Goeking, S. A., and M. A. Windmuller-Campione. 2021. Comparative species assessments of five-needle pines throughout the western United States. *For Ecol Manage* 496:119438.

Goldstein, M. I., B. Woodbridge, M. E. Zaccagnini, and S. B. Canavelli. 1996. An assessment of mortality of Swainson's hawks on wintering grounds in Argentina. *J Raptor Res* 30:106–107.

Goodacre, E. 1933. *Yellowstone National Park Buffalo Ranch Type Map.* Dated August 10 and traced by G. Christensen. Washington, DC: US Department of the Interior, National Park Service.

Goodwin, D. 1986. *Crows of the World*, 2nd ed. Seattle: University of Washington Press.

Grier, J. W. 1982. Ban of DDT and subsequent recovery of reproduction in bald eagles. *Science* 218:1232–1235.

Grilli, M. G., K. L. Bildstein, and S. A. Lambertucci. 2019. Nature's clean-up crew: Quantifying ecosystem services offered by a migratory avian scavenger on a continental scale. *Ecosyst Serv* 39:100990.

Gunther, K. A., R. A. Renkin, J. C. Halfpenny, S. M. Gunther, T. Davis, P. Schullery, and L. Whittlesey. 2009. Presence and distribution of white-tailed jackrabbits in Yellowstone National Park. *Yellowstone Science* 17:24–32.

GYE Committee. 2022. *Greater Yellowstone Ecosystem Common Loon Conservation Plan.* Laramie: Wyoming Department of Game and Fish.

Hackworth, Z. J., J. J. Cox, J. M. Felch, and M. D. Weegman. 2019. A growing conspiracy: Recolonization of common ravens (*Corvus corax*) in central and southern Appalachia, USA. *Southeast Nat* 18:281–296.

Haines, D. B. 2020. Golden Eagle Resource Selection and Environmental Drivers of Reproduction in the Northern Range of Yellowstone National Park. MS thesis, University of Montana.

Hall, L. S., P. R. Krausman, and M. L. Morrison. 1997. The habitat concept and a plea for standard terminology. *Wildl Soc Bull (1973–2006)*, 25:173–182.

Hansen, A. J., and L. Phillips. 2018. Trends in vital signs for greater Yellowstone: Application of a wildland health index. *Ecosphere* 9 (8):e02380.

Hansen, H. A. 1973. Trumpeter swan management. *Wildfowl* 24:27–32.
Harju, S. M., C. V. Olson, J. E. Hess, and B. Bedrosian. 2018. Common raven movement and space use: Influence of anthropogenic subsidies within greater sage-grouse nesting habitat. *Ecosphere* 9:e02348.

Harju, S. M., C. V. Olson, J. E. Hess, and B. Bedrosian. 2018. Common raven movement and space use: influence of anthropogenic subsidies within greater sage-grouse nesting habitat. *Ecosphere* 9(7), e02348.

Harmata, A. 1982. What is the function of undulating flight display in golden eagles? *J Raptor Res* 16:103–109.

Harrington, F. H. 1978. Ravens attracted to wolf howling. *Condor* 80:236–237.

Hayward, G. D., and P. H. Hayward. 2020. Boreal owl (*Aegolius funereus*), version 1.0. In *Birds of the World*, edited by S. M. Billerman. Ithaca, NY: Cornell Lab of Ornithology. doi.org/10.2173/bow.borowl.01.

Heinrich, B. 1988. Why do ravens fear their food? *Condor* 90:950–952.

———. 1994. Dominance and weight changes in the common raven, *Corvus corax. Anim Behav* 48:1463–1465.

———. 1999. *Mind of the Raven: Investigations and Adventures with Wolf-Birds.* New York: Harper Perennial.

Heinrich, B., and J. M. Marzluff. 1991. Do ravens yell because they want to attract others? *Behav Ecol and Sociobiol* 28:13–21.

Heinrich, B., and J. M. Marzluff. 1995. Why ravens share. *Am Sci* 83:342–349.

Heinrich, B., J. M. Marzluff, and W. Adams. 1995. Fear and food recognition in naive common ravens. *Auk* 112:499–503.

Heinrich, B., J. M. Marzluff, and C. S. Marzluff. 1993. Ravens are attracted to the appeasement calls of discoverers when they are attacked at defended food. *Auk* 110:247–254.

Henny, C. J., and H. M. Wight. 1972. Population ecology and environmental pollution: Red-tailed and Cooper's hawks. *Wildl Res Report* 2:229–250.

Herring, G., C. A. Eagles-Smith, and J. Buck. 2017. Characterizing golden eagle risk to lead and anticoagulant rodenticide exposure: A review. *J Raptor Res* 51:273–292.

Higuera, P. E., B. N. Shuman, and K. D. Wolf. 2021. Rocky Mountain subalpine forests now burning more than any time in recent millennia. *PNAS* 118:e2103135118.

Hodges, K. E., L. S. Mills, and K. M. Murphy. 2009. Distribution and abundance of snowshoe hares in Yellowstone National Park. *J Mammal* 90:870–878.

Hoffman, S., and J. Smith. 2003. Population trends of migratory raptors in western North America, 1977–2001. *Condor* 105:397–419.

Hollenbeck, J. P., and W. J. Ripple. 2008. Aspen snag dynamics, cavity-nesting birds, and trophic cascades in Yellowstone's northern range. *For Ecol Manage* 255:1095–1103.

Holt, D. W., and J. L. Petersen. 2020. Northern pygmy-owl (*Glaucidium gnoma*), version 1.0. In *Birds of the World*, edited by A. F. Poole and F. B. Gill. Ithaca, NY: Cornell Lab of Ornithology. doi.org/10.2173/bow.nopowl.01.

Houston, D. B. 1982. *The Northern Yellowstone Elk: Ecology and Management.* New York: Macmillan Publishing.

Hunt, G. 2002. *Golden Eagles in a Perilous Landscape: Predicting the Effects of Mitigation for Wind Turbine Blade-strike Mortality.* Consultant report to California Energy Commission under contract P500-02-043F. Sacramento: California Energy Commission.

Hutchins, H. H., and R. M. Lanner. 1982. The central role of Clark's nutcracker in the dispersal and establishment of whitebark pine. *Oecologia* 55:192–201.

IGBST (Interagency Grizzly Bear Study Team). 2015. *Yellowstone Grizzly Bear Investigations 2014: Report of the Interagency Grizzly Bear Study Team.* US Department of the Interior and US Geological Survey.

Jackson, S. G. 1992. Relationships among Birds, Willows, and Native Ungulates in and around Northern Yellowstone National Park. MS thesis, Utah State University.

Janes, S. W. 1984. Influences of territory composition and interspecific competition on red-tailed hawk reproductive success. *Ecology* 65:862–870.

Johnsgard, P. A. 1968. *Waterfowl: Their Biology and Natural History.* Lincoln: University of Nebraska Press.

———. 1975. *Waterfowl of North America.* Bloomington, IN, and London: Indiana University Press.

———. 2009. *Birds of the Rocky Mountains, with Particular Reference to National Parks in the Northern Rocky Mountain Region.* Lincoln: University of Nebraska Press.

———. 2013. *Yellowstone Wildlife: Ecology and Natural History of the Greater Yellowstone Ecosystem.* Boulder: University Press of Colorado.

Johnson, M. D. 2007. Measuring habitat quality: A review. *Condor* 109:489–504.

Kaeding, L. R., G. D. Boltz, and D. G. Carty. 1996. Lake trout discovered in Yellowstone Lake threaten native cutthroat trout. *Fisheries* 21:16–20.

Kamil, A. C., and R. P. Balda. 1985. Cache recovery and spatial memory in Clark's nutcracker (*Nucifraga columbiana*). *J Exp Psychol Anim Behav Process* 11:95–111. doi.org/10.1037/0097-7403.11.1.95.

Katzner, T. E., M. N. Kochert, K. Steenhof, C. L. McIntyre, E. H. Craig, and T. A. Miller. 2020. Golden eagle (*Aquila chrysaetos*), version 2.0. In *Birds of the World*, edited by P. G. Rodewald and B. K. Keeney. Ithaca, NY: Cornell Lab of Ornithology. doi.org/10.2173/bow.goleag.02.

Katzner, T. E., M. J. Stuber, V. A. Slabe, J. T. Anderson, J. L. Cooper, L. L. Rhea, and B. A. Millsap. 2018. Origins of lead in populations of raptors. *Anim Conserv* 21:232–240.

Kaufman, K. 1996. *Lives of North American Birds (Peterson Natural History Companions).* Boston: Houghton Mifflin Company.

Kear, J., ed. 2005. *Ducks, Geese and Swans: Species Accounts (Cairina to Mergus),* vol 2. Oxford, England: Oxford University Press.

Kemsies, E. 1930. Birds of the Yellowstone National Park, with some recent additions. *Wilson Bull* 42:198–210.

———. 1935. Changes in the list of birds of Yellowstone National Park. *Wilson Bull* 47: 68–70.

Kirk, D. A., and M. J. Mossman. 2020. Turkey vulture (*Cathartes aura*), version 1.0. In *Birds of the World*, edited by A. F. Poole and F. B. Gill. Ithaca, NY: Cornell Lab of Ornithology. doi.org/10.2173/bow.turvul.01.

Kneeland, M. R., E. Berman, T. Grade, J. Cooley, H. Vogel, N. Schoch, C. Cray, V. Stout, D. Evers, and M. Pokras. 2020. Plasma biochemistry and protein electrophoresis reference intervals of the common loon (*Gavia immer*). *J Zoo Wildl Med* 51:561–570.

Knight, W. C. 1902. *The Birds of Wyoming.* Wyoming Experiment Station Bulletin No. 55. Laramie: University of Wyoming, Agricultural College Department.

Kochert, M. N., and K. Steenhof. 2002. Golden eagles in the U.S. and Canada: Status, trends, and conservation challenges. *J Raptor Res* 36:32–40.

Kochert, M. N., K. Steenhof, C. L. McIntyre, and E. H. Craig. 2002. Golden eagle (*Aquila chrysaetos*). Account 684 In *The Birds of North America*, edited by A. Poole and F. Gill. Philadelphia: Academy of Natural Sciences, and Washington, DC: American Ornithologists' Union.

Koel, T. M., J. L. Arnold, P. E. Bigelow, C. R. Detjens, P. D. Doepke, B. D. Ertel, and D. J. MacDonald. 2019a. *Yellowstone National Park Native Fish Conservation Program Report 2015–2018.* YCR-2019–04. Yellowstone National Park, WY: National Park Service.

Koel, T. M., P. E. Bigelow, P. D. Doepke, B. D. Ertel, and D. L. Mahony. 2005. Nonnative lake trout result in Yellowstone cutthroat trout decline and impacts to bears and anglers. *Fisheries* 30:10–19.

Koel, T. M., L. M. Tronstad, J. L. Arnold, K. A. Gunther, D. W. Smith, J. M. Syslo, and P. J. White. 2019b. Predatory fish invasion induces within and across ecosystem effects in Yellowstone National Park. *Sci Adv* 5:eaav1139.

Kohl, M. T., T. K. Ruth, M. C. Metz, D. R. Stahler, D. W. Smith, P. J. White, and D. R. MacNulty. 2019. Do prey select for vacant hunting domains to minimize a multi-predator threat? *Ecology Lett* 22:1724–1733.

Krugman, S. L., and J. L. Jenkinson. 1974. Pinus L. Pine. In *Seeds of Woody Plants in the United States*, edited by C. S. Schopmeyer. USDA Forest Service, Agriculture Handbook 450, Washington, DC.

Langford, N. P. 1905. Diary of the Washburn expedition to the Yellowstone and Firehole rivers in the year 1870. St. Paul, MN: J. E. Haynes.

Lanner, R. M. 1980. Avian seed dispersal as a factor in the ecology and evolution of limber and whitebark pines. In *Proceedings—Sixth North American Forest Biology Workshop*, edited by B. P. Dancik and K. O. Higginbotham. Edmonton: University of Alberta.

Lanner, R. M., and S. B. Vander Wall. 1980. Dispersal of limber pine seed by Clark's nutcracker. *Journal For Res* 78:637–639.

Larison, B., A. R. Lindsay, C. Bossu, M. D. Sorenson, J. D. Kaplan, D. C. Evers, J. Paruk, J. M. DaCosta, T. B. Smith, and K. Ruegg. 2021. Leveraging genomics to understand threats to migratory birds. *Evol Appl* 14:1646–1658.

Latta, B. C., D. E. Driscoll, J. L. Linthicum, R. E. Jackman, and G. Doney. 2005. Capture and translocation of golden eagles from the California Channel Islands to mitigate depredation of endemic island foxes. In *Proceedings—Sixth California Islands Symposium*, edited by D. K. Garcelon and C. A. Schwemm. Arcata, CA: Institute for Wildlife Studies.

Laundré, J. W., L. Hernández, and K. B. Altendorf. 2001. Wolves, elk, and bison: Reestablishing the "landscape of fear" in Yellowstone National Park, U.S.A. *Can J Zoo* 79:1401–1409.

Lemke, T. O., J. A. Mack, and D. B. Houston. 1998. Winter range expansion by the northern Yellowstone elk herd. *Intermountain Journal of Sciences* 4:1–9.

Lenard, S., J. Carlson, J. Ellis, C. Jones, and C. Tilly, eds. 2003. *P. D. Skaar's Montana Bird Distribution*, 6th ed. Helena: Montana Audubon.

Lewis, M., W. Clark, and Members of the Corps of Discovery. 2002. *The Journals of the Lewis and Clark Expedition*. Lincoln: University of Nebraska Press.

Linnell, J.D.C., R. Aanes, J. E. Swenson, J. Odden, and M. E. Smith. 1997. Translocation of carnivores as a method for managing problem animals: A review. *Biodiversity and Conserv* 6:1245–1257.

Lorenz, K. 1935. Der Kumpan in der Umwelt des Vogels. *Journal für Ornithologie* 83:137–215, 289–413.

Lorenz, T. J., K. A. Sullivan, A. V. Bakian, and C. A. Aubrey. 2011. Cache-site selection in Clark's nutcracker (*Nucifraga columbiana*). *Auk* 128:237–247.

Loretto, M., S. Reimann, R. Schuster, D. M. Graulich, and T. Bugnyar. 2016a. Shared space, individually used: Spatial behaviour of non-breeding ravens (*Corvus corax*) close to a permanent anthropogenic food source. *J Ornithol* 157:439–450.

Loretto, M.-C., R. Schuster, and T. Bugnyar. 2016b. GPS tracking of non-breeding ravens reveals the importance of anthropogenic food sources during their dispersal in the eastern Alps. *Curr Zool*, 62 (4):337–344. doi.org/10.1093/cz/zow016.

Loretto, M.-C., R. Schuster, C. Itty, P. Marchand, F. Genero, and T. Bugnyar. 2017. Fission-fusion dynamics over large distances in raven non-breeders. *Sci Rep* 7:380. doi.org/10.1038/s41598-017-00404-4.

Lwiwski, T. C., N. Koper, and D. C. Henderson. 2015. Stocking rates and vegetation structure, heterogeneity, and community in a northern mixed-grass prairie. *Rangeland Ecol Manage* 68:322–331.

Mack, D. E., and W. Yong. 2020. Swainson's thrush (*Catharus ustulatus*), version 1.0. In *Birds of the World*, edited by A. F. Poole and F. B. Gill. Ithaca, NY: Cornell Lab of Ornithology. doi.org/10.2173/bow. swathr.01.

MacNulty, D. R., D. R. Stahler, T. Wyman, J. Ruprecht, L. M. Smith, M. T. Kohl, and D. W. Smith. 2020. Population dynamics of northern Yellowstone elk after wolf reintroduction. In *Yellowstone Wolves: Science and Discovery in the World's First National Park*, edited by D. W. Smith, D. R. MacNulty, and D. R. Stahler. Chicago: University of Chicago Press.

Madden, E. M., R. K. Murphy, A. J. Hansen, and L. Murray. 2000. Models for guiding management of prairie bird habitat in northwestern North Dakota. *Am Midl Nat* 144:377–392.

Marchand, P., M. C. Loretto, P. Y. Henry, O. Duriez, F. Jiguet, T. Bugnyar, and C. Itty. 2018. Relocations and one-time disturbance fail to sustainably disperse non-breeding common ravens *Corvus corax* due to homing behaviour and extensive home ranges. *Eur J Wildl Res* 64:57. doi.org/10.1007/s10344-018-1217-7.

Marks, J. S., D. L. Evans, and D. W. Holt. 2020. Long-eared owl (*Asio otus*), version 1.0. In *Birds of the World*, edited by S. M. Billerman. Ithaca, NY: Cornell Lab of Ornithology. doi.org/10.2173/bow.loeowl.01.

Marzluff, J. M., and T. Angell. 2005. *In the Company of Crows and Ravens*. New Haven, CT: Yale University Press.

Marzluff, J. M., R. B. Boone, and G. W. Cox. 1994. Historical changes in populations and perceptions of native pest bird species in the West. *Stud Avian Biol* 15:202–220.

Marzluff, J. M., and B. Heinrich. 1991. Foraging of vagrant common ravens in the presence and absence of territorial adults: An experimental analysis of social foraging. *Anim Behav* 42:755–770.

Marzluff, J. M., B. Heinrich, and C. S. Marzluff. 1996. Communal roosts of common ravens are mobile information centers. *Anim Behav* 51:89–103.

Marzluff, J. M., M. Loretto, C. K. Ho, G. W. Coleman, and M. Restani. 2022. Thinking like a raven: Restoring integrity, stability, and beauty to western ecosystems. *Hum-Wildl Interact* 15 (3):22.

Marzluff, J. M., and C. Marzluff. 2011. *Dog Days, Raven Nights*. New Haven, CT: Yale University Press.

Marzluff, J. M., and E. Neatherlin. 2006. Corvid response to human settlements and campgrounds: Causes, consequences, and challenges for conservation. *Biol Conserv* 130:301–314.

Matchett, M. R., and B. W. O'Gara. 1987. Methods of controlling golden eagle depredation on domestic sheep in southwestern Montana. *J Raptor Res* 21:85–94.

McClure, C.J.W., R. Van Buskirk, B. P. Pauli, and J. A. Heath. 2017. Commentary: Research recommendations for understanding the decline of American kestrels (*Falco sparverius*) across much of North America. *J Raptor Res* 51:455–464.

McClure, C.J.W., and S. E. Schulwitz. 2022. Historical accounts provide inference into population dynamics of American kestrels (*Falco sparverius*) in the Northeastern USA. *J Raptor Res* 56:89–94.

McColley, S. D., D. B. Tyers, and B. F. Sowell. 2012. Aspen and willow restoration using beaver on the northern Yellowstone winter range. *Restor Ecol* 20:450–455.

McEneaney, T. 1986. *Movements and Habitat Use Patterns of the Centennial Valley Trumpeter Swan Population (Montana) as Determined by Radio Telemetry Data*. Lima, MT: Red Rocks National Wildlife Refuge.

———. 1988. *Birds of Yellowstone*. Boulder, CO: Roberts Rinehart Publishers.

———. 2004. A whooper swan (*Cygnus cygnus*) at Yellowstone National Park, Wyoming, with comments on North American reports of the species. *North American Birds* 58:301–308.

———. 2006. *Yellowstone Bird Report, 2005*. YCR-20062. Yellowstone National Park, WY: National Park Service.

McEneaney, T., B. Heinrich, and B. Oakleaf. 1998. Greater Yellowstone peregrine falcons: Their trials, tribulations, and triumphs. *Yellowstone Sci* 6:16–21.

McEneaney, T., and R. Sjostrom. 1986. *Trumpeter Swan Movements and Seasonal Use, Centennial Valley (Montana): An Analysis of Neck-banding Data*. Lima, MT: Red Rock Lakes National Wildlife Refuge.

McGarvey, L., and C. Geremia. 2022. How do Yellowstone's ungulates make a home on the range? *Yellowstone Science* 28:16–24.

McIntyre, C. L., and J. H. Schmidt. 2012. Ecological and environmental correlates of territory occupancy and breeding performance of migratory golden eagles *Aquila chrysaetos* in interior Alaska. *Ibis* 154:124–135.

McKinney, S. T., C. E. Fiedler, and D. F. Tomback. 2009. Invasive pathogen threatens bird-pine mutualism: Implications for sustaining a high-elevation ecosystem. *Ecol Appl* 19:597–607.

McLaren, T. H. 2022. Clark's Nutcracker Forest Community Use in Yellowstone National Park and Analysis for Long-term Monitoring Efforts. MS thesis, University of Colorado.

Mearns, E. A. 1902. Carded notes on the birds observed in the Yellowstone National Park and vicinity, July 26–July 31, 1889, and April 19–December 16, 1902. Accessed January 16, 2021. www.biodiversitylibrary.org/item/269229.

Mech, L. D. 1966. *The Wolves of Isle Royale*. Washington, DC: US National Park Service and US Government Printing Office.

———. 1970. *The Wolf: The Ecology and Behavior of an Endangered Species*. Minneapolis: University of Minnesota Press.

Mewaldt, L. R. 1952. Reproduction and Molt in Clark's nutcracker *Nucifraga columbiana* Wilson. PhD diss., State College of Washington.

———. 1956. Nesting behavior of the Clark nutcracker. *Condor* 58:3–23.

Millsap, B. A., G. S. Zimmerman, W. L. Kendall, J. G. Barnes, M. A. Braham, B. E. Bedrosian, D. A. Bell, P. H. Bloom, R. H. Crandall, R. Domenech, D. Driscoll, A. E. Duerr, R. Gerhardt, S.E.J. Gibbs, A. R. Harmata, K. Jacobson, T. E. Katzner, R. N. Knight, J. M. Lockhart, C. McIntyre, R. K. Murphy, S. S. Slater, B. W. Smith, J. P. Smith, D. W. Stahlecker, and J. W. Watson. 2022. Age-specific survival rates, causes of death, and allowable take of golden eagles in the western United States. *Ecol Appl* 32 (3):e2544.

Millsap, B. A., G. S. Zimmerman, J. R. Sauer, R. M. Nielson, M. Otto, E. Bjerre, and R. Murphy. 2013. Golden eagle population trends in the western United States: 1968–2010. *J Wildl Manage* 77:1436–1448.

Mitchell, C. D., and M. W. Eichholz. 2020. Trumpeter swan (*Cygnus buccinator*), version 1.0. In *Birds of the World*, edited by P. G. Rodewald. Ithaca, NY: Cornell Lab of Ornithology. doi.org/10.2173/bow.truswa.01.

Mitro, M. G., D. C. Evers, M. W. Meyer, and W. H. Piper. 2008. Common loon survival rates and mercury in New England and Wisconsin. *J Wildl Manage* 72:665–673.

Mowbray, T. B., C. R. Ely, J. S. Sedinger, and R. E. Trost. 2020. Canada goose (*Branta canadensis*), version 1.0. In *Birds of the World*, edited by A. F. Poole and F. B. Gill. Ithaca, NY: Cornell Lab of Ornithology. doi.org/10.2173/bow.cangoo.01.

Munro, A. R., T. E. McMahon, and J. R. Ruzycki. 2005. Natural chemical markers identify source and date of introduction of an exotic species: Lake trout (*Salvelinus namaycush*) in Yellowstone Lake. *Can J Fish Aquat Sci* 62:79–87.

Murie, A. 1940. Ecology of the coyote in the Yellowstone. In *Fauna of the National Parks of the United States, Fauna Series No. 4, 1940, Conservation Bulletin No. 4*. Washington, DC: US Government Printing Office.

Murphy, J. R. 1960. *Ecology of the Bald Eagle in Yellowstone National Park: Progress Report—Summer of 1960*. Unpublished report, Yellowstone National Park, WY.

———. 1961. *Ecology of the Bald Eagle in Yellowstone National Park: Progress Report no. 2—1961*. Unpublished report, Yellowstone National Park, WY.

National Audubon Society. 2002. The Christmas Bird Count Historical Results. www.christmasbirdcount.org [January 13, 2022].

National Audubon Society. 2005. The Christmas Bird Count Historical Results. www.christmasbirdcount.org [January 13, 2022].

Nemeth, N., D. Gould, R. Bowen, and N. Komar. 2006. Natural and experimental West Nile virus infection in five raptor species. *J Wildl Dis* 42:1–13.

Nette, T., D. Burles, and M. Hoefs. 1984. Observations of golden eagle, *Aquila chrysaetos*, predation on Dall sheep, *Ovis dalli*, lambs. *Can Field-Nat* 98:252–254.

Newton, I. 1979. *Population Ecology of Raptors*. Berkhamsted, England: T & AD Poyser.

———. 1998. *Population Limitation in Birds*. London: Academic Press.

Ng, J., M. D. Giovanni, M. J. Bechard, J. K. Schmutz, and P. Pyle. 2020. Ferruginous hawk (*Buteo regalis*), version 1.0. In *Birds of the World*, edited by P. G. Rodewald. Ithaca, NY: Cornell Lab of Ornithology. doi.org/10.2173/bow.ferhaw.01.

Nichols, J. D., J. E. Hines, J. R. Sauer, F. W. Fallon, J. E. Fallon, and P. J. Heglund. 2000. A double-observer approach for estimating detection probability and abundance from point counts. *Auk* 2:393–408.

BIBLIOGRAPHY

Nielson, R. M., L. McManus, T. Rintz, L. L. McDonald, R. K. Murphy, W. H. Howe, and R. E. Good. 2014. Monitoring abundance of golden eagles in the western United States. *J Wildl Manage* 78:721–730.

Niemeyer, C. 1976. Montana golden eagle removal and translocation project, final report. US Fish and Wildlife Service Area Office, Billings, MT.

———. 1977. Montana golden eagle removal and translocation project, final report. US Fish and Wildlife Service Area Office, Billings, MT.

———. 1978. Montana golden eagle removal and translocation project, final report. US Fish and Wildlife Service Area Office, Billings, MT.

———. 1979. Montana golden eagle removal and translocation project, final report. US Fish and Wildlife Service Area Office, Billings, MT.

———. 1980. Montana golden eagle removal and translocation project, final report. US Fish and Wildlife Service Area Office, Billings, MT.

———. 1981. Montana golden eagle removal and translocation project, final report. US Fish and Wildlife Service Area Office, Billings, MT.

———. 1982. Montana golden eagle removal and translocation project, final report. US Fish and Wildlife Service Area Office, Billings, MT.

Oakleaf, R., and G. R. Craig. 2003. Peregrine restoration from a state biologist's perspective. In *Return of the Peregrine: A North American Saga of Tenacity and Teamwork*, edited by T. J. Cade, W. A. Burnham, and P. Burnham. Boise, ID: The Peregrine Fund.

Ohmart, R. D. 1994. The effects of human-induced changes in the avifauna of western riparian habitats. *Stud Avian Biol* 15:273–285.

Olechnowski, B.F.M., and D. M. Debinski. 2008. Response of songbirds to riparian willow habitat structure in the Greater Yellowstone Ecosystem. *Wilson J Ornith* 120:830–839.

Owen, M., and J. M. Black. 1983. *Waterfowl Ecology*. New York: Chapman and Hall.

Owen, M., and J. M. Black. 1990. *Waterfowl Ecology*. Glasgow: Blackie.

Oyler-McCance, S. J., F. A. Ransler, L. K. Berkman, and T. W. Quinn. 2007. A rangewide population genetic study of trumpeter swans. *Conserv Genet* 8:1339–1353.

Pacific Flyway Council. 2017. *Pacific Flyway Management Plan for the Rocky Mountain Population of Trumpeter Swans*. Vancouver, WA: Pacific Flyway Council, care of US Fish and Wildlife Service, and Division of Migratory Bird Management.

Pagel, J. E., K. J. Kritz, B. A. Millsap, R. K. Murphy, E. L. Kershner, and S. Covington. 2013. Bald eagle and golden eagle mortalities at wind energy facilities in the contiguous United States. *J Raptor Res* 47:311–315.

Pagel, J. E., and N. J. Schmitt. 2013. American marten prey remains found within peregrine falcon prey sample in Yellowstone National Park. *J Raptor Res* 47:419–420.

Palmer, R. S., ed. 1976a. *Handbook of North American Birds: Waterfowl*, vol. 2, part 1. New Haven, CT, and London: Yale University Press.

———, ed. 1976b. *Handbook of North American Birds: Waterfowl*, vol. 3, part 2. New Haven, CT, and London: Yale University Press.

———, ed. 1988. *Handbook of North American birds: Diurnal Raptors*, vol. 5, part 2. New Haven, CT, and London: Yale University Press.

Parker, P. G., T. A. Waite, B. Heinrich, and J. M. Marzluff. 1994. Do common ravens share food bonanzas with kin? DNA fingerprinting evidence. *Anim Behav* 48:1085–1093.

Paruk, J. D., D. C. Evers, J. W. McIntyre, J. F. Barr, J. Mager, and W. H. Piper. 2021. Common loon (*Gavia immer*), version 2.0. In *Birds of the World*, edited by P. G. Rodewald and B. K. Keeney. Ithaca, NY: Cornell Lab of Ornithology. doi.org/10.2173/bow.comloo.02.

Peakall, D. B. 1976. The peregrine falcon (*Falco peregrinus*) and pesticides. *Can Field-Nat* 90:301–307.

Persons, N. W., P. A. Hosner, K. A. Meiklejohn, E. L. Braun, and R. T. Kimball. 2016. Sorting out relationships among the grouse and ptarmigan using intron, mitochondrial, and ultra-conserved element sequences. *Mol Phylogenet Evol* 98:123–132.

Phillips, R. L., J. L. Cummings, and J. D. Berry. 1991. Responses of breeding golden eagles to relocation. *Wild Soc Bull* 19:430–434.

Preston, C. R., and R. D. Beane. 2020. Red-tailed hawk (*Buteo jamaicensis*), version 1.0. In *Birds of the World*, edited by A. F. Poole. Ithaca, NY: Cornell Lab of Ornithology. doi.org/10.2173/bow.rethaw.01.

Preston, C. R., R. E. Jones, and N. S. Horton. 2017. Golden eagle diet breadth and reproduction in relation to fluctuations in primary prey abundance in Wyoming's Bighorn Basin. *J Raptor Res* 51:334–346.

Pritchard, J. A. 1999. *Preserving Yellowstone's Natural Conditions: Science and the Perception of Nature*. Lincoln: University of Nebraska Press.

Proffitt, K. M., T. P. McEneaney, P. J. White, and R. A. Garrott. 2009. Trumpeter swan abundance and growth rates in Yellowstone National Park. *J Wildl Manage* 73:728–736.

Proffitt, K. M., T. P. McEneaney, P. J. White, and R. A. Garrott. 2010. Productivity and fledging success of trumpeter swans in Yellowstone National Park, 1987–2007. *Waterbirds* 33:341–348.

QGIS. 2020. QGIS Geographic Information System. www.qgis.org.

Rasmussen, J. L., S. G. Sealy, and R. J. Cannings. 2020. Northern saw-whet owl (*Aegolius acadicus*), version 1.0. In *Birds of the World*, edited by A. F. Poole. Ithaca, NY: Cornell Lab of Ornithology. doi.org/10.2173/bow.nswowl.01.

Ratcliffe, D. A. 1967. Decrease in eggshell weight in certain birds of prey. *Nature* 215:208–210.

———. 1997. *The Raven: A Natural History in Britain and Ireland*. Calton, Staffordshire, UK: T & AD Poyser.

Rattner, B. A., K. E. Horak, R. S. Lazarus, S. L. Schultz, B. Knowles, B. G. Abbo, and S. F. Volker. 2015. Toxicity reference values for chlorophacinone and their application for assessing anticoagulant rodenticide risk to raptors. *Ecotoxicology* 24:720–734.

Reid, R., P. F. Haworth, A. H. Fielding, and D. P. Whitfield. 2019. Spatial distribution of undulating flight displays of territorial golden eagles *Aquila chrysaetos* in Lewis, Scotland. *Bird Study* 66:407–412.

Reiswig, B. 1986. Western mute swan population status and agency attitudes. In *Proceedings—Ninth Trumpeter Swan Society Conference*, edited by D. Compton. Plymouth, MN: The Trumpeter Swan Society.

Renkin, R. A., and D. G. Despain. 1992. Fuel moisture, forest type, and lightning-caused fire in Yellowstone National Park. *Can J For Res* 22:37–45.

Restani, M. 1991. Resource partitioning among three *Buteo* species in the Centennial Valley, Montana. *Condor* 93:1007–1010.

Restani, M., J. M. Marzluff, and R. E. Yates. 2001. Effects of anthropogenic food sources on movements, survivorship, and sociality of common ravens in the Arctic. *Condor* 103:399–404.

Ringelman, J. K. 1991. 13.4.7. Managing beaver to benefit waterfowl. In *Waterfowl Management Handbook, Fish and Wildlife Leaflet 13*, compiled by D. H. Cross. Washington, DC: US Department of the Interior and US Fish and Wildlife Service.

———. 1992. 13.3.6. Ecology of montane wetlands. In *Waterfowl Management Handbook, Fish and Wildlife Leaflet 13*, compiled by D. H. Cross. Washington, DC: US Department of the Interior, US Fish and Wildlife Service.

Ripple, W. J., and R. L. Beschta. 2007. Restoring Yellowstone's aspen with wolves. *Biol Conserv* 138:514–519.

Ripple, W. J., E. J. Larsen, R. A. Renkin, and D. W. Smith. 2001. Trophic cascades among wolves, elk and aspen on Yellowstone National Park's northern range. *Biol Conserv* 102:227–234.

Risely, K., D. G. Noble, and S. R. Baillie. 2008. *The Breeding Bird Survey 2007*. BTO Research Report 508. Thetford, England: British Trust for Ornithology.

Rocca, M. E., P. M. Brown, L. H. MacDonald, and C. M. Carrico. 2014. Climate change impacts on fire regimes and key ecosystem services in Rocky Mountain Forests. *For Ecol Manage* 327:290–305.

Rosenberg, K. V., A. M. Dokter, P. J. Blancher, J. R. Sauer, A. C. Smith, P. A. Smith, J. C. Stanton, A. Panjabi, L. Helft, M. Parr, and P. P. Marra. 2019. Decline of the North American avifauna. *Science* 366:120–124.

Rosenfield, R. N., K. K. Madden, J. Bielefeldt, and O. E. Curtis. 2020. Cooper's hawk (*Accipiter cooperii*), version 1.0. In *Birds of the World*, edited by P. G. Rodewald. Ithaca, NY: Cornell Lab of Ornithology. doi.org/10.2173/bow.coohaw.01.

Rotenberry, J. T., and J. A. Wiens. 1980. Habitat structure, patchiness, and avian communities in North American steppe vegetation: A multivariate analysis. *Ecology* 61:1228–1250.

Sawyer, E. J. 1928. The courtship behavior of Barrow's golden-eye. *Wilson Bull* 40:4–17.

Schaming, T. D. 2016. Clark's nutcracker breeding season space use and foraging behavior. *PLOS One* 11:e0149116. doi.org/10.1371/journal.pone.0149116.

Schmidt, J. H., E. A. Rexstad, C. A. Roland, C. L. McIntyre, M. C. MacCluskie, and M. J. Flamme. 2017. Weather-driven change in primary productivity explains variation in the amplitude of two herbivore population cycles in a boreal system. *Oecologia* 186:435–446.

Schmutz, J. K., S. M. Schmutz, and D. A. Boag. 1980. Coexistence of three species of hawks (*Buteo* spp.) in the prairie-parkland ecotone. *Can J Zool* 58:1075–1089.

Schullery, P., and J. D. Varley. 1995. Cutthroat trout and the Yellowstone Lake ecosystem. In *The Yellowstone Lake Crisis: Confronting a Lake Trout Invasion: A Report to the Director of the National Park Service*, edited by J. D. Varley and P. Schullery. Yellowstone National Park, WY: National Park Service.

Schullery, P., and L. H. Whittlesey. 1999. Greater Yellowstone carnivores: A history of changing attitudes. In *Carnivores in Ecosystems: The Yellowstone Experience*, edited by T. W. Clark, A. P. Curlee, S. C. Minta, and P. M. Kareiva. New Haven, CT, and London: Yale University Press.

Shanahan, E., K. M. Irvine, D. Thoma, S. Wilmoth, A. Ray, K. Legg, and H. Shovic. 2016. Whitebark pine mortality related to white pine blister rust, mountain pine beetle outbreak, and water availability. *Ecosphere* 7:e01610.

Shanahan, E., K. Legg, and R. Daley. 2017. *Status of Whitebark Pine in the Greater Yellowstone Ecosystem: A Step-trend Analysis with Comparisons from 2004 to 2015*. Natural Resource Report NPS/GRYN/NRR—2017/1445. Fort Collins, CO: National Park Service.

Shea, R. 1979. The Ecology of the Trumpeter Swan in Yellowstone National Park and Vicinity. MS thesis, University of Montana.

Shields, E. M. 2021. Retrospective Analysis of a Declining Trumpeter Swan (*Cygnus buccinator*) Population in Yellowstone National Park. MS thesis, Montana State University.

Siders, M. S., and P. L. Kennedy. 1996. Forest structural characteristics of accipiter nesting habitat: Is there an allometric relationship? *Condor* 98:123–132.

Sim, I.M.W., R. D. Gregory, M. H. Hancock, and A. F. Brown. 2005. Recent changes in the abundance of British upland breeding birds. *Bird Study* 52:261–275.

Skinner, M. P. 1916. The nutcrackers of Yellowstone Park. *Condor* 18:62–64.

———. 1917. The birds of Molly Island, Yellowstone National Park. *Condor* 19:177–182.

———. 1921. Yellowstone Park, Wyoming. Bird-Lore's twenty-first Christmas Census. *Bird Lore* 23:29.

———. 1925. *The Birds of Yellowstone National Park*. Syracuse: New York State College of Forestry.

———. 1927. The predatory and fur-bearing animals of the Yellowstone National Park. *Roosevelt Wildlife Bulletin* 4:1–284.

———. 1928a. Yellowstone's winter birds. *Condor* 30:237–242.

BIBLIOGRAPHY

———. 1928b. The Canada goose in Yellowstone National Park. *Wilson Bull* 40:139–149.

———. 1937. Barrow's golden-eye in the Yellowstone National Park. *Wilson Bull* 49:3–10.

Slabe, V. A., J. T. Anderson, B. A. Millsap, J. L. Cooper, A. R. Harmata, M. Restani, R. H. Crandall, et al. 2022. Demographic implications of lead poisoning for eagles across North America. *Science* 375:779–782.

Smallwood, J. A., and D. M. Bird. 2020. American kestrel (*Falco sparverius*), version 1.0. In *Birds of the World*, edited by A. F. Poole and F. B. Gill. Ithaca, NY: Cornell Lab of Ornithology. doi.org/10.2173/bow.amekes.01.

Smallwood, J. A., M. F. Causey, D. H. Mossop, J. R. Klucsarits, B. Robertson, S. Robertson, J. Mason, M. J. Maurer, R. J. Melvin, R. D. Dawson, G. R. Bortolotti, J. W. Parrish, T. F. Breen, and K. Boyd. 2009. Why are American kestrel (*Falco sparverius*) populations declining in North America? Evidence from nest-box programs. *J Raptor Res* 43:274–282.

Smith, D. W., and N. Chambers. 2011. *The Future of Trumpeter Swans in Yellowstone National Park.* Final report summarizing expert workshop, April 26–27, 2011. Yellowstone National Park, WY: National Park Service.

Smith, D. W., R. O. Peterson, and D. B. Houston. 2003. Yellowstone after wolves. *BioScience* 53:330–340.

Smith, D. W., and R. O. Peterson. 2021. Intended and unintended consequences of wolf restoration to Yellowstone and Isle Royale National Parks. *Conserv Sci Pract* 3:e413.

Smith, D. W., D. R. Stahler, and D. R. MacNulty, eds. 2020a. *Yellowstone Wolves: Science and Discovery in the World's First National Park.* Chicago: University of Chicago Press.

Smith, D. W., and D. B. Tyers. 2012. The history and current status and distribution of beavers in Yellowstone National Park. *Northwest Sci* 86:276–288.

Smith, K. G., S. R. Wittenberg, R. B. Macwhirter, and K. L. Bildstein. 2020b. Northern harrier (*Circus hudsonius*), version 1.0. In *Birds of the World*, edited by P. G. Rodewald. Ithaca, NY: Cornell Lab of Ornithology. doi.org/10.2173/bow.norhar2.01.

Smith, R. N., S. L. Cain, S. H. Anderson, and J. R. Dunk. 1998. Blackfly-induced mortality of nestling red-tailed hawks. *Auk* 115:386–375.

Squires, J. R., R. T. Reynolds, J. Orta, and J. S. Marks. 2020. Northern goshawk (*Accipiter gentilis*), version 1.0. In *Birds of the World*, edited by S. M. Billerman. Ithaca, NY: Cornell Lab of Ornitholology. doi.org/10.2173/bow.norgos.01.

Stahler, D. R. 2000. Interspecific Interactions between the Common Raven (*Corvus corax*) and the Gray Wolf (*Canis lupus*) in Yellowstone National Park, Wyoming: Investigations of a Predator and Scavenger Relationship. MS thesis, University of Vermont.

Stahler, D., B. Heinrich, and D. Smith. 2002. Common ravens, *Corvus corax*, preferentially associate with gray wolves, *Canis lupus*, as a foraging strategy in winter. *Anim Behav* 64:283–290.

Steenhof, K. 2020. Prairie falcon (*Falco mexicanus*), version 1.0. In *Birds of the World*, edited by A. F. Poole. Ithaca, NY: Cornell Lab of Ornithology. doi.org/10.2173/bow.prafal.01.

Steenhof, K., M. N. Kochert, and T. L. McDonald. 1997. Interactive effects of prey and weather on golden eagle reproduction. *J Anim Ecol* 66:350–362.

Steenhof, K., and B. E. Peterson. 2009. American kestrel reproduction in southwestern Idaho: Annual variation and long-term trends. *J Raptor Res* 43:283–290.

Stinson, C. H. 1980. Weather-dependent foraging success and sibling aggression in red-tailed hawks in central Washington. *Condor* 82:76–80.

Sullivan, B. L., C. L. Wood, M. J. Iliff, R. E. Bonney, D. Fink, and S. Kelling. 2009. eBird: A citizen-based bird observation network in the biological sciences. *Biol Conserv* 142:2282–2292.

Swanson, D. L., and J. S. Palmer. 2009. Spring migration phenology of birds in the northern prairie region is correlated with local climate change. *J Field Ornithol* 80:351–363.

Swenson, J. E. 1975. Ecology of the Bald Eagle and Osprey in Yellowstone National Park. MS thesis, Montana State University.

———. 1978. Prey and foraging behavior of ospreys on Yellowstone Lake, Wyoming. *J Wildl Manage* 42:87–90.

Swenson, J. E., K. L. Alt, and R. L. Eng. 1986. Ecology of bald eagles in the Greater Yellowstone Ecosystem. *Wildl Monogr* 95:3–46.

Tack, J. D., and B. C. Fedy. 2015. Landscapes for energy and wildlife: Conservation prioritization for golden eagles across large spatial scales. *PLOS ONE* 10:e0134781.

Tack, J. D., B. R. Noon, Z. H. Bowen, L. Strybos, and B. C. Fedy. 2017. No substitute for survival: Perturbation analyses using a golden eagle population model reveal limits to managing for take. *J Raptor Res* 51:258–272.

Takats, D. L., C. M. Francis, G. L. Holroyd, J. R. Duncan, K. M. Mazur, R. J. Cannings, W. Harris, and D. Holt. 2001. *Guidelines for Nocturnal Owl Monitoring in North America.* Edmonton, Alberta: Beaverhill Bird Observatory and Bird Studies Canada.

Tastad, A. C. 2013. The Relative Effects of Grazing by Bison and Cattle on Plant Community Heterogeneity in Northern Mixed Prairie. PhD diss., University of Manitoba.

Tercek, M. T., R. Stottlemyer, and R. Renkin. 2010. Bottom-up factors influencing riparian willow recovery in Yellowstone National Park. *West N Am Nat* 70:387–399.

Tomback, D. F. 1978. Foraging strategies of Clark's nutcracker. *Living Bird* 16:123–161.

———. 1980. How nutcrackers find their seed stores. *Condor* 82:10–19.

———. 1982. Dispersal of whitebark pine seeds by Clark's nutcracker: A mutualism hypothesis. *J Anim Ecol* 51:451–467.

———. 2001. Clark's nutcracker: Agent of regeneration. In *Whitebark Pine Communities: Ecology and Restoration,* edited by D. F. Tomback, S. F. Arno, R. E., Keane. Washington, DC: Island Press.

———. 2020. Clark's nutcracker (*Nucifraga columbiana*), version 1.0. In *Birds of the World,* edited by A. F. Poole and F. B. Gill. Ithaca, NY: Cornell Lab of Ornithology. doi.org/10.2173/bow.clanut.01

Tomback, D. F., and P. Achuff. 2010. Blister rust and western forest biodiversity: Ecology, values and outlook for white pines. *For Pathol* 40:186–225.

Tomback, D. F., P. Achuff, A. W. Schoettle, J. W. Schwandt, and R. J. Mastrogiuseppe. 2011. The magnificent high-elevation five-needle white pines: Ecological roles and future outlook. In *The Future of High-elevation, Five-needle White Pines in Western North America: Proceedings of the High Five Symposium,* edited by R. E. Keane, D. F. Tomback, M. P. Murray, and C. M. Smith. Proceedings RMRS-P-63. Fort Collins, CO: Rocky Mountain Research Station, US Department of Agriculture, and US Forest Service.

Tomback, D. F., A. J. Anderies, K. S. Carsey, M. L. Powell, and S. Mellman-Brown. 2001a. Delayed seed germination in whitebark pine and regeneration patterns following the Yellowstone fires. *Ecology* 82:2587–2600.

Tomback, D. F., S. F. Arno, and R. E. Keane. 2001b. The compelling case for management intervention. In *Whitebark Pine Communities: Ecology and Restoration.* edited by D. F. Tomback, S. F. Arno, and R. E., Keane. Washington, DC: Island Press.

Tomback, D. F., and K. C. Kendall. 2001. Biodiversity losses: The downward spiral. In *Whitebark Pine Communities: Ecology and Restoration,* edited by D. F. Tomback, S. F. Arno, R. E., Keane. Washington, DC: Island Press, 243–262.

Tomback, D. F., and Y. B. Linhart. 1990. The evolution of bird-dispersed pines. *Evol Ecol* 4:185–219.

Touchstone, T. H. 1997. Golden Eagle Recovery Techniques and Success in the Southern Appalachian Region. MS thesis, State University of West Georgia.

Turner, J. F. 1968. Preliminary report on the osprey, *Pandion haliaetus,* in northwestern Wyoming. Unpublished report, University of Michigan, Ann Arbor.

Turner, M. G., K. H. Braziunas, W. D. Hansen, and B. J. Harvey. 2019. Short-interval severe fire erodes the resilience of subalpine lodgepole pine forests. *PNAS* 116:11319–11328.

US Shorebird Conservation Partnership. 2022. Assessment of the conservation status of shorebirds. www.shorebirdplan.org/science/assessment-conservation-status-shorebirds/.

USFWS (US Fish and Wildlife Service). 1994. The North American trumpeter swan status report—1990. Unpublished report. Canadian Wildlife Service and the Trumpeter Swan Society.

———. 2008. *Birds of Conservation Concern 2008.* Arlington, VA: US Department of the Interior, US Fish and Wildlife Service, and Division of Migratory Bird Management.

———. 2016. *Bald and Golden Eagles: Population Demographics and Estimation of Sustainable Take in the United States, 2016 update.* Washington, DC: US Department of the Interior, US Fish and Wildlife Service, and Division of Migratory Bird Management.

———. 2020. *Waterfowl Population Status, 2019.* Washington, DC: US Department of the Interior.

———. 2021. *Birds of Conservation Concern 2021.* US Department of the Interior, US Fish and Wildlife Service, and Division of Migratory Bird Management, Arlington, VA.

USGS (US Geological Survey). 2018. North American Breeding Bird Survey Home. Last modified March 27, 2018. www.pwrc.usgs.gov/bbs/.

Van Horne, B. 1983. Density as a misleading indicator of habitat quality. *J Wildl Manage* 47:893–901.

Vander Wall, S. B. 1982. An experimental analysis of cache recovery in Clark's nutcracker. *Anim Behav* 30:84–94.

Vander Wall, S. B., and R. P. Balda. 1977. Coadaptations of the Clark's nutcracker and piñon pine for efficient seed harvest and dispersal. *Ecol Monogr* 47:89–111.

VanDerWal, J., H. T. Murphy, A. S. Kutt, G. C. Perkins, B. L. Bateman, J. J. Perry, and A. E. Reside. 2013. Focus on poleward shifts in species' distribution underestimates the fingerprint of climate change. *Nat Clim Change* 3:239–243.

Vickery, P. D., P. L. Tubaro, J. M. Cardosa da Silva, B. G. Peterjohn, J. R. Herkert, and R. B. Cavalcanti. 1999. Conservation of grassland birds in the Western Hemisphere. *Stud Avian Biol* 19:2–26.

Walker, L. E., L. M. Baril, D. B. Haines, and D. W. Smith. 2019. Reproductive characteristics of red-tailed hawks in Yellowstone National Park, an intact temperate landscape. *J Raptor Res* 53:309–318.

Walker, L. E., J. M. Marzluff, M. C. Metz, A. J. Wirsing, L. M. Moskal, D. R. Stahler, and D. W. Smith. 2018. Population responses of common ravens to reintroduced gray wolves. *Ecol Evol* 8:11158–11168.

Walker, L. E., D. W. Smith, M. E. Albrechtsen, H. Weinberg, K. Burke, and K. Duffy. 2020. *Yellowstone Bird Program 2019 Annual Report.* YCR-2020–02. Yellowstone National Park, WY: National Park Service.

Warkentin, I. G., N. S. Sodhi, R.H.M. Espie, A. F. Poole, L. W. Oliphant, and P. C. James. 2020. Merlin (*Falco columbarius*), version 1.0. In *Birds of the World,* edited by S. M. Billerman. Ithaca, NY: Cornell Lab of Ornithology. doi.org/10.2173/bow.merlin.01.

Watson, J. 2010. *The Golden Eagle.* Calton, England: T & AD Poyser.

Weaver, T., and F. Forcella. 1986. Cone production in *Pinus albicaulis* forests. In *Proceedings—Conifer Tree Seed in the Inland Mountain Symposium,* compiled by R. C. Shearer. Ogden, UT: US Department of Agriculture, US Forest Service, and Intermountain Research Station.

Webb, W. C., J. M. Marzluff, and J. Hepinstall. 2011. Linking resource use with demography in a synanthropic population of common ravens. *Biol Conserv* 144:2264–2273.

BIBLIOGRAPHY

Webb, W. C., J. M. Marzluff, and J. Hepinstall-Cymerman. 2012. Differences in space use by common ravens in relation to sex, breeding status, and kinship. *Condor* 114:584–594.

Weidensaul, S. 2018. Losing ground: What's behind the worldwide decline of shorebirds? *Living Bird*, September 19, 2018. www.allaboutbirds.org/losingground-whats-behind-the-worldwide-decline-of-shorebirds/.

Weller, M. W. 1981. *Freshwater Marshes: Ecology and Wildlife Management*, 3rd ed. Minneapolis: University of Minnesota Press.

Westerling, A. L., M. G. Turner, E.A.H. Smithwick, W. H. Romme, and M. G. Ryan. 2011. Continued warming could transform Greater Yellowstone fire regimes by mid-21st century. *PNAS* 108:13165–13170.

Western Hemisphere Shorebird Reserve Network. 2109. whsrn.org/.

WGFD (Wyoming Game and Fish Department). 2017. *Wyoming State Wildlife Action Plan*. Cheyenne: Wyoming Game and Fish Department.

Wheeler, M. 2014. The Genetics of Conservation Translocations: A Comparison of North American Golden Eagles (*Aquila chrysaetos canadensis*) and Bald Eagles (*Haliaeetus leucocephalus*). PhD diss., Duquesne University.

White, C. M., N. J. Clum, T. J. Cade, and W. G. Hunt. 2020. Peregrine falcon (*Falco peregrinus*), version 1.0. In *Birds of the World*, edited by S. M. Billerman. Ithaca, NY: Cornell Lab of Ornithology. doi.org/10.2173/bow.perfal.01.

White, E. B. 1970. *The Trumpet of the Swan*. New York: Harper and Row.

White, P. J., K. A. Gunther, and F. T. Van Manen, eds. 2017. *Yellowstone Grizzly Bears: Ecology and Conservation of an Icon of Wildness*. Bozeman, MT: Yellowstone Forever.

Whitfield, D. P., A. H. Fielding, D.R.A. Mcleod, and P. F. Haworth. 2004. Modelling the effects of persecution on the population dynamics of golden eagles in Scotland. *Biol Conserv* 119:319–333.

Whittlesey, L. H. 1995. "They're going to build a railroad!": Cinnabar, Stephens Creek, and the Game Ranch addition to Yellowstone National Park. Gardiner, MT: Yellowstone National Park, Heritage Research Center.

Whittlesey, L. H., and S. Bone. 2018. *The History of Mammals in the Greater Yellowstone Ecosystem, 1796–1881: A Cross-disciplinary Analysis of Thousands of Historical Observations*. Yellowstone National Park, WY: National Park Service.

Wiens, J. D., P. S. Kolar, W. G. Hunt, T. Hunt, M. R. Fuller, and D. A. Bell. 2018. Spatial patterns in occupancy and reproduction of golden eagles during drought: Prospects for conservation in changing environments. *Condor* 120:106–124.

Wiens, J. D., N. H. Schumaker, R. D. Inman, T. C. Esque, K. M. Longshore, and K. E. Nussear. 2017. Spatial demographic models to inform conservation planning of golden eagles in renewable energy landscapes. *J Raptor Res* 51:234–257.

Williams, J. R., C. S. Guy, P. E. Bigelow, and T. M. Koel. 2022. Quantifying the spatial structure of invasive lake trout in Yellowstone Lake to improve suppression efficacy. *N Am J Fish Manage* 42:50–62.

Williams, T. J., D. F. Tomback, N. Grevstad, and K. Broms. 2020. Temporal and energetic drivers of seed resource use by Clark's nutcracker, keystone seed disperser of coniferous forests. *Ecosphere* 11:e03085.

Wilmers, C. C., D. R. Stahler, R. L. Crabtree, D. W. Smith, and W. M. Getz. 2003. Resource dispersion and consumer dominance: Scavenging at wolf- and hunter-killed carcasses in Greater Yellowstone, USA. *Ecol Lett* 6:996–1003.

Windell, J. T., B. E. Willard, D. J. Cooper, S. Q. Foster, C. F. Knud-Hansen, L. P. Rink, and G. N. Kiladis. 1986. An ecological characterization of Rocky Mountain montane and subalpine wetlands. *US Fish and Wildlife Service Biological Report* 86 (11):298.

Woodbridge, B. 1991. Habitat Selection by Nesting Swainson's Hawks: A Hierarchical Approach. MS thesis, Oregon State University.

Woodbridge, B., K. K. Finley, and S. T. Seager. 1995. An investigation of the Swainson's hawk in Argentina. *J Raptor Res* 29:202–204.

Wright, G. M. 1930–31. Field Notes. Journal v1720. Museum of Vertebrate Zoology, University of California, Berkeley.

———. 1934. The primitive persists in bird life of Yellowstone National Park. *Condor* 36:145–153.

Wu, J. X., C. B. Wilsey, L. Taylor, and G. W. Schuurman. 2018. Projected avifaunal responses to climate change across the US National Park System. *PLOS ONE* 13:e0190557.

Yeats, W. B. 1919. The wild swans at Coole. In *The Wild Swans at Coole*. New York,: MacMillan.

Yellowstone National Park. 2021. *Yellowstone Resources and Issues Handbook: 2021*. Yellowstone National Park, WY: National Park Service.

Yellowstone National Park Bird Program. Birds of Yellowstone Checklist. 2020. www.nps.gov/yell/learn/nature/upload/274-BirdChecklist_2020-web.pdf.

———. Birds of Yellowstone Checklist. 2021. www.nps.gov/yell/learn/nature/upload/274-Bird-Checklist-2021.pdf.

Zimmerman, T. 2009. Wildland fire management policy. Learning from the past and present and responding to future challenges. *Yellowstone Sci* 17:31–34.

CONTRIBUTORS

ALBRECHTSEN, MARY BETH
National Park Service, Yellowstone National Park, Wyoming.
P.O. Box 168, Yellowstone National Park, WY 82190
mbalbrechtsen@gmail.com

CASSIDY, KIRA A.
Yellowstone Forever, Yellowstone National Park, Wyoming.
P.O. Box 168, Yellowstone National Park, WY 82190
kira.a.cassidy@gmail.com

COLEMAN, GEORGIA
School of Environmental and Forestry Science,
University of Washington–Seattle, Seattle, Washington.
3715 West Stevens Way NE, Seattle, WA 98195
gwc083@gmail.com

DELEHANTY, DAVID J.
Idaho State University, Pocatello, Idaho.
Idaho State University, 921 S. 8th Ave.,
Mail Stop 8007, Pocatello, ID 83209
deledavi@isu.edu

DREITZ, VICTORIA J.
Avian Science Center, W. A. Franke College of Forestry and
Conservation, University of Montana, Missoula, Montana.
32 Campus Dr., Missoula, MT 59812
victoria.dreitz@mso.umt.edu

DUFFY, KATHARINE E.
National Park Service (retired), Yellowstone National Park, Wyoming.
P.O. Box 168, Yellowstone National Park, WY 82190
owlpals@wyellowstone.com

DUNCAN, DAYTON
daytonduncan49@gmail.com

EVERS, DAVID
Biodiversity Research Institute, Portland, Maine.
276 Canco Rd., Portland, ME 04103
david.evers@briwildlife.org

FAGRE, DANIELLE A.
Avian Science Center, W. A. Franke College of Forestry and
Conservation, University of Montana, Missoula, Montana.
32 Campus Dr., Missoula, MT 59812
danielle.fagre@gmail.com

FAIR, JEFF
Palmer, Alaska.
fairwinds@briloon.org

HAINES, DAVID B.
National Park Service, Yellowstone National Park, Wyoming.
P.O. Box 168, Yellowstone National Park, WY 82190
david_b_haines@nps.gov

HO, CAMERON
School of Environmental and Forestry Science,
University of Washington–Seattle, Seattle, Washington.
3715 West Stevens Way NE, Seattle, WA 98195
ho.cameron98@gmail.com

JANOUSEK, WILLIAM M.
Avian Science Center, W. A. Franke College of Forestry and
Conservation, University of Montana, Missoula, Montana.
32 Campus Dr., Missoula, MT 59812
janousek12@gmail.com

KATZNER, TODD E.
United States Geological Survey, Boise, Idaho.
230 N. Collins Rd., Boise, ID 83702
tkatzner@usgs.gov

LEAVITT, ARCATA
Ricketts Conservation Foundation, Bondurant, Wyoming.
aleavitt@rickettsconservation.org

LONG, WILLIAM
Wyoming Wetland Society, Jackson, Wyoming.
bill@wyomingwetlandssociety.org

CONTRIBUTORS

LORETTO, MATTHIAS
Technical University of Munich, Munich, Germany.
matthias.loretto@gmail.com

MARZLUFF, JOHN M.
School of Environmental and Forestry Science, University of Washington–Seattle, Seattle, Washington.
3715 West Stevens Way NE, Seattle, WA 98195
corvid@uw.edu

McINTYRE, CAROL L.
National Park Service, Denali National Park and Preserve, Alaska.
P.O. Box 9, Denali Park, AK 99755
carol_mcintyre@nps.gov

McLAREN, THOMAS H.
University of Colorado–Denver, Integrative Biology, Denver, Colorado.
P.O. Box 173364, Denver, CO 80217
thomas.mclaren@ucdenver.edu

MITCHELL, CARL D.
United States Fish and Wildlife Service (retired), Grays Lake, Idaho.
mitch@silverstar.com

PARKER, JOHN
National Park Service (retired), Yellowstone National Park, Wyoming.
P.O. Box 168, Yellowstone National Park, WY 82190
conundrumjp@gmail.com

PATLA, SUSAN
Wyoming Game and Fish Department (retired).
susan_patla@hotmail.com

ROTELLA, JAY
Montana State University, Department of Ecology, Bozeman, Montana.
P.O. Box 173460, Bozeman, MT 59717
rotella@montana.edu

SAVOY, LUCAS
Biodiversity Research Institute, Portland, Maine.
276 Canco Rd., Portland, ME 04103
lucas.savoy@briwildlife.org

SHEA, RUTH
Northern Rockies Conservation Cooperative, Jackson Hole, Wyoming.
ruth@nrcooperative.org

SHIELDS, EVAN M.
Montana State University, Department of Ecology, Bozeman, Montana.
P.O. Box 173460, Bozeman, MT 59717
emshields4@gmail.com

SMITH, DOUGLAS W.
National Park Service (retired), Yellowstone National Park, Wyoming.
P.O. Box 168, Yellowstone National Park, WY 82190
dwsmith0526@gmail.com

SPAGNUOLO, VINCENT
Ricketts Conservation Foundation (retired), Bondurant, Wyoming.
vincent.spagnuolo@gmail.com

STAHLER, DANIEL
National Park Service, Yellowstone National Park, Wyoming.
P.O. Box 168, Yellowstone National Park, WY 82190
dan_stahler@nps.gov

TALIAFERRO, JOHN
Pray, Montana.
johntalia@att.net

TOMBACK, DIANA F.
University of Colorado–Denver, Integrative Biology, Denver, Colorado.
P.O. Box 173364, Denver, CO 80217
diana.tomback@ucdenver.edu

WALKER, LAUREN E.
National Park Service, Yellowstone National Park, Wyoming.
P.O. Box 168, Yellowstone National Park, WY 82190
lauren.seckel@gmail.com

WEINBERG, HOWARD J.
National Park Service, Yellowstone National Park, Wyoming.
P.O. Box 168, Yellowstone National Park, WY 82190
howard_weinberg@nps.gov

WEHTJE, WALTER G.
Ricketts Conservation Foundation, Bondurant, Wyoming.
wwehtje@rickettsconservation.org

INDEX

Italic pagination refers to figures and tables

INDEX